Fate of Pharmaceuticals in the Environment and in Water Treatment Systems

Fate of Pharmaceuticals in the Environment and in Water Treatment Systems

Edited by

Diana S. Aga

CRC Press
Taylor & Francis Group
Boca Raton London New York

CRC Press is an imprint of the
Taylor & Francis Group, an **informa** business

CRC Press
Taylor & Francis Group
6000 Broken Sound Parkway NW, Suite 300
Boca Raton, FL 33487-2742

First issued in paperback 2019

ISBN-13: 978-1-4200-5232-9 (hbk)
ISBN-13: 978-0-367-38787-7 (pbk)

Library of Congress Cataloging-in-Publication Data

Fate of pharmaceuticals in the environment and in water treatment systems /
   edited by Diana S. Aga.
      p. cm.
   Includes bibliographical references and index.
   ISBN 978-1-4200-5232-9 (alk. paper)
      1. Drugs--Environmental aspects. 2. Water--Pollution. 3. Water--Purification.
   I. Aga, Diana S., 1967- II. Title.

TD196.D78F38 2008
628.5'2--dc22   2007029935

Visit the Taylor & Francis Web site at
http://www.taylorandfrancis.com

and the CRC Press Web site at
http://www.crcpress.com

Contents

PART I Occurrence and Analysis of Pharmaceuticals in the Environment

PART II Environmental Fate and Transformations of Veterinary Pharmaceuticals

PART III Treatment of Pharmaceuticals in Drinking Water and Wastewater

Preface

Recent advances in analytical instrumentation have been mirrored by our increased ability to detect and quantify organic contaminants at trace levels, even in highly complex matrices such as wastewater, manure, and soil. In contrast to the hydrophobic persistent organic pollutants (for example, PCBs, DDT) that are often found in the environment at parts-per-million or parts-per-billion concentrations, pharmaceutical compounds are generally present at the parts-per-trillion or low parts-per-billion range and are mostly polar. Consequently, the detection of pharmaceutical residues in the environment remained elusive until modern instruments such as liquid chromatography/mass spectrometry (LC/MS) became commonplace in many environmental laboratories. As a result, many scientists have now documented the occurrence of residues of pharmaceuticals and personal-care products, which have been termed "emerging contaminants" in various environmental compartments. Questions regarding persistence and long-term adverse effects of pharmaceuticals in the environment have been raised because there have been reports that very low drug concentrations (for example, ng/L) in the environment can have undesirable ecological and potentially human health effects.

The first section of this book, "Occurrence and Analysis of Pharmaceuticals in the Environment," includes a chapter prepared by leading researchers from the U.S. Environmental Protection Agency and the U.S. Geological Survey, which provides an overview of the momentous publications that have been instrumental in the recognition of emerging contaminants. A compilation of the most current (2004 through 2006) literature on the presence and concentrations of pharmaceuticals in the environment is also presented. This information is complemented by the subsequent review chapters on the recent advances in instrumentation and sample preparation techniques in environmental analysis that have played a critical role in the advancement of our knowledge on the environmental fate of pharmaceuticals. Finally, an example of how risk assessment is conducted to investigate the fate and effects of pharmaceutical contaminants is included in this first section.

An important source of pharmaceutical contaminants is through land application of livestock manure to fertilize crops. Many animal operations generate manure that contains antibiotics since animals receive these drugs in feed rations, either as growth promoters or as therapeutic agents. Therefore, the second section of this book, "Environmental Fate and Transformations of Veterinary Pharmaceuticals," is dedicated to chapters that explore the behavior of pharmaceuticals in soil and the potential effects of antibiotics on plants after uptake.

Pharmaceuticals are also introduced into the environment via wastewater treatment plants (WWTPs), which are currently not designed and operated to remove low concentrations of organic contaminants. Because increasing amounts of treated wastewater are recycled for industrial and domestic use, it is important to improve treatment technologies for both wastewater and drinking water sources. Therefore, the third section of this book, "Treatment of Pharmaceuticals in Drinking Water and

Wastewater," includes chapters that examine various treatment processes that can be employed to reduce the concentrations of pharmaceuticals at the source.

This book covers important issues regarding the analysis, occurrence, persistence, treatment, and transformations of pharmaceuticals in the environment. Topics range from field studies documenting the occurrence of pharmaceuticals in several environmental compartments to laboratory studies determining the degradation kinetics and formation of byproducts during treatment. This book will provide information that will help scientists, regulators, and engineers understand the factors that affect the environmental fate of pharmaceuticals in soil and water to facilitate the development of best management practices and optimize treatment systems for effective removal of these compounds in the environment.

Diana S. Aga
Department of Chemistry
University at Buffalo
Buffalo, New York

Editor

Diana Aga, Ph.D., is an associate professor of chemistry at the University at Buffalo, The State University of New York. Her current research involves the investigations on the fate, transport, and ecotoxicological effects of pharmaceuticals, endocrine-disrupting chemicals, and persistent organic pollutants in the environment. A major focus of her research is to identify unknown transformation products of pharmaceuticals in various environmental matrices (for example, manure, plants, soil, wastewater) using a combination of novel strategies in sample preparation, bioassays, and modern mass spectrometric techniques. Dr. Aga received her B.S. in agricultural chemistry at the University of the Philippines at Los Baños (1988) and her Ph.D. in environmental and analytical chemistry at the University of Kansas (1995). She was a research assistant at the U.S. Geological Survey, Lawrence, Kansas (1993–1996), and a postdoctoral fellow at the Swiss Federal Institute of Aquatic Science and Technology (EAWAG), Switzerland (1996–1998). Dr. Aga is recipient of various research awards, such as the National Science Foundation CAREER Award, the North Atlantic Treaty Organization Scientific and Environmental Affairs Fellowship, and the Alexander von Humboldt Research Fellowship.

Contributors

Craig D. Adams
Department of Civil, Architectural,
and Environmental Engineering
University of Missouri at Rolla
Rolla, Missouri

Damià Barceló
Department of Environmental
Chemistry
IIQAB-CSIC
Barcelona, Spain

Angela Batt
National Exposure Research
Laboratories
Office of Research and Development
U.S. Environmental Protection
Agency
Cincinnati, Ohio

James O. Berry
Biology Department
University at Buffalo
Buffalo, New York

Alistair B.A. Boxall
Central Science Laboratory
University of York
Sand Hutton, York
United Kingdom

Nadia Carmosini
Department of Agronomy
Purdue University
West Lafayette, Indiana

Joel R. Coats
Department of Entomology
Iowa State University
Ames, Iowa

Michael C. Dodd
Swiss Federal Institute of Aquatic
Science and Technology (EAWAG)
Duebendorf, Switzerland

Michael H. Farkas
Biology Department
University at Buffalo
Buffalo, New York

Tamara Floyd-Smith
Department of Chemical Engineering
Tuskegee University
Tuskegee, Alabama

Michael J. Focazio
U.S. Geological Survey
Reston, Virginia

Edward T. Furlong
U.S. Geological Survey
Denver, Colorado

Susan T. Glassmeyer
National Exposure Research
Laboratory
Office of Research and Development
U.S. Environmental Protection
Agency
Cincinnati, Ohio

Willie F. Harper, Jr.
Department of Civil Engineering
Auburn University
Auburn, Alabama

Keri L. Henderson
Department of Entomology
Iowa State University
Ames, Iowa

Ching-Hua Huang
School of Civil and Environmental
Engineering
Georgia Institute of Technology
Atlanta, Georgia

Sungpyo Kim
Department of Earth and
Environmental Engineering
Columbia University
New York

Christine Klein
Department of Chemistry
University at Buffalo
Buffalo, New York

K.F. Knowlton
Department of Dairy Science
Virginia Polytechnic Institute and
State University
Blacksburg, Virginia

Dana W. Kolpin
U.S. Geological Survey
Iowa City, Iowa

Reinhard Länge
Bayer Schering Pharma
Nonclinical Drug Safety
Berlin, Germany

Linda S. Lee
Department of Agronomy
Purdue University
West Lafayette, Indiana

Hongxia Lei
Water Quality Research and
Development Division
Southern Nevada Water Authority
Las Vegas, Nevada

Jonas Locke
Department of Chemistry
University at Buffalo
Buffalo, New York

N.G. Love
Department of Civil and
Environmental Engineering
Virginia Polytechnic Institute and
State University
Blacksburg, Virginia

Thomas B. Moorman
U.S. Department of Agriculture–
Agricultural Research Service
(ARS)
National Soil Tilth Laboratory
Ames, Iowa

Claudia Neubert
Bayer Schering Pharma
Nonclinical Drug Safety
Berlin, Germany

Seamus O'Connor
Department of Chemistry
University at Buffalo
Buffalo, New York

Vanessa J. Pereira
Instituto de Biologia Experimental e
Tecnólogica
Oeiras, Portugal

Sandra Pérez
Department of Environmental
Chemistry
Consejo Superior de Investigaciones
Científicas (CSIC)
Barcelona, Spain

Rudolf J. Schneider
Department of Analytical Chemistry:
Reference Materials
Federal Institute for Materials
Research and Testing
Berlin, Germany

Amisha D. Shah
School of Civil and Environmental
Engineering
Georgia Institute of Technology
Atlanta, Georgia

Shane A. Snyder
Water Quality Research and
  Development Division
Southern Nevada Water Authority
Las Vegas, Nevada

Thomas Steger-Hartmann
Bayer Schering Pharma
Nonclinical Drug Safety
Berlin, Germany

A. Scott Weber
Department of Civil, Structural, and
  Environmental Engineering
University at Buffalo
Buffalo, New York

Howard S. Weinberg
Department of Environmental
  Sciences and Engineering
University of North Carolina
Chapel Hill, North Carolina

Eric C. Wert
Water Quality Research and
  Development Division
Southern Nevada Water Authority
Las Vegas, Nevada

Zhengqi Ye
Enthalpy Analytical, Inc.
Durham, North Carolina

Taewoo Yi
Department of Civil Engineering
Auburn University
Auburn, Alabama

Z. Zhao
Department of Dairy Science
Virginia Polytechnic Institute and
  State University
Blacksburg, Virginia

Part I

---

*Occurrence and Analysis
of Pharmaceuticals
in the Environment*

1 Environmental Presence and Persistence of Pharmaceuticals
An Overview

Susan T. Glassmeyer, Dana W. Kolpin,
Edward T. Furlong, and Michael J. Focazio

Contents

1.1 INTRODUCTION

Emerging contaminants (ECs) in the environment—that is, chemicals with domestic, municipal, industrial, or agricultural sources that are not commonly monitored but may have the potential for adverse environmental effects—is a rapidly growing field of research. The use of "emerging" is not intended to infer that the presence of these compounds in the environment is new. These chemicals have been released into the environment as long as they have been in production or, in the case of hormones and other endogenous compounds, since the rise of animal life. What is emerging is the interest by the scientific and lay communities in the presence of these chemicals in the environment, the analytical capabilities required for detection, and the subtle effects that very small concentrations of these chemicals appear to have on aquatic biota. In December 2006, *Environmental Science & Technology* devoted an entire special issue (volume 40, number 23) to the topic of ECs, illustrating the increased interest in the subject. Within the ECs, one particular class that has seen a substantial increase in research over the past 10 years is pharmaceuticals and personal-care products (PPCPs). This increased research interest can be demonstrated by several means, including requests for proposals from funding agencies, but the clearest indication of a focused effort to understand the introduction, transformation, and potential health and environmental effects of PPCPs and ECs, in general, is the number of published reports. This increase can be shown by examining six environmental journals that regularly publish PPCP-related papers—*Chemosphere, Environmental Science & Technology, Environmental Toxicology and Chemistry, Science of the*

3

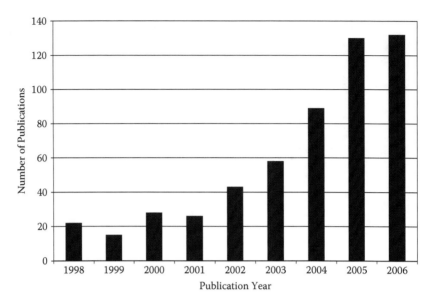

FIGURE 1.1   Cumulative yearly number of pharmaceutical-related papers published in six environmental journals from 1998 to 2006.

Total Environment, Water Research, and *Water Science and Technology.* In 1998 there were 22 papers published on pharmaceuticals, antibiotics, or drugs in these 6 journals; by 2006, this number increased sixfold to 132 papers (Figure 1.1).

This growth can be attributed to a number of factors. The presence of pharmaceuticals in surface-water samples from Europe and the United States was documented in several sentinel papers.[1–4] These ground-breaking works encouraged other scientists to examine the rivers, streams, lakes, and reservoirs in their regions for such chemicals. In addition the intense public attention paid to news reports on the environmental detections of these chemicals and possible effects on aquatic life has made this issue visible to the wastewater-treatment, drinking-water treatment, and regulatory communities. This has driven the funding bodies associated with these communities to fund studies or request proposals that address the presence, fate, and effects of PPCPs in aquatic systems. The release of the first comprehensive reconnaissance of pharmaceuticals and other wastewater contaminants in the United States[2] provides an example of the intense media interest in this topic. Within 6 days of online publication of this study, 72 newspapers across the United States had published articles describing the results, either locally written or based on international media syndicate reports. There also was substantial concurrent coverage by local and national radio and television outlets, including the Cable News Network, *ABC World News Tonight*, and National Public Radio. A substantial fraction of these news stories may be attributable to press releases and media briefings prior to publication. However, the interest by television and print journalists in reporting the results of a peer-reviewed journal article to the general public was motivated by the recognition that describing the presence of PPCPs in water supplies would be of interest to the public. To better convey the results of the study published by Kolpin et al.[2] to the public, a separate general-interest fact sheet was published to summarize the important points of the study.[5] Because PPCPs are commonly and widely used by individuals, there is likely a preexisting, personal identification with these compounds that does not occur for the wide range of other organic and inorganic contaminants whose presence in the environment has previously been described. This greater public "name recognition" makes itself known through the media to the regulatory and technical community and has prompted interest in sponsoring research that defines the composition and concentrations of PPCPs in potential sources and their fate and effects following release into the environment.

Independent of the drivers that potentially fuel the interest in studies of PPCPs, it is clear that PPCP research has grown beyond surface-water studies to examine issues such as:

- Presence in other matrices, such as groundwater,[6–11] landfill leachates,[12–15] sediments,[16,17] and biosolids.[18,19]
- Environmental transport and fate in surface water,[20–23] groundwater,[8,9,24–30] and soils amended with reclaimed water[31,32] or biosolids.[33–35]
- PPCP source elucidation, such as wastewater treatment plant (WWTP) effluents,[20,23,36–40] confined animal feeding operations (CAFOs),[41,42] and aquaculture.[43,44]
- Removal during wastewater[23,45–53] and drinking-water[10,54–61] treatment.
- Effects on aquatic ecosystems,[62–66] terrestrial ecosystems,[67,68] and human health.[69,70]

The chapters in this book provide an extensive examination of current environmental pharmaceutical research and are divided into three sections: "Occurrence and Analysis of Pharmaceuticals in the Environment," "Environmental Fate and Transformations of Veterinary Pharmaceuticals," and "Treatment of Pharmaceuticals in Drinking Water and Wastewater." The purpose of this introductory overview chapter is to outline current (2004–2006) knowledge about the presence and concentration of PPCPs as described in the published literature. Previous reviews[1,71–73] should be consulted for discussions on pre-2004 publications. Those reviews will provide the reader with a comprehensive introduction to the topic of PPCPs in the environment. This chapter describes the sources of PPCPs and other organic contaminants often associated with human wastewater into the environment, the range of concentrations present in various environmental compartments, and the potential routes of removal/sequestration. An overview of the sources and fate of veterinary pharmaceuticals will be discussed in Chapter 5, "Fate and Transport of Veterinary Medicines in the Soil Environment."

1.2 OVERVIEW OF RECENT LITERATURE

Between January 2004 and July 2006, more than 80 papers were published discussing the worldwide presence of PPCPs in different environmental compartments, such as raw and treated wastewater and surface water (Table 1.1), biosolids and sludges, sediments, groundwater, and drinking water (Table 1.2). Figure 1.2 is an analysis of the papers listed in Table 1.1 and Table 1.2, categorized by sample location.

Although the United States had the most recently published papers of any individual country, the total number of publications from Europe slightly exceeds those from North America. This result likely reflects the earlier attention paid to the environmental presence of PPCPs in Europe, particularly in Germany, Switzerland, Italy, and the United Kingdom. The global nature of this issue is illustrated by the number of countries reporting studies. Whereas PPCPs once were only found on Western European and North American research agendas, the importance of the issue has resulted in the expansion of studies into other parts of the world, including Eastern Europe and Asia.[74–78] The presence of PPCPs also has been reported for some of the countries missing from this list, such as Australia and Brazil, in articles that predate the time addressed in this review. Their absence likely reflects the timing of publication and the publication in journals that were missed in the searches used for this review. However, in many other regions of the world where detection of these compounds would be expected, such as urbanized watersheds in Latin America, Africa, the Middle East, and China, no concentrations have been reported in the literature.[79] There may be decreased access to PPCPs in some of the remote parts of these regions; however, wastewater and drinking-water treatment in these same regions is also likely to be minimal. Therefore, detectable, and perhaps substantial, concentrations would be expected. Studies are necessary to assess the concentration and composition of PPCPs present in the aquatic environments of these regions to determine the potential for the environmental effect of PPCPs in regions of the globe with rapidly growing populations and less advanced water treatment.

In the 80 papers cited in Table 1.1 and Table 1.2, detections and concentrations for more than 120 chemicals were reported. It should be noted that "nondetections"

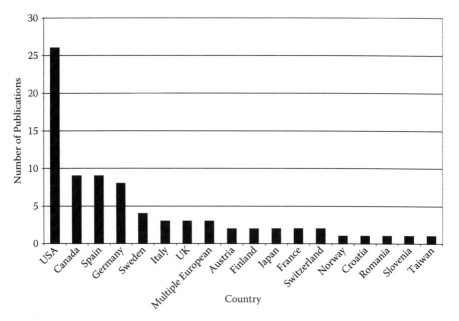

FIGURE 1.2 Distribution of recent publications on pharmaceuticals and personal-care products by location sampled.

TABLE 1.1

Concentrations of Pharmaceuticals and Other Emerging Contaminants in Wastewater Influents, Effluents, and Surface Water, Reported in the Literature Since 2004

Compound	CAS Number	WWTP Influent Concentration µg/L	Reference	WWTP Effluent Concentration µg/L	Reference	Surface-Water Concentration µg/L	Reference
1,4-dichlorobenzene	106-46-7			nd[a]-0.91 (0.11)	20	nd-0.28 (nd)	20
						0.027-0.098	116
						0.095 max[b]	102
1,7-dimethylxanthine	611-59-6			nd-8.55 (nd)	20	nd-1.76 (nd)	20
						0.294 max	102
17-alpha-ethynyl estradiol	57-63-6					0.1-0.13	117
17-beta-estradoil	50-28-2	0.003-0.022 (0.009)	45	nd-0.002 (0.002)	45	0.1 only[d]	117
		0.0081 med[c]	118			nd-0.005	119
1-methylnaphthalene	90-12-0			nd-0.1 (nd)	20	nd-0.095 (nd)	20
2-methylnaphthalene	91-57-6			nd-0.06 (nd)	20	nd-0.061 (nd)	20
						0.056 max	102
3,4-dichlorophenyl isocyanate	102-36-3			0.047-0.32 (0.15)	20	nd-0.28 (0.047)	20
						0.055-0.23	112

[a] nd = not detected
[b] max = maximum concentration reported
[c] med = median concentration reported
[d] only = only measured concentration reported
[e] ave = average concentration reported

Note: Entries in **bold** are concentrations reported for multiple related compartments in individual studies. For concentration ranges, the median (if available) is presented in parentheses.

TABLE 1.1
(Continued)

Compound	CAS Number	WWTP Influent Concentration μg/L	Reference	WWTP Effluent Concentration μg/L	Reference	Surface-Water Concentration μg/L	Reference
4-cumylphenol	599-64-4					nd-0.76	116
4-tert-oct-lphenol	140-66-9	0.38-3.56 (3.08)	45	0.01-0.47 (0.06)	45	0.02-0.19	117
				nd-1.1 (nd)	20	nd-0.29 (nd)	20
						nd-0.18	116
						0.22 max	102
5-methyl-1h-benzotriazole	136-85-6			nd-1.7 (0.82)	20	nd-1.1 (nd)	20
						0.27 max	102
acetaminophen	103-90-2	0.13-26.09	23	nd-5.99	23	nd-0.25	23
		5.529-69.57	53	nd-1.06 (0.006)	20	nd-1.78 (nd)	20
				nd-9	40	nd-3.6	40
				0.5-29	120	nd-0.066	121
						0.555 max	103
						0.025-0.065	119
						1.95 max	102
						nd-0.014	112
acetophenone	98-86-2			nd-0.26(nd)	20	nd-0.78 (nd)	20
						0.22 max	102
albuterol	18559-94-9			nd-0.0327 (nd)	20	nd-0.343 (nd)	20
				0.009 med	36	0.001 med	36
						0.003 max	122
						0.268 max	103
alkylpheno	68555-24-8			nd-6.2	37	nd-1.3	37
anthraquinone	84-65-1			nd-0.096	20	nd-0.58 (nd)	20
						0.066 max	102
						nd-0.073	112

Compound	CAS Number	WWTP Influent Concentration µg/L	Reference	WWTP Effluent Concentration µg/L	Reference	Surface-Water Concentration µg/L	Reference
aspirin	50-78-2					nd-0.037	77
atenolol	29122-68-7	nd-0.74	23	nd-1.15	23	nd-0.25	23
		0.03 only	50	0.16 only	50	0.06 max	50
				0.466 med	36	0.017 med	36
				0.19 med	123	0.042 max	122
				0.1-122	120		
azithromycin	83905-01-5	nd-0.3	23	0.05-0.21	23	nd-0.02	23
		0.09-0.38 (0.17)	87	0.08-0.4 (0.16)	87		
		0.26ave^e	78	0.085-0.255	124		
				0.015-0.066	113		
bezafibrate	41859-67-0	nd-0.05	23	nd-0.01	23	nd-0.01	23
		2.775 ave	46	0.565 ave	46	0.847 ave	46
		1.738 ave	47	0.018 ave	47	nd-0.088	121
		2.2 ave	125	0.14 ave	125	0.003 max	122
		2.6 ave	48	0.24 ave	48	0.78 max	104
		1.55-7.6	49	nd-4.8	49		
		1.9 only	126	0.0548 med	36	0.002 med	36
				nd-0.81	40	nd-0.47	40
				0.004-0.024	38	nd-0.004	38
benzophenone	119-61-9			0.081-0.61 (0.2)	20	nd-0.51 (0.06)	20
						nd-0.16	116
						0.11 max	102
						nd-0.17	112

(Continued)

TABLE 1.1
(Continued)

Compound	CAS Number	WWTP Influent Concentration μg/L	Reference	WWTP Effluent Concentration μg/L	Reference	Surface-Water Concentration μg/L	Reference
bha (butylated hydroxyanisole)	25013-16-5			**nd-0.32 (nd)**	20	**nd-0.23 (nd)**	20
bht (butylated hydroxytoluene)	128-37-0	**2.53 only**	50	**0.61 only**	50	**0.47 max**	50
bisphenol A	80-05-7	**0.72-2.376**	49	**0.026-1.53**	49	nd-0.114	121
		0.21-2.4 (1.28)	45	**0.02-0.45 (0.18)**	45	0.01-0.02	117
				nd-0.31 (0.12)	20	**nd-0.3 (nd)**	20
						nd-0.147	119
						0.7 max	102
						nd-0.23	112
bromacil	314-40-9			**nd-0.69 (nd)**	20	**nd-0.34 (nd)**	20
						nd-0.79	116
						0.39 max	102
bromoform	75-25-2			**nd-0.22 (nd)**	20	**nd-0.62 (nd)**	20
						nd-0.041	116
						0.16 max	102
						nd-0.071	112
caffeine	58-08-2	**3.69 only**	50	**0.22 only**	50	**0.11 max**	50
		16.3 max	51	**4.52 max**	51	nd-0.88	116
		42.4-43.8	127	**0.013-0.036**	127	nd-0.038	119
				nd-7.99 (0.0532)	20	**nd-2.6 (0.0458)**	20
				nd-9.9	22	**nd-0.31**	22
				1.742-8.132	39	**nd-1.59**	39
				0.46-1.56	128	0.428-9.7	77
				0.036 ave	129	1.39 max	102
						0.021-0.055	112

Compound	CAS Number	WWTP Influent Concentration μg/L	Reference	WWTP Effluent Concentration μg/L	Reference	Surface-Water Concentration μg/L	Reference
camphor	76-22-2			nd-0.13 (nd)	20	nd-0.084 (nd)	20
						0.084	102
carbamazepine	298-46-4	1.68 only	50	1.18 only	50	0.5 max	50
		nd-0.95	23	nd-0.63	23	nd-0.11	23
		2.1 max	51	0.75 max	51	nd-7.1	121
		1.45 med	130	1.65 med	130	0.043 max	10
		0.356 ave	52	0.251 ave	52	1.15 max	104
		0.325-1.85	49	0.465-1.594	49	0.043-0.114	119
				nd-0.27 (0.0802)	20	nd-0.186 (0.03)	20
				0.291 med	36	0.023 med	36
				nd-0.24	40	nd-0.17	40
				nd-0.059	37	nd-0.075	77
				0.42 (only)	75	nd-0.024	131
				0.44 med	123	0.263 max	102
				0.03-0.07	120	0.0002-0.016	132
						0.044-0.13	112
carbaryl	63-25-2					nd-0.22 (nd)	20
						nd-0.076	116
chlortetracycline	64-72-2	0.26	133			nd-0.16	133
						0.192 only	131
						0.1 max	102
cholesterol	57-88-5			nd-8.7 (2)	20	nd-8 (0.89)	20
						nd-1.6	116
						4.3 max	102

(Continued)

TABLE 1.1
(Continued)

Compound	CAS Number	WWTP Influent Concentration µg/L	Reference	WWTP Effluent Concentration µg/L	Reference	Surface-Water Concentration µg/L	Reference
cimetidine	51481-61-9			nd-0.426 (nd)	20	nd-0.354 (nd)	20
						0.338 max	102
ciprofloxacin	85721-33-1	0.09-0.3	134	nd-0.06	134	0.026 max	122
		0.228 max	100	0.054 max	100	nd-0.039	116
		nd-0.21	135	nd-0.14	135	0.03 max	102
		3.6-101	136	0.251 med	36	0.014 only	36
				0.091-5.6	22	0.031-0.36	22
				nd-0.37	137		
clarithromycin	81103-11-9	0.33-0.6 (0.38)	87	0.11-0.35 (0.24)	87	0.02 max	122
		0.647	78	0.018 med	36	0.002 med	36
				0.22-0.329	124		
clindamycin	18323-44-9			nd-1	22	nd-0.14	22
clofibric acid	882-09-7	nd-0.36	23	0.02-0.03	23	0.01-0.02	23
		0.163 ave	46	0.109 ave	46	0.279 ave	46
		0.098 ave	47	0.023 ave	47	nd-0.022	121
		nd-0.651	53	nd-0.044	53	0.003-0.027	119
		0.34 only	126	nd-0.038	40		
clotrimazole	23593-75-1	0.023-0.033	53	0.01-0.027	53	0.006-0.034	53
codeine	76-57-3			nd-0.73 (0.139)	20	nd-0.217 (0.0092)	20
				0.01-5.7	120	nd-0.054	77
						0.119 max	102
coprostanol	360-68-9			nd-5.9 (1.3)	20	nd-5.6 (0.26)	20
						nd-1.5	116
						1.3 max	102

Compound	CAS Number	WWTP Influent Concentration μg/L	Reference	WWTP Effluent Concentration μg/L	Reference	Surface-Water Concentration μg/L	Reference
cotinine	486-56-6			nd-1.03 (0.024)	20	nd-0.481 (0.0183)	20
						nd-0.052	116
						0.528	102
						0.013-0.024	112
dehydronifedipine	67035-22-7			nd-0.0214 (0.0112)	20	nd-0.0216 (0.0026)	20
						0.01 max	102
						0.0051-0.018	112
demeclocyclin	127-33-3	1.14 only	133	0.09 only	133	nd-0.53	133
dextropropoxyphene	469-62-5	0.022-0.033	53	0.037-0.064	53	nd-0.098	53
				nd-0.585 (0.195)	21	nd-0.682 (0.058)	21
diazepam	439-14-5	0.31 only	126			nd-0.034	77
dibutyl phthalate	84-74-2	0.15 only	50	0.03 only	50	0.06 max	50
						0.11-6.58	117
diisobutyl phthalate	84-69-5	0.04 only	50	0.01 only	50	0.02 max	50
diclofenac	15307-86-5	0.16 only	50	0.12 only	50	0.12 max	50
		0.05-0.54	23	nd-0.39	23	nd-0.06	23
		2.333 ave	46	1.561 ave	46	0.272 ave	46
		1.532 ave	47	0.437 ave	47	nd-0.069	121
		0.46 ave	125	0.4 ave	125	nd-0.282 (nd)	74
		0.28 ave	48	1.9 ave	48		
		0.905-4.11	49	0.78-1.68	49		
		0.901-1.036	53	0.261-0.598	53		
		0.14 med	118	0.14 med	118		
		0.05-2.45 (0.17)	45	0.07-0.25 (0.11)	45		

(Continued)

TABLE 1.1
(Continued)

Compound	CAS Number	WWTP Influent Concentration µg/L	Reference	WWTP Effluent Concentration µg/L	Reference	Surface-Water Concentration µg/L	Reference
diclofenac *(cont'd)*		0.33–0.49	127	nd-2.349 (0.424)	21	nd-0.568 (nd)	21
		4.1 only	126	nd-0.5	40	nd-0.089	40
				0.011-0.04	38	nd-0.003	38
				0.032-0.457	39		
				0.29 med	123		
				0.06-1.9	120		
diethyl phthalate	84-66-2	0.19 only	50	0.02	50	0.03 max	50
				nd-0.71 (nd)	20		
diethylhexyl phthalate	117-81-7	0.27 only	50	0.02	50	0.04 max	50
				nd-27 (nd)	20	nd-7.5 (nd)	20
						nd-12.74	117
diltiazem	42399-41-7			nd-0.146 (0.0491)	20	nd-0.0736 (0.0042)	20
						0.106 max	102
diphenhydramine	58-73-1			nd-0.387 (0.0784)	20	nd-0.273 (nd)	20
						0.023 max	102
						nd-0.0058	112
d-limonene	5989-27-5					0.029 only	116
						0.94 max	102
doxycycline	564-25-0	0.22 only	133	0.09 only	133	nd-0.08	133
		nd-2.21	134	nd-0.88	134	nd-0.073	131
		0.6-6.7	136				

Compound	CAS Number	WWTP Influent Concentration µg/L	Reference	WWTP Effluent Concentration µg/L	Reference	Surface-Water Concentration µg/L	Reference
enalaprilat	76420-72-9					0.0005 max	122
						0.0001 med	36
enrofloxacin	93106-60-6					0.01 max	102
erythromycin	114-07-8	0.71-0.141	53	0.1 max	128	nd-0.07	53
				0.145-0.29	53	0.003 med	36
				0.047med	36	nd-1.022 (nd)	21
				nd-1.842 (nd)	21	0.016 max	122
				0.01-0.03	120	nd-0.175	116
						nd-0.051	131
						nd-0.03	23
						nd-0.007	132
erythromycin-H2O (anyhydro-erythromycin)	114-07-8	0.2 ave	138	0.08 ave	138	0.17 only	138
		0.06-0.19 (0.07)	87	0.06-0.11 (0.07)	87	nd-1.209	116
		nd-1.2	135	nd-0.3	135	0.22 max	102
		0.09-0.35	139	0.04-0.12	139		
				nd-0.48 (0.15)	20	nd-0.61 (nd)	20
				0.055-0.075	124		
estrone	53-16-7	0.032 med	118	0.013 med	118	nd-0.022 (0.004)	7
		0.008-0.052 (0.016)	45	nd-0.054 (0.005)	45	nd-0.005	119
		2.4 max	140	4.4 max	140		
estrone-3-sulfate	481-97-0					nd-0.007 (0.006)	7
ethyl citrate	77-93-0			0.11-0.52 (0.27)	20	nd-0.4 (0.072)	20
						nd-0.27	116
						0.17 max	102
fluoxetine	54910-89-3	0.0004-0.0024	141	nd-0.0013	141	nd-0.0212 (nd)	20

(Continued)

TABLE 1.1
(Continued)

Compound	CAS Number	WWTP Influent Concentration μg/L	Reference	WWTP Effluent Concentration μg/L	Reference	Surface-Water Concentration μg/L	Reference
furosemide	54-31-9			**0.585 med**	36	**0.0035 med**	36
						0.067 max	122
gadolinium	7440-54-2			**0.0023-0.14 (0.094)**	142	**0.026-0.076**	142
galaxolide (HHCB)*	1222-05-5	**0.79 only**	50	**1.08 only**	50	**0.23 max**	50
		0.83-4.443	49	**0.451-0.87**	49	0.106 only	10
		1.701 med	118	**0.876 med**	118	nd-1.4	116
		2.1-3.4	140	**0.49-0.6**	140	0.172-0.313	77
				0.66-2.6 (1)	20	**nd-2.1 (0.14)**	20
						1.2 max	102
						0.43-1.1	112
gemfibrozil	25812-30-0	**0.71 only**	50	**0.18 only**	50	**0.17 max**	50
		nd-0.36	23	**nd-0.32**	23	**nd-0.06**	23
		0.418 med	118	**0.255 med**	118	nd-0.027	121
		0.12-36.53 (0.26)	45	**0.08-2.09 (0.19)**	45	0.014 ave	131
		0.19 only	126	**0.11-1.4**	40	**nd-0.58**	40
				nd-0.158	37	**nd-0.18**	37
				0.08-0.478	39	**nd-0.035**	39
				0.56 med	123	nd-0.013	132
hydrochlorothiazide	58-93-5			**0.439 med**	36	**0.005 med**	36
						0.024 max	122

*The original USGS publications (References 20, 102, 112, and 116) mislabeled galaxolide and tonalide. The concentrations presented here are correct. See http://nwql.usgs.gov/Public/tech_memos/nwql.07-03.html (accessed August 13, 2007) for a detailed explanation.

Compound	CAS Number	WWTP Influent Concentration µg/L	Reference	WWTP Effluent Concentration µg/L	Reference	Surface-Water Concentration µg/L	Reference
ibuprofen	15687-27-1	**3.59 only**	50	**0.15 only**	50	**0.22 max**	50
		nd-0.9	23	**0.04-0.8**	23	**nd-0.15**	23
		7.741-33.764	53	**1.979-4.239**	53	**0.144-2.37**	53
		143 max	51	**10.1 max**	51	nd-0.146	121
		23.4 ave	125	**0.04 ave**	125	0.01 max	122
		9.5-14.7	127	**0.01-0.022**	127	0.014 only	10
		5.7 ave	48	**0.18 ave**	48	3.08 max	103
		1.2-2.679	49	**nd-2.4**	49	5.6 max	104
		8.84 med	118	**0.353 med**	118	nd-0.034	119
		4.1-10.21 (6.77)	45	**0.11-2.17 (0.31)**	45	nd-0.115	77
		2.64-5.7	140	0.91-2.1	140		
		5.518 ave	47	**0.121 med**	36	**0.013 med**	36
		5.533 ave	46	**nd-27.256 (3.086)**	21	**nd-5.044 (0.826)**	21
		3.8 only	126	**nd-22**	40	**nd-6.4**	40
				0.005-0.425	37	**0.003-0.25**	37
				2.235-6.718	39	**nd-0.0095**	39
				0.0035-0.064	38	**nd-0.014**	38
				0.018 ave	129		
				0.11 med	123		
				1.1-151	120		
				0.03 only	75		
indole	120-72-9			nd-0.2 (nd)	20	nd-0.026	116

(Continued)

TABLE 1.1
(Continued)

Compound	CAS Number	WWTP Influent Concentration µg/L	Reference	WWTP Effluent Concentration µg/L	Reference	Surface-Water Concentration µg/L	Reference
indomethacin	53-86-1	**0.151 ave**	46	**0.091 ave**	46	**0.066 ave**	46
		0.196 med	118	**0.149 med**	118	nd-0.01	23
		0.03-0.43 (0.28)	45	**0.04-0.49 (0.18)**	45		
		0.22 only	126	**nd-0.31**	40	**nd-0.15**	40
iopromide	73334-07-3	**nd-3.84**	49	**nd-5.06**	49		
		6.6 max	140	**9.3 max**	140	0.011 max	10
		0.13 only	126				
isopropylphenazone	479-92-5					nd-0.033	121
ketoprofen	22071-15-4	**0.94 only**	50	**0.33 only**	50	**0.07 max**	50
		0.321 ave	46	**0.146 ave**	46	**0.329 ave**	46
		2.1 max	51	**1.76 max**	51		
		2.9 ave	125	**0.23 ave**	125		
		0.41-0.52	127	**0.08-0.023**	127		
		0.47 ave	48	**0.18 ave**	48		
		0.16-0.97	23	**0.13-0.62**	23		
		0.136 med	118	**0.114 med**	118		
		0.06-0.15 (0.08)	45	**0.04-0.09 (0.05)**	45		
				nd-0.31	40	**nd-0.079**	40
				nd-0.029	37	**nd-0.01**	37
				nd-0.039	38	**nd-0.023**	38
				0.008-0.351	39		
				0.023 ave	129		

Compound	CAS Number	WWTP Influent Concentration µg/L	Reference	WWTP Effluent Concentration µg/L	Reference	Surface-Water Concentration µg/L	Reference
lincomycin	154-21-2			**0.031 med**	36	**0.033 med**	36
						0.249 max	122
						nd-0.355	131
						0.01 max	102
						nd-0.046	132
loratadine	79794-75-5					nd-0.02	23
mefenamic acid	61-68-7	**nd-0.005**	23	**nd-0.01**	23	**nd-0.003**	23
		0.136-0.363	53	0.29-0.396	53	0.242 max	103
metformin	657-24-9			**nd-1.44 (0.133)**	21	**nd-0.366 (0.062)**	21
methyl salicylate	119-36-8			**nd-0.698 (nd)**	20	**nd-0.112 (nd)**	20
						nd-0.099 (nd)	20
						0.19 max	102
metoprolol	37350-58-6	**0.16 only**	50	**0.19 only**	50	**0.07 max**	50
				0.08 med	123		
monensin	22373-78-0					nd-0.22	132
n,n-diethyl-m-toluamide (DEET)	134-62-3			**nd-2.1 (0.18)**	20	**nd-0.64 (0.088)**	20
						1.13 max	143
						nd-0.083	116
						0.13 max	102
						0.051-0.099	112
n4-acetyl-sulfamethoxazole	21312-10-7	**0.85-1.6 (1.4)**	87	**0.21-0.88 (0.4)**	87		
				0.071-0.082	124		
naphthalene	91-20-3			nd-0.15 (nd)	20	nd-0.16 (nd)	20
						nd-0.02	116
						0.082 max	102

(Continued)

TABLE 1.1
(Continued)

Compound	CAS Number	WWTP Influent Concentration µg/L	Reference	WWTP Effluent Concentration µg/L	Reference	Surface-Water Concentration µg/L	Reference
naproxen	22204-53-1	**3.65 only**	50	**0.25 only**	50	**0.25 max**	50
		nd-0.19	23	**nd-0.16**	23	**nd-0.05**	23
		11.4 max	51	**3.12 max**	51	nd-0.032	121
		8.6 ave	125	**0.42 ave**	125	nd-0.313 (0.021)	74
		10.3-12.8	127	**0.012-0.038**	127	nd-0.135	119
		0.95 ave	48	**0.27 ave**	48	0.042 ave	131
		5.22 med	118	**0.351 med**	118	nd-0.045	38
		0.732 ave	46	**0.261 ave**	46	nd-0.041	132
		1.73-6.03 (2.76)	45	**0.36-2.54 (0.82)**	45		
		1.79-4.6	140	**0.8-2.6**	140		
		0.806 ave	47	**0.17 only**	75	**0.03 only**	75
		0.54 only	126	**nd-14**	40	**nd-4.5**	40
				nd-0.172	37	**nd-0.105**	37
				0.633-7.962	39	**nd-0.271**	39
				0.031 ave	129		
				0.41 med	123		
				0.017-0.057	38		
norfloxaxin	70458-96-7	**0.319 max**	100	**0.071 max**	100	0.03 max	102
		0.066-0.174	134	**nd-0.037**	134		
nonylphenol monoethoxylate		**4.06-7.299**	49	nd-2.58	49		
				nd-18 (0.88)	20	**nd-12 (0.35)**	20
nonylphenol diethoxylate		**0.6-4.645**	49	**nd-1.36**	49	nd-7.4	116
				nd-38 (2.2)	20	**nd-15 (0.56)**	20
						2.5 max	102

Compound	CAS Number	WWTP Influent Concentration µg/L	Reference	WWTP Effluent Concentration µg/L	Reference	Surface-Water Concentration µg/L	Reference
ofloxacin	82419-36-1	nd-0.287	134	nd-0.052	134	nd-0.109	116
		0.2-7.6	136	0.6 med	36	0.0331 med	36
				0.0175-0.186	76		
				(0.0518)	144		
				0.1 only	137		
				nd-0.35			
octylphenol monoethoxylate		0.042-0.66	49	nd-0.47	49	nd-1.1	116
				nd-1.9 (nd)	20		
octylphenol diethoxylate		nd-0.114	49	nd-0.15	49	nd-0.68	116
				nd-0.36 (0.12)	20	nd-0.34 (nd)	20
oxytetracycline	79-57-2	0.33 only	133			nd-0.13	133
palmitic acid	57-10-3	35.91 only	50	0.71 only	50	0.78 max	50
para-nonylphenol-total	84852-15-3	1.14 only	50	0.34 only	50	0.2 max	50
		1.28-4.031	49	0.285-0.482	49	nd-0.18	117
		2.72-25 (14.6)	45	0.32-3.21 (0.07)	45	nd-1.2	116
				nd-22 (nd)	20	nd-12 (nd)	20
						0.88 max	102
phenazone	60-80-0	0.35 med	130	0.33 med	130	nd-0.085	121
propranolol	525-66-6	0.05 only	50	0.03 only	50	0.01 max	50
		0.06-0.119	53	0.195-0.373	53	0.037-0.107	53
		nd-0.25	145	0.003-0.16	145	nd-0.032	145
		0.08-0.29	23	0.1-0.47	23		

(Continued)

TABLE 1.1
(Continued)

Compound	CAS Number	WWTP Influent Concentration µg/L	Reference	WWTP Effluent Concentration µg/L	Reference	Surface-Water Concentration µg/L	Reference
propranolol *(cont'd)*				0.016-0.284 (0.076)	21	nd-0.215 (0.029)	21
				0.2-6.5	120		
propyphenazone	479-92-5	0.12 med	130	0.13 med	130		
ranitidine	66357-35-5	nd-0.29	23	nd-0.2	23	nd-0.01	23
				nd-0.295 (nd)	20	nd-0.16(nd)	20
				0.288 med	36	0.001 med	36
				0.4-1.7	120	0.004 max	122
						0.027 max	102
roxithromycin	80214-83-1	0.01-0.04 (0.02)	87	0.01-0.03 (0.01)	87	0.036 max	10
		0.025-0.117	49	0.036-0.069	49	nd-0.002	131
				0.01-0.023	124	0.06 only	138
salicylic acid	69-72-7	0.861 ave	46	0.092 ave	46	1.098 ave	46
		14.1 med	118	0.104 med	118		
		2.82-12.7 (6.86)	45	0.01-0.32 (0.14)	45		
sitosterol	83-46-5			nd-2.9 (1.1)	20	nd-2.9 (0.71)	20
						2.9 max	102
sotalol	3930-20-9	0.12-0.2	23	nd-0.21	23	nd-0.07	23
spiramycin	8025-81-8			0.075 med	36	0.01 med	36
						0.044 max	122
stearic acid	57-11-4	41 only	50	0.8 only	50	0.84 max	50
stigmastanol	19466-47-8			nd-0.81(nd)	20	nd-1.2 (nd)	20
						nd-1.6	116

Compound	CAS Number	WWTP Influent Concentration µg/L	Reference	WWTP Effluent Concentration µg/L	Reference	Surface-Water Concentration µg/L	Reference
sulfachloropyradizine	80-32-0					nd-0.007	131
						nd-0.06	133
						nd-0.02	132
sulfadiazine	68-35-9					0.236 only	146
sulfadimethoxine	122-11-2	0.34 only	133			nd-0.09	133
						0.028 only	146
						nd-0.004	116
						nd-0.056	131
						0.01 max	102
sulfamerazine	127-79-7						
sulfamethazine	57-68-1	0.68 only	133	nd-0.019	124	nd-0.19	133
		nd-0.21	135			nd-0.22	133
						nd-0.408	131
						nd-0.038	132
sulfamethoxazole	723-46-6	0.02 only	50	0.07 only	50	0.01 max	50
		0.52 only	133	0.13 only	133	nd-0.12	133
		0.23-0.57 (0.43)	87	0.211-0.86 (0.29)	87	0.402 only	146
		nd-1.25	135	nd-0.37	135	0.202 max	10
		nd-0.145	49	nd-0.091	49	nd-0.5	116
		nd-0.87	23	nd-0.82	23	nd-0.009	131
		nd-0.674	134	nd-0.304	134	0.07 max	102
		0.58 max	140	0.25 max	140	nd-0.034	112
		0.4-12.8	136	nd-0.589 (0.15)	20	nd-0.763 (0.0279)	20
				0.37-6	22	0.043-0.45	22

(Continued)

TABLE 1.1
(Continued)

Compound	CAS Number	WWTP Influent Concentration µg/L	Reference	WWTP Effluent Concentration µg/L	Reference	Surface-Water Concentration µg/L	Reference
sulfamethoxazole (*cont'd*)				0.352 max	124		
				1.3 max	128		
				0.127 med	36		
				nd-0.132 (nd)	21		
				nd-1.6	137		
sulfapyridine	144-83-2	**0.06-0.15 (0.09)**	87	**0.04-0.35 (0.09)**	87	0.121 only	146
sulfathiazole	72-14-0					nd-0.016	131
tamoxifen	54965-24-1	**0.143-0.215**	53	**0.146-0.369**	53	**0.027-0.212**	53
				nd-0.042 (nd)	21		
tetracycline	60-54-6	**0.45 only**	133	**0.19 only**	133	**nd-0.14**	133
		nd-1.2	135	**nd-0.85**	135	0.03 max	102
				0.62 max	128		
				nd-0.56	22		
thiabendazole	148-79-8			**nd-0.0063 (nd)**	20	**nd-0.515 (nd)**	20
tonalide (AHTN)*	1506-02-1	**0.21-1.106**	49	**0.092-0.17**	49	0.01 max	10
		0.687 med	118	**0.298 med**	118	nd-0.24	116
		0.9-1.69	140	**0.15-0.2**	140	0.081-0.106	77
				0.11-0.53	20	**nd-0.35 (nd)**	20
				(0.28)		0.56 max	102
						0.012-0.32	112

*The original USGS publications (References 20, 102, 112, and 116) mislabeled galaxolide and tonalide. The concentrations presented here are correct. See http://nwql.usgs.gov/Public/tech_memos/nwql.07-03.html (accessed August 13, 2007) for a detailed explanation.

Compound	CAS Number	WWTP Influent Concentration µg/L	Reference	WWTP Effluent Concentration µg/L	Reference	Surface-Water Concentration µg/L	Reference
tri(2-chloroethyl)phosphate	115-96-8			0.13-0.43 (0.33)	20	**nd-0.48 (0.1)**	20
						nd-0.7	116
						0.25 max	102
						0.089-0.28	112
tri(dichlorisopropyl) phosphate	13674-87-8			**0.2-0.48 (0.3)**	20	**nd-0.39 (0.075)**	20
						nd-0.4	116
						0.4	102
						0.11-0.32	112
tributylphosphate	126-73-8			**0.074-0.47 (0.18)**	20	**nd-0.34 (0.083)**	20
						nd-0.56	116
						0.14	102
						nd-0.26	112
triclosan	3380-34-5	0.38 only	50	**0.16 only**	50	**0.07 max**	50
		2.7-26.8	147	**0.03-0.25**	147	**0.03-0.29**	147
		0.121-13.9 (0.382)	148	**0.321 max**	148	nd-0.25	116
		3-3.6	127	**0.028-0.072**	127	0.009-0.035	119
		1.86 med	118	**0.106 med**	118	nd-0.057	77
		0.87-1.83 (1.35)	45	**0.05-0.36 (0.14)**	45	0.14 max	102
				0.097-1.6 (0.25)	21	**nd-1 (nd)**	21
				0.08 only	149	**0.008 max**	149
				0.072 ave	129	0.028-0.12	112
				0.18 med	123		

(Continued)

TABLE 1.1
(Continued)

Compound	CAS Number	WWTP Influent Concentration μg/L	Reference	WWTP Effluent Concentration μg/L	Reference	Surface-Water Concentration μg/L	Reference
trimethoprim	738-70-5	0.08 only	50	0.04 only	50	0.02 max	50
		nd-4.22	23	0.07-0.31	23	nd-0.02	23
		0.213-0.3	53	0.218-0.322	53	0.004-0.019	53
		0.21-0.44 (0.29)	87	0.02-0.31 (0.07)	87	nd-0.19	116
		1.86 max	100	1.88 max	100	nd-0.015	131
		0.05-1.3	135	nd-0.55	135	0.08 max	102
		0.099-1.3	134	0.066-1.34	134	nd-0.002	132
		0.6-7.6	136	nd-0.353 (0.0376)	20	nd-0.414 (0.0068)	20
				nd-1.288 (0.07)	21	nd-0.042 (nd)	21
				nd-0.53	22	nd-0.31	22
				0.068-0.081	124		
				0.12-0.16	128		
				0.03-1.22	137		
				0.05 med	123		
				0.01-0.03	120		

Compound	CAS Number	WWTP Influent Concentration µg/L	Reference	WWTP Effluent Concentration µg/L	Reference	Surface-Water Concentration µg/L	Reference
triphenyl phosphate	115-86-6			nd-0.18 (0.072)	20	nd-0.096 (nd)	20
						nd-0.03	116
						0.12 max	102
						0.012-0.066	112
tris(2-butoxyethyl) phosphate	78-51-3	9.44 only	50	1.89 only	50	1.21 max	50
						0.87 max	102
						nd-0.46	112
tris(2-chloroisopropyl) phosphate	13674-84-5	2.79 only	50	2.26 only	50	1.13 max	50
tylosin	1401-69-0	1.15 ave	138	0.06 ave	138	nd-0.13	138
		0.06-0.18	139	nd-0.05	139	nd-0.012	116

TABLE 1.2

Concentrations of Pharmaceuticals and Other Emerging Contaminants in Sludges and Biosolids, Sediments, Groundwater, and Drinking Water, Reported in the Literature Since 2004

Compound	CAS Number	Sludge and Biosolid Concentration mg/kg	References	Sediment Concentration mg/kg	References	Groundwater Concentration µg/L	References	Drinking-Water Concentration µg/L	References
1,7-dimethylxanthine	611-59-6					0.022 max[a]	6	nd[b]-0.0645	31
17-beta-estradiol	50-28-2			0.0089 only[c]	17				
acetaminophen	103-90-2					0.015 max	6	0.0081-0.0653	31
acetophenone	98-86-2								
albuterol	18559-94-9							nd-0.0114	31
alkylphenol	68555-24-8								
anthraquinone	84-65-1							0.072 max	54
azithromycin	83905-01-5	64 ave[d]	87						
		0.052-0.127	150						
bezafibrate	41859-67-0					nd-0.19	8		
						nd-0.01	9		
benzophenone	119-61-9	3.4 only	33					0.02 ave	106
								0.13 max	54
bha (butylated hydroxyanisole)	25013-16-5	0.3 only	33					0.23	106
bht (butylated hydroxytoluene)	128-37-0	4.9 only	33						
bisphenol A	80-05-7	1.090-14.400[e]	19	0.0248 only	17	nd-0.007	7	0.005 only	7
						nd-0.84	12	0.42 max	54
bromoform	314-40-9							21	54

Compound	CAS Number	Sludge and Biosolid Concentration mg/kg	References	Sediment Concentration mg/kg	References	Groundwater Concentration µg/L	References	Drinking-Water Concentration µg/L	References
caffeine	58-08-2					0.12 max	6	0.00197-0.0172	31
								0.119 max	54
camphor	76-22-2								
carbamazepine	298-46-4	69.6 ave	52			0.071 max	10	0.0431-0.0936	31
		0.019 only	151			0.11-1.19	8	0.258 max	54
		0.015-1.20[e]	19			nd-0.5	9		
cholesterol	57-88-5	19.1-402e	19			0.022-0.044	12		
ciprofloxacin	85721-33-1	7.7 max	100						
		0.5-4.8	134						
clarithromycin	81103-11-9	67 ave	87						
		0.027-0.063	150						
clofibric acid	882-09-7	0.087 only	151						
codeine	76-57-3					nd-0.29	11	nd-0.0587	31
						nd-0.125	9		
						0.08 max	6		
coprostanol	360-68-9	8.1-1460[e]	19			nd-0.074	12		
cotinine	486-56-6					0.06 max	6	0.0165-0.0618	31
						nd-0.13	12	0.025 max	54

[a] max = maximum concentration reported
[b] nd = not detected
[c] only = only measured concentration reported
[d] ave = average concentration reported
[e] concentrations were carbon normalized

Note: Entries in **bold** are concentrations reported for multiple related compartments in individual studies. For concentration ranges, the median (if available) is presented in parentheses.

(Continued)

TABLE 1.2
(Continued)

Compound	CAS Number	Sludge and Biosolid Concentration mg/kg	References	Sediment Concentration mg/kg	References	Groundwater Concentration µg/L	References	Drinking-Water Concentration µg/L	References
dehydronifedipine	67035-22-7					0.003 max	6	0.00124-0.00358	31
								0.004 max	54
dibutyl phthalate	84-74-2	93.9	33						
diclofenac	15307-86-5	0.31-7.02	151					0.18	106
diethyl phthalate	84-66-2	2.5	33			nd-0.05	11	0.16	106
diethylhexyl phthalate	117-81-7	3.46-31.7^c	19			nd-0.9	8	0.34	106
diltiazem	42399-41-7					nd-0.035	9	0.00217-0.00743	31
diphenhydramine	58-73-1	0.032-22^c	19					0.01-0.0721	31
doxycycline	564-25-0	nd-1.5	134						
erythromycin	114-07-8							0.154-0.611	31
erythromycin-H2O (anhydro-erythromycin)	114-07-8					0.75 max	6		
estrone	53-16-7			0.0018 only	17				
ethyl citrate	77-93-0							0.062 max	54
fluoxetine	54910-89-3	0.10-4.70^c	19			nd-0.17	8	0.00123-0.0054	31
galaxolide (HHCB)*	1222-05-5	7.84	33			0.041 max	10	0.49 max	54
		13.1-187	151						
		0.281-1340^c	19						

*The original USGS publications (References 19 and 54) mislabeled galaxolide and tonalide. The concentrations presented here are correct. See http://nwql.usgs.gov/Public/tech_memos/nwql.07-03.html (accessed August 13, 2007) for a detailed explanation.

Compound	CAS Number	Sludge and Biosolid Concentration mg/kg	References	Sediment Concentration mg/kg	References	Groundwater Concentration µg/L	References	Drinking-Water Concentration µg/L	References
gemfibrozil	25812-30-0							nd-0.0936	31
ibuprofen	15687-27-1	0.12 only	151			0.129 max	6	0.009 max	125
iopromide	73334-07-3					nd-0.012	8	0.12 ave	106
ketoprofen	22071-15-4					nd-0.053	8	0.008	125
lincomycin	154-21-2					nd-0.1	12		
n,n-diethyl-m-toluamide (DEET)	134-62-3					nd-13	12	0.066 max	54
naphthalene	91-20-3								
norfloxaxin	70458-96-7	5.8 max 0.1-0.31	100 134			nd-0.09	12		
nonylphenol monoethoxylate		3.96-79.4e	19			nd-7	12		
nonylphenol diethoxylate		0.793-89.0e	19			nd-10	12		
ofloxacin	82419-36-1	nd-2	134						
octylphenol monoethoxylate						nd-1	12		
octylphenol diethoxylate						nd-0.3	12		
para-nonylphenol-total	84852-15-3	193 only 2.18-1520e	33 19			nd-3	12		
primadone	125-33-7					nd-0.11	9		

(Continued)

TABLE 1.2
(Continued)

Compound	CAS Number	Sludge and Biosolid Concentration mg/kg	References	Sediment Concentration mg/kg	References	Groundwater Concentration µg/L	References	Drinking-Water Concentration µg/L	References
propyphenazone	479-92-5					nd-0.25	11		
						0.005-0.95	9		
roxithromycin	80214-83-1	nd-0.131	150			nd-0.096	8		
stigmastanol	19466-47-8					nd-2	12		
sulfamethoxazole	723-46-6	68 ave	87			0.012 max	10	nd-0.0592	31
		0.041-0.113	150			0.15 max	6		
sulfapyridine	144-83-2	28 ave	87						
		0.011-0.16	150						
thiabendazole	148-79-8							0.00567-0.0727	31
tonalide (AHTN)*	1506-02-1	10.2-183	151			0.026 max	10	0.49 max	54
		0.047-554^e	19			nd-0.073	8	0.082 max	54

*The original USGS publications (References 19 and 54) mislabeled galaxolide and tonalide. The concentrations presented here are correct. See http://nwql.usgs.gov/Public/tech_memos/nwql.07-03.html (accessed August 13, 2007) for a detailed explanation.

Compound	CAS Number	Sludge and Biosolid Concentration mg/kg	References	Sediment Concentration mg/kg	References	Groundwater Concentration µg/L	References	Drinking-Water Concentration µg/L	References
tri(2-chloroethyl)-phosphate	115-96-8					nd-0.74	12	0.099 max	54
tri(dichlorisopropyl) phosphate	13674-87-8							0.25 max	54
tributylphosphate	126-73-8							0.1 max	54
triclosan	3380-34-5	1.17-32.9c	19	0.0152 only 0.0027-0.1307	16 16	nd-0.21	12	0.049 ave	106
trimethoprim	21411-53-0	41 ave	87			0.58 max	6	0.00177-0.042	31
warfarin	81-81-2	0.021-0.107	150			0.009 max	6	nd-0.0734	31

of chemicals reported in these papers generally were not assimilated into the tables (the exception was when a range of concentrations was reported, with nondetections being the lower boundary). The number of compounds found in each environmental compartment varied and likely reflects multiple factors, such as variation in the total number of compounds potentially present, the total number of compounds determined by the analytical methods used in these studies, the varying presence of interferences in the different media sampled, and the total number of studies addressing any one environmental compartment or sample matrix. Thus, the total number of compounds mentioned for any environmental compartment or sample matrix is not easily comparable, and the differences in compounds between related or connected environmental compartments should not be used to infer removal or release processes. Sixty-seven chemicals were found in wastewater influents, 105 were detected in the WWTP effluents, and 22 were present in biosolids. In surface-water samples, 124 compounds were measured, with 35 found in groundwater, 4 in sediments, and 35 PPCPs measured in drinking water (Table 1.1 and Table 1.2). Of the 126 compounds measured in at least 1 matrix, 18 were detected in more than 10 different studies. These compounds include:

Analgesics:
- Acetaminophen/Paracetamol—an analgesic and antipyretic used to relieve minor aches, pains, and fevers
- Diclofenac—a nonsteroidal antiinflammatory drug (NSAID) used to reduce pain, inflammation, and stiffness from arthritis, menstrual cramps, and similar conditions
- Ibuprofen—an NSAID used to reduce fever, pain, and inflammation from headaches, toothaches, backaches, and similar conditions
- Ketoprofen—an NSAID used to reduce pain, inflammation, and stiffness from rheumatoid arthritis, osteoarthritis, menstrual cramps, and similar conditions
- Naproxen—an NSAID used to reduce pain, inflammation, and stiffness from rheumatoid arthritis, osteoarthritis, tendonitis, bursitis, and similar conditions

Antibiotics/Antimicrobials:
- Ciprofloxacin—a fluroquinolone antibiotic used to treat infections by gram-positive and gram-negative bacteria
- Erythromycin plus erythromycin-H_2O (anhydro-erythromycin)—a macrolide antibiotic (and its facilely formed dehydration product) primarily used to treat infections from gram-positive bacteria
- Sulfamethoxazole—a sulfonamide antibiotic often prescribed in tandem with trimethoprim and commonly used to treat urinary tract infections, bronchitis, and ear infections
- Trimethoprim—an antibiotic that potentiates the effects of sulfonamide antibiotics
- Triclosan—an antimicrobial compound commonly found in soaps, toothpastes, and other consumer products, such as trash bags and kitchen utensils

Antihyperlipidemics: All are fibrate drugs that decrease the production of cholesterol and triglycerides and increase the production of high-density lipoproteins ("good" cholesterol)
  - Bezafibrate
  - Clofibric acid
  - Gemfibrozil

Fragrances: Synthetic polycyclic musks used to scent detergents, cosmetics, and lotions
  - Galaxolide (HHCB)
  - Tonalide (AHTN)

Other Compounds:
  - Bisphenol A—a precursor to polycarbonate plastics and a component of epoxy resins and other plastics
  - Caffeine—a natural stimulant in coffee, tea, and guarana and added to other beverages or as an over-the-counter stimulant
  - Carbamazepine—an antiepileptic used in the treatment of psychomotor and grand mal seizures, trigeminal neuralgia, schizophrenia, and bipolar disorder

1.3   INTRODUCTION INTO THE ENVIRONMENT

The primary route of entry for the human use of PPCPs into the environment is through wastewater point sources. In the United States about 75% of households are connected to municipal sewers.[80,81] The remaining households discharge their waste through on-site wastewater-treatment (septic) systems. PPCPs enter the wastewater system either through excretion of unmetabolized products, the rinsing off of dermally applied products, the use of surfactants and fragrances for bodily or household cleansing, or though the use of the wastewater system to dispose of excess medications. While PPCPs can enter the environment at pharmaceutical manufacturing and packaging facilities, the presence of these chemicals in typical households allows for more numerous and widespread routes of entry than traditional pollutants.[82] Additionally, the chemical contaminants in wastewater have been shown to enter surface water through combined sewer overflow during wet weather conditions.[83–85] Thus, combined sewer overflows may be an additional route of entry for pharmaceuticals into the environment.

Table 1.1 lists the concentrations of pharmaceuticals and other emerging contaminants in wastewater influents and effluents. Some studies have investigated the removal of PPCPs in much greater detail, but the clearest indicator of total removal efficiency during wastewater treatment is concentration differences between the influent and effluent. Removal efficiency typically refers to removal from the liquid waste stream; biosolids produced from wastewater treatment may be the compartment in which pharmaceuticals removed from the liquid waste stream are sequestered but may not be included in the calculation of removal efficiency. Thus, biosolids have the potential to be sources in the environment.[19,52,86] Removal efficiencies that are discussed in studies need to be evaluated to determine whether a complete mass balance change from influent to effluent has been calculated and that removal efficiency determinations include solid and liquid components of the waste stream in order to accurately assess transport and transformation of pharmaceuticals from source to ultimate sink.

Within this conceptual framework, it is not surprising that the concentrations of chemicals in the influent (untreated) wastewater tend to be higher than any other environmental compartment. Numerous pharmaceuticals are in the range of 1 µg/L (parts-per-billion), and several exceed 10 µg/L; ibuprofen has been measured at more than 140 µg/L.[51] Although wastewater treatment is designed to remove pathogens and nutrients from sewage, pharmaceuticals and other chemicals found in wastewater can be incidentally removed, but the elimination is variable. Some compounds, such as ibuprofen and naproxen, show decreases of an order of magnitude or more resulting from wastewater treatment. Other chemicals, such as carbamazepine and diclofenac, are reduced only minimally. The processes that mediate the removal of pharmaceuticals in wastewater treatment will be discussed in further detail in Chapter 13—"Hormones in Waste from Concentrated Animal Feeding Operations;" Chapter 14—"Treatment of Antibiotics in Swine Wastewater;" Chapter 15—"Removal of Pharmaceuticals in Biological Wastewater Treatment Plants"; and Chapter 16—"Chemical Processes during Wastewater Treatment."

Although treatment may eliminate many chemicals in the liquid effluent discharged from wastewater treatment facilities, it may not completely remove the potential for future environmental entry. Chemicals are moderated in wastewater treatment either through dilution, oxidation by disinfectants, biological degradation, or sorption to the solid materials settled out of the waste stream.[86] Although the concentrations of pharmaceuticals in liquid effluents typically are in the range of low parts-per-billion [low micrograms-per-liter (µg/L)] to high parts-per-trillion [nanograms-per-liter (ng/L)], they are found in biosolids at parts-per-million [milligrams-per-kilogram (mg/kg)] concentrations (Table 1.2). For example, azithromycin was detected by Göbel et al.[87] at median concentrations of 0.17 µg/L and 0.16 µg/L in WWTP influents and effluents, respectively; in biosolids, the concentration was 64 mg/kg, a greater than 375,000-fold difference. The mean concentration of carbamazepine detected by Miao et al.[52] was 0.356 µg/L in untreated wastewater and 0.251 µg/L in WWTP effluents; in biosolids, the concentration was 69.6 mg/kg. Harrison et al.[18] recently reviewed the literature to determine the range of compound classes detected in biosolids. Most of the individual studies evaluated in that review only determined a limited set of PPCPs in the biosolids studied, but in aggregate, hundreds of anthropogenically derived chemicals were identified in one or more studies. Of the pharmaceuticals identified, the antibiotic ciprofloxacin was present at concentrations as high as 4.8 mg/kg, triclosan as high as 15.6 mg/kg, acetaminophen as high as 4.5 mg/kg, ibuprofen as high as 4.0 mg/kg, and gemfibrozil as high as 1.2 mg/kg. Kinney et al.[19] assessed the presence and concentration of 87 pharmaceuticals and other wastewater compounds in 9 different biosolids and biosolid products, detecting 55 in any one biosolid, with 25 compounds present in all 9 biosolids. The carbon normalized concentrations of the pharmaceuticals carbmazepine and fluoxetine ranged between 0.015 and 1.20 mg/kg and 0.100 and 4.70 mg/kg, respectively. The median concentrations in the biosolids were 0.068 mg/kg for carbamazepine and 0.370 for fluoxetine. The fragrances galaxolide and tonalide were detected at carbon normalized concentrations that ranged between 0.281 and 1340 mg/kg and 0.047 and 554 mg/kg, respectively.* The median concentrations were 11.6 mg/kg for galaxolide and 3.9 mg/kg for tonalide.

About 50% of the biosolids produced during wastewater treatment are spread on agricultural lands as a soil amendment, with the remainder disposed of in landfills or destroyed through incineration.[88,89] Pharmaceuticals can be leached by precipitation from these biosolids into surface and groundwater.[86] Veterinary pharmaceuticals have a similar mode of entry into the environment, either from manures directly deposited onto the ground, or seepage from manure lagoons. Several chapters in this book will explore the processes that govern the transferring of pharmaceuticals from solid sources to the environment in further detail. The factors that affect mobility and transport of veterinary pharmaceuticals in soil will be reviewed in Chapter 6— "Sorption and Degradation of Selected Pharmaceuticals in Soil and Manure"; and a case study is presented in Chapter 7—"Mobility of Tylosin and Enteric Bacteria in Soil Columns." Recently, evidences of plant uptake of pharmaceuticals from soil and transformations by crop plants have been reported. Two chapters in this book will address these issues: Chapter 8—"Plant Uptake of Pharmaceuticals from Soil— Determined by ELISA"; and Chapter 9—"Antibiotic Transformations in Plants via Glutathione Conjugation."

1.4   ENVIRONMENTAL PRESENCE AND FATE

Regardless of their route of entry, a pharmaceutical's concentration and persistence in water systems are guided by several physicochemical processes.[90] As shown in Figure 1.3, the fate of PPCPs can be divided into three categories: transport, sequestration, and degradation. The least disruptive category is simple transport of the dissolved and particle-based chemicals from the point of entry. Thus dispersion and any additional dilution from influent tributaries, and groundwater discharge can reduce PPCP concentration in the water column. Turbulent mixing and aeration as a water parcel is transported may further reduce PPCP concentrations for the relatively more volatile compounds, such as the polycyclic musks.

The next fate classification is the sequestration of PPCPs in other environmental compartments. Pharmaceuticals may be transferred without degradation and stored, at least temporarily, in other matrices or compartments through processes such as bioconcentration, sorption, and deposition of particles. It is important to emphasize the potentially transient storage of PPCPs, because it is possible for the chemicals to reenter the water column through the depuration of aquatic life, desorption from sediments, as well as through reintroduction from the atmosphere by particulate deposition or gas exchange. The processes most likely to transform or mineralize PPCPs, and more permanently remove them from the aquatic environment, include photolysis,[91–96] biodegradation,[30,48,97–100] and hydrolysis.[48] In these processes, the

---

* The original USGS publications (Reference 19) mislabeled galaxolide and tonalide. The concentrations presented here are correct. See http://nwql.usgs.gov/Public/tech_memos/nwql.07-03.html (accessed August 13, 2007) for a detailed explanation.

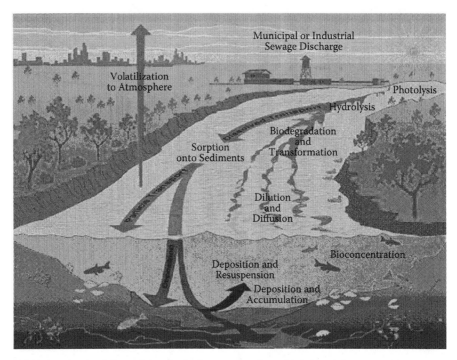

FIGURE 1.3   Depiction of the physical processes that affect the environmental concentrations of PPCPs and other wastewater-derived chemicals. (From Barber et al., U.S. Geological Survey, Reston, VA, 1995.[152] With permission.)

pharmaceuticals either are completely remineralized or are transformed into degradates that may be subject to subsequent remineralization and thus have the potential to be permanently removed from the system. However, these degradates also may be more mobile and persistent than their corresponding parent compound; therefore, they may be more frequently detected in the environment.[101] This potential environmental presence, coupled with similar (or possibly even greater) toxicity than the parent compounds, indicates that degradates should be included as analytes in studies to determine the total environmental impact of PPCPs.

Even after removal, albeit incomplete, from wastewater treatment and the possible additional removal after discharge, pharmaceuticals have been frequently detected in the environment. Table 1.1 lists the concentrations of 124 compounds found in surface-water samples. Ninety-three of the chemicals were found in at least 1 of the 17 studies that concurrently sampled surface water and WWTP influents or effluents, or both (indicated by entries in bold text). These studies show a decrease in the concentration of PPCPs between the WWTP effluents and the surface water, often by an order of magnitude. The resulting surface-water concentrations are typically at sub-part-per-billion concentrations, but some compounds, such as caffeine[20,39,77,102] and ibuprofen,[21,40,103,104] were found in excess of 1 µg/L. Fewer studies have examined the concentrations of pharmaceuticals in groundwater (Table 1.2). The concentrations in groundwater typically are even lower than those found in surface water, usually by about a factor of 10. Nevertheless, a number of PPCPs were detected in groundwater systems sufficiently downgradient from sources to suggest persistence in this environmental compartment.

Groundwater and surface water downstream from wastewater effluent discharges can be the source water used for the genesis of drinking water. The processes involved in the treatment of surface water for drinking water—sedimentation, flocculation, carbon bed filtration, and disinfection—provide additional possible removal[54,55,105,106] or transformation pathways[107–109] for chemicals, but pharmaceuticals still have been detected in finished drinking water (Table 1.2). The infrequent detection of PPCPs in finished drinking water may reflect the relative efficiency of drinking-water treatment technologies to reduce compound concentrations to less-than-detectable levels but also may reflect the paucity of studies performed on drinking water. Methodologies that determine the transformation products of PPCPs during drinking-water treatment at ambient environmental concentrations are needed to fully resolve the effects of drinking-water treatment processes on pharmaceuticals present in source water. Further discussions on drinking-water treatment can be found in Chapter 10—"Drugs in Drinking Water: Treatment Options"; Chapter 11—"Removal of Endocrine Disruptors and Pharmaceuticals during Water Treatment"; and Chapter 12—"Reaction and Transformation of Antibacterial Agents with Aqueous Chlorine under Relevant Water Treatment Conditions."

Several trends are identifiable from this brief overview, and from the related chapters in this volume, that point to future developments in the study of PPCPs in the environment. These developments suggest that new measurement and monitoring approaches need to be used more widely in future studies. For example, one important trend is that many studies have looked at similar suites of compounds, which may reflect similar PPCP use globally, or the tendency to look in more detail at compounds previously identified as frequently detected in other large studies. Also, the simplicity of using existing published methods to begin studies of the presence of pharmaceuticals in the environment may have constrained the variety of compounds detected. However, pharmaceutical prescribing trends change as new products are approved for use, particularly in North America and Europe. The website http://www.rxlist.com provides, by year, a ranking of the top 200 to 300 prescribed pharmaceuticals dispensed in the United States and provides access to results for the last 10 years. A year-by-year examination of the top 20 compounds prescribed in any 1 year illustrates the changes in prescription trends. Some compounds, such as albuterol or furosemide, are consistently present, whereas new pharmaceuticals, such as various antidepressants, may appear in the top 20 as they are introduced and rapidly gain acceptance. Conversely, other compounds, such as premarin and rofecoxib (Vioxx™), drop from the lists of commonly used drugs as adverse health effects are identified that result from their use or as newer, more efficacious compounds are introduced. The dynamics of these prescribing trends suggest the need for ongoing research to improve the analytical methods used to determine dilute concentrations of commonly prescribed compounds and their primary degradates and for the development of methods for newly introduced compounds whose environmental effect may not be known. Current trends in the analysis and sample preparation for pharmaceuticals in aqueous and solid samples are discussed in Chapter 2—"Advances in the Analysis of Pharmaceuticals in the Aquatic Environment" and in Chapter 3—"Sample Preparation and Analysis of Solid-Bound Pharmaceuticals." Methods to evaluate ecotoxicological effects of pharmaceuticals are discussed in Chapter 4—"Gadolinium Containing Contrast Agents for MRI: Investigations on the Environmental Fate and Effects."

Another trend reflected in this review, and in the chapters comprising this volume, is the emphasis on multiple media assessments of PPCPs. A comprehensive mass-balance approach to hydrogeologic, wastewater treatment, and drinking-water treatment systems is necessary to understand whether biotic and abiotic processes that may act upon PPCPs result in true removal or simply transfer them between environmental compartments. Although not explicitly reviewed herein, the determination of PPCPs in biota is a critical matrix that needs to be included in multiple media assessments. Intimately linked to multimedia assessments, and reflected in this review, is the need to monitor not just parent PPCPs but also to determine the degradation or transformation products. This comprehensive precursor-product assessment is necessary if the efficiency of treatment or environmental transformation processes are to be accurately assessed, and it is particularly important if the degraded or transformed PPCP retains pharmacological activity or becomes toxicologically significant by transformation.[110]

Although many of the landmark studies have attempted to define the boundaries of the emerging subdiscipline of the environmental chemistry of PPCPs, these studies provide guidance but do not preclude the need for improvements to analytical methodologies and ongoing monitoring. As noted above, new pharmaceuticals and other PPCPs are being constantly introduced or removed from human use, which reflects a dynamic that will be a continuous hallmark of this research area. As new methods are developed for these nascent chemicals and methods for established chemicals are revised, it would be useful if the analytical methodologies are standardized to allow easier intercomparison between laboratories. International interlaboratory method comparisons also need to be developed to establish a benchmark for existing methods. Monitoring programs that use discrete sample collection will continue to provide valuable information about the discharge and fate of PPCPs into the environment, and the emerging focus on specific biologically active classes will offer insight into their effects. Deployment of integrating samplers into environments likely to result in human or ecosystem exposure, such as drinking-water sources or aquatic habitats known or suspected to contain species sensitive to PPCPs, may be the monitoring strategy that best links the presence of pharmaceuticals in aquatic systems with focused studies of aquatic health effects. Semipermeable membrane devices (SPMDs)[111] and Polar Integrating Chemical Samplers (POCIS)[111–113] are complementary, related integrating sampler technologies that have been demonstrated to concentrate a wide range of environmental contaminants, including pharmaceuticals, endocrine-disrupting compounds, and other PPCP constituents in streams. These samplers exclude particulate-sorbed compounds, but if the assumption is valid that bioavailability to human or aquatic life is primarily through the dissolved phase,[114,115] then POCIS and SPMDs can provide a measure not just of concentration/load but exposure. In addition, transient exposures from events such as the discharge of less-well-treated sewage from combined storm-water overflow systems will be captured by integrating samplers that would not be detected by discrete sample monitoring programs. Substantial research remains to be conducted on how well POCIS and SPMDs work for monitoring PPCPs, particularly the development of uptake rate constants that allow the conversion of sampler PPCP contents to average concentrations during deployment. However, as the studies reviewed herein show, the POCIS and SPMDs are collecting detectable amounts of PPCPs, and better environmental calibration will only expand the use of these samplers.

This review and the subsequent chapters in this book demonstrate that research on the environmental chemistry of PPCPs has rapidly expanded from initial exploratory studies to comprehensive assessments of the presence and concentrations of these compounds. These studies of the presence and concentrations of PPCPs have led to multiple focused studies that address the need to understand the transformation and fate of PPCPs as they are released, move through wastewater-treatment processes, are discharged and transported through the aquatic environment, and ultimately result in exposure to human and aquatic life. As the results of these diverse studies converge, and a more detailed understanding of the processes affecting the concentrations of PPCPs and their degradates develops, it can be envisioned that well-designed comprehensive monitoring approaches will result that integrate exposure to the dynamic environmental PPCP signature and provide water managers and other decision makers with the information to accurately assess and, if necessary, mitigate aquatic and human exposure to PPCPs.

Disclaimer: This paper has been reviewed in accordance with the peer and administrative review policies of the U.S. Environmental Protection Agency and the U.S. Geological Survey. The use of trade, product, or firm names is for descriptive purposes only and does not imply endorsement by the U.S. Government.

REFERENCES

1. Daughton, C.G. and Ternes, T.A., Pharmaceuticals and personal care products in the environment: Agents of subtle change? *Environmental Health Perspectives* 107, 907–938, 1999.
2. Kolpin, D.W., Furlong, E.T., Meyer, M.T., Thurman, E.M., Zaugg, S.D., Barber, L.B., and Buxton, H.T., Pharmaceuticals, hormones, and other organic wastewater contaminants in U.S. streams, 1999–2000: A national reconnaissance, *Environmental Science & Technology* 36 (6), 1202–1211, 2002.
3. Heberer, T., Schmidt-Baumler, K., and Stan, H.J., Occurrence and distribution of organic contaminants in the aquatic system in Berlin. Part 1: Drug residues and other polar contaminants in Berlin surface and groundwater, *Acta Hydrochimica Et Hydrobiologica* 26 (5), 272–278, 1998.
4. Ternes, T.A., Occurrence of drugs in German sewage treatment plants and rivers, *Water Research* 32 (11), 3245–3260, 1998.
5. Buxton, H.T. and Kolpin, D.W., Pharmaceuticals, hormones, and other organic wastewater contaminants in U.S. streams: U.S. Geological Survey Fact Sheet FS-027-02, p. 2, 2002.
6. Verstraeten, I.M., Fetterman, G.S., Meyer, M.T., Bullen, T., and Sebree, S.K., Use of tracers and isotopes to evaluate vulnerability of water in domestic wells to septic waste, *Ground Water Monitoring and Remediation* 25 (2), 107–117, 2005.

7. Rodriguez-Mozaz, S., de Alda, M. J.L., and Barcelo, D., Monitoring of estrogens, pesticides and bisphenol A in natural waters and drinking water treatment plants by solid-phase extraction-liquid chromatography-mass spectrometry, *Journal of Chromatography A* 1045 (1–2), 85–92, 2004.
8. Kreuzinger, N., Clara, M., Strenn, B., and Vogel, B., Investigation on the behaviour of selected pharmaceuticals in the groundwater after infiltration of treated wastewater, *Water Science and Technology* 50 (2), 221–228, 2004.
9. Heberer, T., Mechlinski, A., Fanck, B., Knappe, A., Massmann, G., Pekdeger, A., and Fritz, B., Field studies on the fate and transport of pharmaceutical residues in bank filtration, *Ground Water Monitoring and Remediation* 24 (2), 70–77, 2004.
10. Bruchet, A., Hochereau, C., Picard, C., Decottignies, V., Rodrigues, J.M., and Janex-Habibi, M.L., Analysis of drugs and personal care products in French source and drinking waters: The analytical challenge and examples of application, *Water Science and Technology* 52 (8), 53–61, 2005.
11. Scheytt, T., Mersmann, P., Leidig, M., Pekdeger, A., and Heberer, T., Transport of pharmaceutically active compounds in saturated laboratory columns, *Ground Water* 42 (5), 767–773, 2004.
12. Barnes, K.K., Christenson, S.C., Kolpin, D.W., Focazio, M., Furlong, E.T., Zaugg, S.D., Meyer, M.T., and Barber, L.B., Pharmaceuticals and other organic waste water contaminants within a leachate plume downgradient of a municipal landfill, *Ground Water Monitoring and Remediation* 24 (2), 119–126, 2004.
13. Bound, J.P. and Voulvoulis, N., Household disposal of pharmaceuticals as a pathway for aquatic contamination in the United Kingdom, *Environmental Health Perspectives* 113 (12), 1705–1711, 2005.
14. Slack, R.J., Gronow, J., and Vulvulis, N., Household hazardous waste in municipal landfills: Contaminants in leachate, *Science of the Total Environment* 337 (1–3), 119–137, 2005.
15. Slack, R.J., Zerva, P., Gronow, J.R., and Voulvoulis, N., Assessing quantities and disposal routes for household hazardous products in the United Kingdom, *Environmental Science & Technology* 39 (6), 1912–1919, 2005.
16. Aguera, A., Fernandez-Alba, A.R., Piedra, L., Mezcua, M., and Gomez, M.J., Evaluation of triclosan and biphenylol in marine sediments and urban wastewaters by pressurized liquid extraction and solid phase extraction followed by gas chromatography mass spectrometry and liquid chromatography mass spectrometry, *Analytica Chimica Acta* 480 (2), 193–205, 2003.
17. Morales-Munoz, S., Luque-Garcia, J.L., Ramos, M.J., Fernandez-Alba, A., and de Castro, M.D.L., Sequential superheated liquid extraction of pesticides, pharmaceutical and personal care products with different polarity from marine sediments followed by gas chromatography mass spectrometry detection, *Analytica Chimica Acta* 552 (1–2), 50–59, 2005.
18. Harrison, E.Z., Oakes, S.R., Hysell, M., and Hay, A., Organic chemicals in sewage sludges, *Science of the Total Environment* 367 (2–3), 481–497, 2006.
19. Kinney, C.A., Furlong, E.T., Zaugg, S.D., Burkhardt, M.R., Werner, S.L., Cahill, J.D., and Jorgensen, G.R., Survey of organic wastewater contaminants in biosolids destined for land application, *Environmental Science & Technology* 40 (23), 7207–7215, 2006.
20. Glassmeyer, S.T., Furlong, E.T., Kolpin, D.W., Cahill, J.D., Zaugg, S.D., Werner, S.L., Meyer, M.T., and Kryak, D.D., Transport of chemical and microbial compounds from known wastewater discharges: Potential for use as indicators of human fecal contamination, *Environmental Science & Technology* 39 (14), 5157–5169, 2005.
21. Ashton, D., Hilton, M., and Thomas, K.V., Investigating the environmental transport of human pharmaceuticals to streams in the United Kingdom, *Science of the Total Environment* 333 (1–3), 167–184, 2004.

22. Batt, A.L., Bruce, I.B., and Aga, D.S., Evaluating the vulnerability of surface waters to antibiotic contamination from varying wastewater treatment plant discharges, *Environmental Pollution* 142 (2), 295–302, 2006.

23. Gros, M., Petrovic, M., and Barcelo, D., Development of a multi-residue analytical methodology based on liquid chromatography-tandem mass spectrometry (LC-MS/MS) for screening and trace level determination of pharmaceuticals in surface and wastewaters., *Talanta* 70, 678–690, 2006.

24. Masters, R.W., Verstraeten, I.M., and Heberer, T., Fate and transport of pharmaceuticals and endocrine disrupting compounds during ground water recharge, *Ground Water Monitoring and Remediation* 24 (2), 54–57, 2004.

25. Drewes, J.E., Heberer, T., Rauch, T., and Reddersen, K., Fate of pharmaceuticals during ground water recharge, *Ground Water Monitoring and Remediation* 23, 64–72, 2003.

26. Asano, T. and Cotruvo, J.A., Groundwater recharge with reclaimed municipal wastewater: Health and regulatory considerations, *Water Research* 38, 1941–1951, 2004.

27. Cordy, G.E., Duran, N.L., Bouwer, H., Rice, R.C., Furlong, E.T., Zaugg, S.D., Meyer, M.T., Barber, L.B., and Kolpin, D.W., Do pharmaceuticals, pathogens, and other organic waste water compounds persist when waste water is used for recharge? *Ground Water Monitoring and Remediation* 24 (2), 58–69, 2004.

28. Matamoros, V., Garcia, J., and Bayona, J.M., Behavior of selected pharmaceuticals in subsurface flow constructed wetlands: A pilot-scale study, *Environmental Science & Technology* 39 (14), 5449–5454, 2005.

29. Zuehlke, S., Duennbier, U., Heberer, T., and Fritz, B., Analysis of endocrine disrupting steroids: Investigation of their release into the environment and their behavior during bank filtration, *Ground Water Monitoring and Remediation* 24 (2), 78–85, 2004.

30. Snyder, S.A., Leising, J., Westerhoff, P., Yoon, Y., Mash, H., and Vanderford, B., Biological and physical attenuation of endocrine disruptors and pharmaceuticals: Implications for water reuse, *Ground Water Monitoring and Remediation* 24 (2), 108–118, 2004.

31. Kinney, C.A., Furlong, E.T., Werner, S.L., and Cahill, J.D., Presence and distribution of wastewater-derived pharmaceuticals in soil irrigated with reclaimed water, *Environmental Toxicology and Chemistry* 25 (2), 317–326, 2006.

32. Pedersen, J.A., Soliman, M., and Suffet, I.H., Human pharmaceuticals, hormones, and personal care product ingredients in runoff from agricultural fields irrigated with treated wastewater, *Journal of Agricultural and Food Chemistry* 53 (5), 1625–1632, 2005.

33. Buyuksonmez, F. and Sekeroglu, S., Presence of pharmaceuticals and personal care products (PPCPs) in biosolids and their degradation during composting, *Journal of Residuals Science & Technology* 2 (1), 31–40, 2005.

34. Overcash, M., Sims, R.C., Sims, J.L., and Nieman, J.K.C., Beneficial reuse and sustainability: The fate of organic compounds in land-applied waste, *Journal of Environmental Quality* 34 (1), 29–41, 2005.

35. Xia, K., Bhandari, A., Das, K., and Pillar, G., Occurrence and fate of pharmaceuticals and personal care products (PPCPs) in biosolids, *Journal of Environmental Quality* 34 (1), 91–104, 2005.

36. Zuccato, E., Castiglioni, S., and Fanelli, R., Identification of the pharmaceuticals for human use contaminating the Italian aquatic environment, *Journal of Hazardous Materials* 122 (3), 205–209, 2005.

37. Gross, B., Montgomery-Brown, J., Naumann, A., and Reinhard, M., Occurrence and fate of pharmaceuticals and alkylphenol ethoxylate metabolites in an effluent-dominated river and wetland, *Environmental Toxicology and Chemistry* 23 (9), 2074–2083, 2004.

38. Lindqvist, N., Tuhkanen, T., and Kronberg, L., Occurrence of acidic pharmaceuticals in raw and treated sewages and in receiving waters, *Water Research* 39 (11), 2219–2228, 2005.
39. Verenitch, S.S., Lowe, C.J., and Mazumder, A., Determination of acidic drugs and caffeine in municipal wastewaters and receiving waters by gas chromatography-ion trap tandem mass spectrometry, *Journal of Chromatography A* 1116 (1–2), 193–203, 2006.
40. Brun, G.L., Bernier, M., Losier, R., Doe, K., Jackman, P., and Lee, H.B., Pharmaceutically active compounds in Atlantic Canadian sewage treatment plant effluents and receiving waters, and potential for environmental effects as measured by acute and chronic aquatic toxicity, *Environmental Toxicology and Chemistry* 25, 2163–2176, 2006.
41. Batt, A.L., Snow, D.D., and Aga, D.S., Occurrence of sulfonamide antimicrobials in private water wells in Washington County, Idaho, USA, *Chemosphere* 64 (11), 1963–1971, 2006.
42. Brown, K.D., Kulis, J., Thomson, B., Chapman, T.H., and Mawhinney, D.B., Occurrence of antibiotics in hospital, residential, and dairy, effluent, municipal wastewater, and the Rio Grande in New Mexico, *Science of the Total Environment* 366 (2–3), 772–783, 2006.
43. Cabello, F.C., Heavy use of prophylactic antibiotics in aquaculture: A growing problem for human and animal health and for the environment, *Environmental Microbiology* 8 (7), 1137–1144, 2006.
44. Dietze, J.E., Scribner, E.A., Meyer, M.T., and Kolpin, D.W., Occurrence of antibiotics in water from 13 fish hatcheries, 2001–2003, *International Journal of Environmental Analytical Chemistry* 85 (15), 1141–1152, 2005.
45. Lee, H.B., Peart, T.E., and Svoboda, M.L., Determination of endocrine-disrupting phenols, acidic pharmaceuticals, and personal-care products in sewage by solid-phase extraction and gas chromatography-mass spectrometry, *Journal of Chromatography A* 1094 (1–2), 122–129, 2005.
46. Quintana, J.B. and Reemtsma, T., Sensitive determination of acidic drugs and triclosan in surface and wastewater by ion-pair reverse-phase liquid chromatography/tandem mass spectrometry, *Rapid Communications in Mass Spectrometry* 18 (7), 765–774, 2004.
47. Quintana, J.B., Rodil, R., and Reemtsma, T., Suitability of hollow fibre liquid-phase microextraction for the determination of acidic pharmaceuticals in wastewater by liquid chromatography-electrospray tandem mass spectrometry without matrix effects, *Journal of Chromatography A* 1061 (1), 19–26, 2004.
48. Quintana, J.B., Weiss, S., and Reemtsma, T., Pathways and metabolites of microbial degradation of selected acidic pharmaceutical and their occurrence in municipal wastewater treated by a membrane bioreactor, *Water Research* 39 (12), 2654–2664, 2005.
49. Clara, M., Strenn, B., Gans, O., Martinez, E., Kreuzinger, N., and Kroiss, H., Removal of selected pharmaceuticals, fragrances and endocrine disrupting compounds in a membrane bioreactor and conventional wastewater treatment plants, *Water Research* 39 (19), 4797–4807, 2005.
50. Bendz, D., Paxeus, N.A., Ginn, T.R., and Loge, F.J., Occurrence and fate of pharmaceutically active compounds in the environment, a case study: Hoje River in Sweden, *Journal of Hazardous Materials* 122 (3), 195–204, 2005.
51. Santos, J.L., Aparicio, I., Alonso, E., and Callejon, M., Simultaneous determination of pharmaceutically active compounds in wastewater samples by solid phase extraction and high-performance liquid chromatography with diode array and fluorescence detectors, *Analytica Chimica Acta* 550 (1–2), 116–122, 2005.

52. Miao, X.S., Yang, J.J., and Metcalfe, C.D., Carbamazepine and its metabolites in wastewater and in biosolids in a municipal wastewater treatment plant, *Environmental Science & Technology* 39 (19), 7469–7475, 2005.
53. Roberts, P.H. and Thomas, K.V., The occurrence of selected pharmaceuticals in wastewater effluent and surface waters of the lower Tyne catchment, *Science of the Total Environment* 356 (1–3), 143–153, 2006.
54. Stackelberg, P.E., Furlong, E.T., Meyer, M.T., Zaugg, S.D., Henderson, A.K., and Reissman, D.B., Persistence of pharmaceutical compounds and other organic wastewater contaminants in a conventional drinking-watertreatment plant, *Science of the Total Environment* 329 (1–3), 99–113, 2004.
55. Westerhoff, P., Yoon, Y., Snyder, S., and Wert, E., Fate of endocrine-disruptor, pharmaceutical, and personal care product chemicals during simulated drinking water treatment processes, *Environmental Science & Technology* 39 (17), 6649–6663, 2005.
56. Jones, O.A., Lester, J.N., and Voulvoulis, N., Pharmaceuticals: A threat to drinking water? *Trends in Biotechnology* 23 (4), 163–167, 2005.
57. Snyder, S.A., Westerhoff, P., Yoon, Y., and Sedlak, D.L., Pharmaceuticals, personal care products, and endocrine disruptors in water: Implications for the water industry, *Environmental Engineering Science* 20 (5), 449–469, 2003.
58. Hernando, M.D., Heath, E., Petrovic, M., and Barcelo, D., Trace-level determination of pharmaceutical residues by LC-MS/MS in natural and treated waters. A pilot-survey study, *Analytical and Bioanalytical Chemistry* 385 (6), 985–991, 2006.
59. Rule, K.L., Ebbett, V.R., and Vikesland, P.J., Formation of chloroform and chlorinated organics by free-chlorine-mediated oxidation of triclosan, *Environmental Science & Technology* 39 (9), 3176–3185, 2005.
60. Dodd, M.C. and Huang, C.H., Transformation of the antibacterial agent sulfamethoxazole in reactions with chlorine: Kinetics mechanisms, and pathways, *Environmental Science & Technology* 38 (21), 5607–5615, 2004.
61. Seitz, W., Jiang, J.Q., Weber, W.H., Lloyd, B.J., Maier, M., and Maier, D., Removal of iodinated X-ray contrast media during drinking water treatment, *Environmental Chemistry* 3 (1), 35–39, 2006.
62. Cunningham, V.L., Buzby, M., Hutchinson, T., Mastrocco, F., Parke, N., and Roden, N., Effects of human pharmaceuticals on aquatic life: Next steps, *Environmental Science & Technology* 40 (11), 3456–3462, 2006.
63. Sanderson, H., Johnson, D.J., Reitsma, T., Brain, R.A., Wilson, C.J., and Solomon, K.R., Ranking and prioritization of environmental risks of pharmaceuticals in surface waters, *Regulatory Toxicology and Pharmacology* 39 (2), 158–183, 2004.
64. Fent, K., Weston, A.A., and Caminada, D., Ecotoxicology of human pharmaceuticals, *Aquatic Toxicology* 76 (2), 122–159, 2006.
65. Richards, S.M., Wilson, C.J., Johnson, D.J., Castle, D.M., Lam, M., Mabury, S.A., Sibley, P.K., and Solomon, K.R., Effects of pharmaceutical mixtures in aquatic microcosms, *Environmental Toxicology and Chemistry* 23 (4), 1035–1042, 2004.
66. Mimeault, C., Woodhouse, A., Miao, X.S., Metcalfe, C.D., Moon, T.W., and Trudeau, V.L., The human lipid regulator, gemfibrozil bioconcentrates and reduces testosterone in the goldfish, Carassius auratus, *Aquatic Toxicology* 73 (1), 44–54, 2005.
67. Swan, G., Naidoo, V., Cuthbert, R., Green, R.E., Pain, D.J., Swarup, D., Prakash, V., Taggart, M., Bekker, L., Das, D., Diekmann, J., Diekmann, M., Killian, E., Meharg, A., Patra, R.C., Saini, M., and Wolter, K., Removing the threat of diclofenac to critically endangered Asian vultures, *Plos Biology* 4 (3), 395–402, 2006.
68. Meteyer, C.U., Rideout, B.A., Gilbert, M., Shivaprasad, H.L., and Oaks, J.L., Pathology and proposed pathophysiology of diclofenac poisoning in free-living and experimentally exposed oriental white-backed vultures (Gyps bengalensis), *Journal of Wildlife Diseases* 41 (4), 707–716, 2005.

69. Pomati, F., Castiglioni, S., Zuccato, E., Fanelli, R., Vigetti, D., Rossetti, C., and Cala-mari, D., Effects of a complex mixture of therapeutic drugs at environmental levels on human embryonic cells, *Environmental Science & Technology* 40 (7), 2442–2447, 2006.

70. Jones, A.H., Voulvoulis, N., and Lester, J.N., Potential ecological and human health risks associated with the presence of pharmaceutically active compounds in the aquatic environment, *Critical Reviews in Toxicology* 34 (4), 335–350, 2004.

71. Halling-Sorensen, B., Nielsen, S.N., Lanzky, P.F., Ingerslev, F., Lutzhoft, H.C.H., and Jorgensen, S.E., Occurrence, fate and effects of pharmaceutical substances in the environment—A review, *Chemosphere* 36 (2), 357–394, 1998.

72. Heberer, T., Occurrence, fate, and removal of pharmaceutical residues in the aquatic environment: A review of recent research data, *Toxicology Letters* 131 (1–2), 5–17, 2002.

73. Diaz-Cruz, S. and Barcelo, D., Occurrence and analysis of selected pharmaceuticals and metabolites as contaminants present in waste waters, sludge and sediments, in *Emerging Organic Pollutants in Waste Waters and Sludge,* 15, 227–260, Part I, 2004.

74. Kosjek, T., Heath, E., and Krbavcic, A., Determination of non-steroidal anti-inflam-matory drug (NSAIDs) residues in water samples, *Environment International* 31 (5), 679–685, 2005.

75. Lin, W.C., Chen, H.C., and Ding, W.H., Determination of pharmaceutical residues in waters by solid-phase extraction and large-volume on-line derivatization with gas chromatography-mass spectrometry, *Journal of Chromatography A* 1065 (2), 279–285, 2005.

76. Mitani, K. and Kataoka, H., Determination of fluoroquinolones in environmental waters by in-tube solid-phase microextraction coupled with liquid chromatography-tandem mass spectrometry, *Analytica Chimica Acta* 562 (1), 16–22, 2006.

77. Moldovan, Z., Occurrences of pharmaceutical and personal care products as micropol-lutants in rivers from Romania, *Chemosphere* 64, 1808–1817, 2006.

78. Yasojima, M., Nakada, N., Komori, K., Suzuki, Y., and Tanaka, H., Occurrence of levofloxacin, clarithromycin and azithromycin in wastewater treatment plant in Japan, *Water Science and Technology* 53 (11), 227–233, 2006.

79. Richardson, B.J., Larn, P.K.S., and Martin, M., Emerging chemicals of concern: Phar-maceuticals and personal care products (PPCPs) in Asia, with particular reference to Southern China, *Marine Pollution Bulletin* 50 (9), 913–920, 2005.

80. USEPA, Onsite Wastewater Treatment Systems Manual, *USEPA* EPA-625-R-00-008, 2002.

81. U.S. Census Bureau, Historical Census of Housing Tables: Sewage Disposal, http://www.census.gov/hhes/www/housing/census/historic/sewage.html, 1990.

82. Kümmerer, K., Pharmaceuticals in the environment-scope of the book and introduc-tion, in *Pharmaceuticals in the Environment—Sources, Fate, Effects, and Risk Second Edition*, Kummerer, K. Springer, Berlin Heidelberg, 2004.

83. Managaki, S., Takada, H., Kim, D.-M., Horiguchi, T., and Shiraishi, H., Three-dimen-sional distributions of sewage markers in Tokyo Bay water-fluorescent whitening agents, *Marine Pollution Bulletin* 52, 281–292, 2006.

84. Eganhouse, R.P. and Sherblom, P.M., Anthropogenic organic contaminants in the efflu-ent of a combined sewer overflow: Impact on Boston Harbor, *Marine Environmental Research* 51, 51–74, 2001.

85. Diez, S., Jover, E., Albaiges, J., and Bayona, J.M., Occurrence and degradation of butyl-tins and wastewater marker compounds in sediments from Barcelona Harbor, Spain, *Environment International* 32, 858–865, 2006.

86. Ternes, T.A., Joss, A., and Siegrist, H., Scrutinizing pharmaceuticals and personal care products in wastewater treatment, *Environmental Science & Technology* 38 (20), 392A–399A, 2004.
87. Göbel, A., Thomsen, A., McArdell, C.S., Joss, A., and Giger, W., Occurrence and sorption behavior of sulfonamides, macrolides, and trimethoprim in activated sludge treatment, *Environmental Science & Technology* 39 (11), 3981–3989, 2005.
88. USEPA, Biosolids generation, use, and disposal in the United States, *USEPA* EPA-530-R-99-009, 1999.
89. USEPA, http:/www.epa.gov/owm/mtb/biosolids/genqa.htm, 2003.
90. Gurr, C.J. and Reinhard, M., Harnessing natural attenuation of pharmaceuticals and hormones in rivers, *Environmental Science & Technology* 40 (9), 2872–2876, 2006.
91. Perez-Estrada, L.A., Malato, S., Gernjak, W., Aguera, A., Thurman, E.M., Ferrer, I., and Fernandez-Alba, A.R., Photo-fenton degradation of diclofenac: Identification of main intermediates and degradation pathway, *Environmental Science & Technology* 39 (21), 8300–8306, 2005.
92. Lin, A.Y.C. and Reinhard, M., Photodegradation of common environmental pharmaceuticals and estrogens in river water, *Environmental Toxicology and Chemistry* 24 (6), 1303–1309, 2005.
93. Andreozzi, R., Raffaele, M., and Nicklas, P., Pharmaceuticals in STP effluents and their solar photodegradation in aquatic environment, *Chemosphere* 50 (10), 1319–1330, 2003.
94. Kwon, J.W. and Armbrust, K.L., Degradation of citalopram by simulated sunlight, *Environmental Toxicology and Chemistry* 24 (7), 1618–1623, 2005.
95. Lam, M.W. and Mabury, S.A., Photodegradation of the pharmaceuticals atorvastatin, carbamazepine, levofloxacin, and sulfamethoxazole in natural waters, *Aquatic Sciences* 67 (2), 177–188, 2005.
96. Packer, J.L., Werner, J.J., Latch, D.E., McNeill, K., and Arnold, W.A., Photochemical fate of pharmaceuticals in the environment: Naproxen, diclofenac, clofibric acid, and ibuprofen, *Aquatic Sciences* 65 (4), 342–351, 2003.
97. Perez, S., Eichhorn, P., and Aga, D.S., Evaluating the biodegradability of sulfamethazine, sulfamethoxazole, sulfathiazole, and trimethoprim at different stages of sewage treatment, *Environmental Toxicology and Chemistry* 24 (6), 1361–1367, 2005.
98. Kim, S., Eichhorn, P., Jensen, J.N., Weber, A.S., and Aga, D.S., Removal of antibiotics in wastewater: Effect of hydraulic and solid retention times on the fate of tetracycline in the activated sludge process, *Environmental Science & Technology* 39 (15), 5816–5823, 2005.
99. Carucci, A., Cappai, G., and Piredda, M., Biodegradability and toxicity of pharmaceuticals in biological wastewater treatment plants, *Journal of Environmental Science and Health Part A—Toxic/Hazardous Substances & Environmental Engineering* 41 (9), 1831–1842, 2006.
100. Lindberg, R.H., Olofsson, U., Rendahl, P., Johansson, M.I., Tysklind, M., and Andersson, B.A.V., Behavior of fluoroquinolones and trimethoprim during mechanical, chemical, and active sludge treatment of sewage water and digestion of sludge, *Environmental Science & Technology* 40 (3), 1042–1048, 2006.
101. Boxall, A.B.A., Sinclair, C.J., Fenner, K., Kolpin, D., and Maud, S.J., When synthetic chemicals degrade in the environment, *Environmental Science & Technology* 38 (19), 368A–375A, 2004.
102. Kolpin, D.W., Skopec, M., Meyer, M.T., Furlong, E.T., and Zaugg, S.D., Urban contribution of pharmaceuticals and other organic wastewater contaminants to streams during differing flow conditions, *Science of the Total Environment* 328 (1–3), 119–130, 2004.

103. Bound, J.P. and Voulvoulis, N., Predicted and measured concentrations for selected pharmaceuticals in U.K. rivers: Implications for risk assessment, *Water Research* 40, 2885–2892, 2006.

104. Comoretto, L. and Chiron, S., Comparing pharmaceutical and pesticide loads into a small Mediterranean river, *Science of the Total Environment* 349 (1–3), 201–210, 2005.

105. Ternes, T.A., Meisenheimer, M., McDowell, D., Sacher, F., Brauch, H.J., Gulde, B.H., Preuss, G., Wilme, U., and Seibert, N.Z., Removal of pharmaceuticals during drinking water treatment, *Environmental Science & Technology* 36 (17), 3855–3863, 2002.

106. Loraine, G.A. and Pettigrove, M.E., Seasonal variations in concentrations of pharmaceuticals and personal care products in drinking water and reclaimed wastewater in Southern California, *Environmental Science & Technology* 40 (3), 687–695, 2006.

107. Glassmeyer, S.T. and Shoemaker, J.A., Effects of chlorination on the persistence of pharmaceuticals in the environment, *Bulletin of Environmental Contamination and Toxicology* 74 (1), 24–31, 2005.

108. Bedner, M. and Maccrehan, W.A., Transformation of acetaminophen by chlorination produces the toxicants 1,4-benzoquinone and N-acetyl-p-benzoquinone imine, *Environmental Science & Technology* 40 (2), 516–522, 2006.

109. Pinkston, K.E. and Sedlak, D.L., Transformation of aromatic ether-and amine-containing pharmaceuticals during chlorine disinfection, *Environmental Science & Technology* 38 (14), 4019–4025, 2004.

110. Naumann, K., Influence of chlorine substituents on biological activity of chemicals: A review (Reprinted from *J Prakt Chem*, Vol. 341, pp. 417–435, 1999), *Pest Management Science* 56 (1), 3–21, 2000.

111. Petty, J.D., Huckins, J.N., Alvarez, D.A., Brumbaugh, W.G., Cranor, W.L., Gale, R.W., Rastall, A.C., Jones-Lepp, T.L., Leiker, T.J., Rostad, C.E., and Furlong, E.T., A holistic passive integrative sampling approach for assessing the presence and potential impacts of waterborne environmental contaminants, *Chemosphere* 54 (6), 695–705, 2004.

112. Alvarez, D.A., Stackelberg, P.E., Petty, J.D., Huckins, J.N., Furlong, E.T., Zaugg, S.D., and Meyer, M.T., Comparison of a novel passive sampler to standard water-column sampling for organic contaminants associated with wastewater effluents entering a New Jersey stream, *Chemosphere* 61 (5), 610–622, 2005.

113. Jones-Lepp, T.L., Alvarez, D.A., Petty, J.D., and Huckins, J.N., Polar organic chemical integrative sampling and liquid chromatography-electrospray/ion-trap mass spectrometry for assessing selected prescription and illicit drugs in treated sewage effluents, *Archives of Environmental Contamination and Toxicology* 47 (4), 427–439, 2004.

114. Gourlay, C., Miege, C., Noir, A., Ravelet, C., Garric, J., and Mouchel, J.M., How accurately do semi-permeable membrane devices measure the bioavailability of polycyclic aromatic hydrocarbons to Daphnia magna? *Chemosphere* 61 (11), 1734–1739, 2005.

115. Voutsas, E., Magoulas, K., and Tassios, D., Prediction of the bioaccumulation of persistent organic pollutants in aquatic food webs, *Chemosphere* 48 (7), 645–651, 2002.

116. Haggard, B.E., Galloway, J.M., Green, W.R., and Meyer, M.T., Pharmaceuticals and other organic chemicals in selected north-central and northwestern Arkansas streams, *Journal of Environmental Quality* 35 (4), 1078–1087, 2006.

117. Brossa, L., Marce, R.A., Borrull, F., and Pocurull, E., Occurrence of twenty-six endocrine-disrupting compounds in environmental water samples from Catalonia, Spain, *Environmental Toxicology and Chemistry* 24 (2), 261–267, 2005.

118. Lishman, L., Smyth, S.A., Sarafin, K., Kleywegt, S., Toito, J., Peart, T., Lee, B., Servos, M., Beland, M., and Seto, P., Occurrence and reductions of pharmaceuticals and personal care products and estrogens by municipal wastewater treatment plants in Ontario, Canada, *Science of the Total Environment* 367 (2–3), 544–558, 2006.

119. Zhang, S., Zhang, Q., Darisaw, S., Ehie, O., and Wang, G., Simultaneous quantification of polycyclic aromatic hydrocarbons (PAHs), polychlorinated biphenyls (PCBs), and pharmaceuticals and personal care products (PPCPs) in Mississippi River water, in New Orleans, Louisiana, *Chemosphere*, 66, 1057–1069, 2007.

120. Gomez, M.J., Petrovic, M., Fernandez-Alba, A.R., and Barcelo, D., Determination of pharmaceuticals of various therapeutic classes by solid-phase extraction and liquid chromatography-tandem mass spectrometry analysis in hospital effluent wastewaters, *Journal of Chromatography A* 1114 (2), 224–233, 2006.

121. Wiegel, S., Aulinger, A., Brockmeyer, R., Harms, H., Loffler, J., Reincke, H., Schmidt, R., Stachel, B., von Tumpling, W., and Wanke, A., Pharmaceuticals in the river Elbe and its tributaries, *Chemosphere* 57 (2), 107–126, 2004.

122. Castiglioni, S., Fanelli, R., Calamari, D., Bagnati, R., and Zuccato, E., Methodological approaches for studying pharmaceuticals in the environment by comparing predicted and measured concentrations in River Po, Italy, *Regulatory Toxicology and Pharmacology* 39 (1), 25–32, 2004.

123. Paxeus, N., Removal of selected non-steroidal anti-inflammatory drugs (NSAIDs), gemfibrozil, carbamazepine, beta-blockers, trimethoprim and triclosan in conventional wastewater treatment plants in five EU countries and their discharge to the aquatic environment, *Water Science and Technology* 50 (5), 253–260, 2004.

124. Gobel, A., McArdell, C.S., Suter, M.J.F., and Giger, W., Trace determination of macrolide and sulfonamide antimicrobials, a human sulfonamide metabolite, and trimethoprim in wastewater using liquid chromatography coupled to electrospray tandem mass spectrometry, *Analytical Chemistry* 76 (16), 4756–4764, 2004.

125. Vieno, N.M., Tuhkanen, T., and Kronberg, L., Seasonal variation in the occurrence of pharmaceuticals in effluents from a sewage treatment plant and in the recipient water, *Environmental Science & Technology* 39 (21), 8220–8226, 2005.

126. Wolf, L., Held, I., Eiswirth, M., and Hotzl, H., Impact of leaky sewers on groundwater quality, *Acta Hydrochimica Et Hydrobiologica* 32 (4–5), 361–373, 2004.

127. Thomas, P.M. and Foster, G.D., Tracking acidic pharmaceuticals, caffeine, and triclosan through the wastewater treatment process, *Environmental Toxicology and Chemistry* 24 (1), 25–30, 2005.

128. Batt, A.L. and Aga, D.S., Simultaneous analysis of multiple classes of antibiotics by ion trap LC/MS/MS for assessing surface water and groundwater contamination, *Analytical Chemistry* 77 (9), 2940–2947, 2005.

129. Thomas, P.M. and Foster, G.D., Determination of nonsteroidal anti-inflammatory drugs, caffeine, and triclosan in wastewater by gas chromatography-mass spectrometry, *Journal of Environmental Science and Health Part A—Toxic/Hazardous Substances & Environmental Engineering* 39 (8), 1969–1978, 2004.

130. Zuehlke, S., Duennbier, U., and Heberer, T., Determination of polar drug residues in sewage and surface water applying liquid chromatography-tandem mass spectrometry, *Analytical Chemistry* 76 (22), 6548–6554, 2004.

131. Lissemore, L., Hao, C.Y., Yang, P., Sibley, P.K., Mabury, S., and Solomon, K.R., An exposure assessment for selected pharmaceuticals within a watershed in southern Ontario, *Chemosphere* 64 (5), 717–729, 2006.

132. Hao, C.Y., Lissemore, L., Nguyen, B., Kleywegt, S., Yang, P., and Solomon, K., Determination of pharmaceuticals in environmental waters by liquid chromatography/electrospray ionization/tandem mass spectrometry, *Analytical and Bioanalytical Chemistry* 384 (2), 505–513, 2006.

133. Yang, S. and Carlson, K., Routine monitoring of antibiotics in water and wastewater with a radioimmunoassay technique, *Water Research* 38 (14–15), 3155–3166, 2004.

134. Lindberg, R.H., Wennberg, P., Johansson, M.I., Tysklind, M., and Andersson, B.A.V., Screening of human antibiotic substances and determination of weekly mass flows in five sewage treatment plants in Sweden, *Environmental Science & Technology* 39 (10), 3421–3429, 2005.

135. Karthikeyan, K.G. and Meyer, M.T., Occurrence of antibiotics in wastewater treatment facilities in Wisconsin, USA, *Science of the Total Environment* 361 (1–3), 196–207, 2006.

136. Lindberg, R., Jarnheimer, P.A., Olsen, B., Johansson, M., and Tysklind, M., Determination of antibiotic substances in hospital sewage water using solid phase extraction and liquid chromatography/mass spectrometry and group analogue internal standards, *Chemosphere* 57 (10), 1479–1488, 2004.

137. Renew, J.E. and Huang, C.H., Simultaneous determination of fluoroquinolone, sulfonamide, and trimethoprim antibiotics in wastewater using tandem solid phase extraction and liquid chromatography-electrospray mass spectrometry, *Journal of Chromatography A* 1042 (1–2), 113–121, 2004.

138. Yang, S. and Carlson, K.H., Solid-phase extraction-high-performance liquid chromatography-ion trap mass spectrometry for analysis of trace concentrations of macrolide antibiotics in natural and waste water matrices, *Journal of Chromatography A* 1038 (1–2), 141–155, 2004.

139. Yang, S.W., Cha, J.M., and Carlson, K., Trace analysis and occurrence of anhydroerythromycin and tylosin in influent and effluent wastewater by liquid chromatography combined with electrospray tandem mass spectrometry, *Analytical and Bioanalytical Chemistry* 385 (3), 623–636, 2006.

140. Carballa, M., Omil, F., Lema, J.M., Llompart, M., Garcia-Jares, C., Rodriguez, I., Gomez, M., and Ternes, T., Behavior of pharmaceuticals, cosmetics and hormones in a sewage treatment plant, *Water Research* 38 (12), 2918–2926, 2004.

141. Vasskog, T., Berger, U., Samuelsen, P.J., Kallenborn, R., and Jensen, E., Selective serotonin reuptake inhibitors in sewage influents and effluents from Tromso, Norway, *Journal of Chromatography A* 1115 (1–2), 187–195, 2006.

142. Verplanck, P.L., Taylor, H.E., Nordstrom, D.K., and Barber, L.B., Aqueous stability of gadolinium in surface waters receiving sewage treatment plant effluent, Boulder Creek, Colorado, *Environmental Science & Technology* 39 (18), 6923–6929, 2005.

143. Sandstrom, M.W., Kolpin, D.W., Thurman, E.M., and Zaugg, S.D., Widespread detection of N,N-diethyl-m-toluamide in US streams: Comparison with concentrations of pesticides, personal care products, and other organic wastewater compounds, *Environmental Toxicology and Chemistry* 24 (5), 1029–1034, 2005.

144. Nakata, H., Kannan, K., Jones, P.D., and Giesy, J.P., Determination of fluoroquinolone antibiotics in wastewater effluents by liquid chromatography-mass spectrometry and fluorescence detection, *Chemosphere* 58 (6), 759–766, 2005.

145. Fono, L.J. and Sedlak, D.L., Use of the chiral pharmaceutical propranolol to identify sewage discharges into surface waters, *Environmental Science & Technology* 39 (23), 9244–9252, 2005.

146. Perret, D., Gentili, A., Marchese, S., Greco, A., and Curini, R., Sulphonamide residues in Italian surface and drinking waters: A small scale reconnaissance, *Chromatographia* 63 (5–6), 225–232, 2006.

147. Waltman, E.L., Venables, B.J., and Waller, W.Z., Triclosan in a North Texas wastewater treatment plant and the influent and effluent of an experimental constructed wetland, *Environmental Toxicology and Chemistry* 25 (2), 367–372, 2006.

148. Canosa, P., Rodriguez, I., Rubi, E., and Cela, R., Optimization of solid-phase microextraction conditions for the determination of triclosan and possible related compounds in water samples, *Journal of Chromatography A* 1072 (1), 107–115, 2005.

149. Hua, W.Y., Bennett, E.R., and Letcher, R.J., Triclosan in waste and surface waters from the upper Detroit River by liquid chromatography-electrospray-tandem quadrupole mass spectrometry, *Environment International* 31 (5), 621–630, 2005.

150. Gobel, A., Thomsen, A., McArdell, C.S., Alder, A.C., Giger, W., Theiss, N., Loffler, D., and Ternes, T.A., Extraction and determination of sulfonamides, macrolides, and trimethoprim in sewage sludge, *Journal of Chromatography A* 1085 (2), 179–189, 2005.

151. Ternes, T.A., Herrmann, N., Bonerz, M., Knacker, T., Siegrist, H., and Joss, A., A rapid method to measure the solid-water distribution coefficient (K-d) for pharmaceuticals and musk fragrances in sewage sludge, *Water Research* 38 (19), 4075–4084, 2004.

152. Barber, L.B., Leenheer, J.A., Pereira, W.E., Noyes, T.I., Brown, G.K., Tabor, C.F., and Writer, J.H., *Organic Contamination of the Mississippi River from Municipal and Industrial Wastewater, in Contaminants in the Mississippi River, US Geological Survey Circular 1133*, Meade, R.H., U.S. Geological Survey, Reston, VA, 1995, http://pubs.usgs.gov/circ/circ1133/.

2 Advances in the Analysis of Pharmaceuticals in the Aquatic Environment

Sandra Pérez and Damià Barceló

Contents

2.1 INTRODUCTION

Recently, the focus of environmental analysis has shifted from the classic contaminants, such as the persistent organic pollutants, toward the "emerging contaminants" detected recently in many environmental compartments.[1] Emerging contaminants are defined as compounds that are not currently covered by existing regulations of water quality, have not been previously studied, and are thought to be potential threats to environmental ecosystems and human health and safety. In particular, the compounds that are being addressed include pharmaceuticals, drugs of abuse, and personal-care products.[2] The high water solubility of these organic compounds makes them mobile in the aquatic media, hence they can potentially infiltrate the soil and then reach groundwater. Eventually, these compounds may find their way into the drinking water supplies.

In recent years the increasing use of drugs in farming, aquaculture, and human health has become a growing public concern because of their potential to cause undesirable ecological and human health effects. The main concern regarding pharmaceuticals is that they are being introduced continuously into water bodies as pollutants, and due to their biological activity this can lead to adverse effects in aquatic ecosystems and potentially impact drinking water supplies.[3] Antibiotics are one of the most problematic groups of pharmaceuticals, since their increasing use for more than four decades has led to the selection of resistant bacteria that can threaten the effectiveness of antibiotics for the treatment of human infections.[4] Another group that has caused environmental concern is contraceptives and other endocrine disruptors; due to their endocrine properties they can induce the feminization or masculinization in aquatic organisms.[5] In general the short- and long-term ecotoxicological effects of pharmaceuticals on wildlife have not yet been studied sufficiently.

Prescription and over-the-counter drugs have probably been in the environment for as long as they have been used, but only recently have analytical methods been developed to detect pharmaceuticals at trace levels.[6] Due to the dilution and possible degradation of these substances in the environment, low levels can be expected. Therefore, an analyte preconcentration procedure is almost always necessary in order to achieve desired levels of analytical sensitivity, often requiring high enrichment factors, between 100 and 10,000. Such enrichment factors for drug analysis are usually achieved using solid-phase extraction (SPE). Sensitive detection methods such as gas chromatography-mass spectrometry (GC-MS), GC-tandem mass spectrometry (GC-MS/MS) or liquid chromatography-mass spectrometry (LC-MS), and LC-tandem mass spectrometry (LC-MS/MS) are also crucial for the analytical determination of drugs in the environment. The main drawback of GC for drug analysis, however, is that this technique is limited to compounds with high vapor pressure. Since most drugs are polar substances, they need to be derivatized prior to injection in the GC. For this reason, the combination of atmospheric pressure ionization-MS (API-MS) with separation techniques such as LC or ultra performance liquid chromatography (UPLC) has become the method of choice in drug analysis. LC with a single quadrupole MS analyzer offers good sensitivity, but when very complex matrices such as raw sewage are investigated, insufficient selectivity often impairs the unequivocal identification of the analytes. Tandem MS affords superior performance in terms of sensitivity and selectivity in comparison with single quadrupole instruments. Liquid chromatographic techniques coupled to tandem MS or hybrid mass spectrometers with distinct analyzers such as triple quadrupole (QqQ), time-of-flight (ToF), quadrupole time-of-flight (QqToF), quadrupole ion trap (IT), and recently the quadrupole linear ion trap (QqLIT) are the most widely used instrumental techniques for drug analysis.[7]

Most of the data on the presence on pharmaceuticals in wastewaters, rivers, and drinking water come primarily from European studies[8,9] followed by those carried out in the United States.[10] These substances that are used in human and veterinary medicine can enter the environment via a number of pathways but mainly from discharges of wastewater treatment plants (WWTPs) or land application of sewage sludge and animal manure, as depicted in Figure 2.1. Most active ingredients of pharmaceuticals are transformed only partially in the body and thus are excreted as a mixture of metabolites and bioactive forms into sewage systems. Therefore, the treatment of wastewaters in WWTPs plays a crucial role in the elimination of pharmaceutical compounds before their discharge into rivers. During the application of

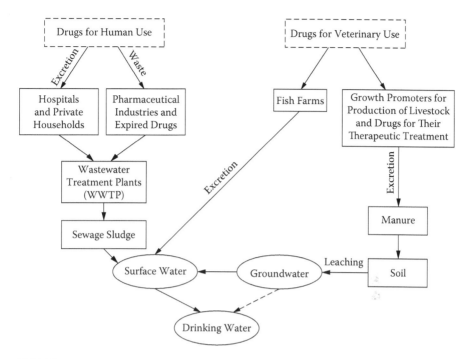

FIGURE 2.1 Pathways of pharmaceuticals and their metabolites in the environment.

primary and secondary treatments, pharmaceutical compounds can be eliminated by sorption onto the sludge or microbially degraded to form metabolites that are usually more polar than the parent drugs.[11] In many cases the high polarity combined with the low biodegradability exhibited by some pharmaceutical compounds results in inefficient elimination in WWTPs. The removal efficiencies vary from plant to plant and depend on the design and operation of the treatment systems.[12,13] Thus, the major source of pharmaceutical residues detectable in surface waters are discharges from WWTPs. Several studies reported the occurrence of pharmaceuticals at levels up to the μg/L range in rivers, streams, lakes, and groundwater.[1,14]

Researchers have yet to determine the occurrence, fate, and possible effects of the most frequently consumed drugs and their main metabolites in the aquatic environment. Exceptionally high levels of drugs have been reported—for example, the occurrence of the antiasthma drug salbutamol in water from the Po River.[15] The researchers concluded that their data reflected the illegal use of salbutamol by local farmers to promote growth in cattle. The determination of pharmaceuticals and drugs of abuse in the environment by applying the principle that what goes in must come out can be a helpful tool to estimate the drug consumption in the investigated areas. For example, Italian researchers measured the levels of benzoylecgonine, the major urinary metabolite of cocaine, in wastewater from several Italian cities.[16] What they found was surprising: cocaine use appeared to be far higher than the public health officials previously thought.

This review provides an overview on analytical protocols used in determining drugs and some of their metabolites in aqueous and solid environmental samples.

Technological progress in the fields of sample extraction and detection by mass spectrometry (MS) techniques (hybrid and tandem mass spectrometers) for analyzing antibiotics, antiinflammatory/analgesics, lipid regulating agents, psychiatric drugs, sedatives, iodated X-ray contrast media, diuretics, drugs of abuse, and some human metabolites in the aquatic environment are discussed.[17–19] The recent trends in multiresidue methodologies for the determination of the drugs and their human metabolites will be reviewed here.

2.2   MULTIRESIDUE METHODS

Many analytical methodologies for the determination of drugs in water bodies focused on selected therapeutic classes. Multiresidue methods, however, are becoming more widespread in response to the need of monitoring a wide range of pharmaceuticals that belong to diverse drug classes in wastewater, surface water, and groundwater. The latter approach offers advantages in terms of providing a more comprehensive picture of the occurrence and fate of the contaminants in the environment. In addition, the simultaneous determination of a large number of analytes by a single method represents a less time-consuming and hence more economical approach as compared with applying class-specific analytical protocols. The multiresidue methods found in the literature are diverse, with target analytes being selected commonly on the basis of their consumption in the country where the study is being conducted, the rate of metabolism of drugs, the environmental occurrence, and persistence in the environment. In this chapter, multiresidue methods for drug analysis in the aquatic environment are reviewed.

Ternes et al.[20] reported the determination of neutral pharmaceuticals: propyphenazone (analgesic), phenylbutazone (antiinflammatory), diazepam (psychiatric drug), omeprazole (antiulcer), nifedipine (calcium antagonist), glibenclamide (antidiabetic), and two human metabolites: 4-aminoantipyridine (metabolite of metamizole) and oxyphenbutazone (metabolite of phenylbutazone) with a multiresidue methodology including a one-step extraction method based on SPE with RP-C_{18} material eluted with methanol. The analysis was performed by LC with detection by electrospray ionization (ESI) tandem MS in multiple reaction monitoring (MRM) mode, which is the acquisition mode providing the best sensitivity and selectivity for quantitative analysis. Low limits of detection and reasonable recoveries for the selected drugs in different matrices were achieved. This work[20] investigated the losses in the recoveries of some drugs due to matrix impurities, which either reduced the sorption efficiencies on the C_{18} material or led to signal suppression in the ESI interface. The authors spiked influent wastewater extracts with the target analytes and found that the recoveries were not appreciably higher in comparison to the recoveries over the total method. Consequently, the signal suppression in ESI played a decisive role in the losses of 4-aminoantipyrine, omeprazole, oxyphenbutazone, phenylbutazone, and propyphenazone. For most of the compounds, compensation for the losses was achieved by addition of the surrogate standard 10,11-dihydrocarbamazepine. However, low corrected recoveries for oxyphenbutazone, phenylbutazone and 4-aminoantipyridine indicated that the determination of these compounds was still rather semiquantitative.[20]

Vanderford et al.[21] developed an analytical method for the determination of 21 pharmaceuticals in water, choosing them based on their occurrence in the environment and their dissimilar structural and physicochemical structures. They used also a one-step extraction method employing SPE with a hydrophilic-lipophilic balance (HLB). The separation and detection was performed with LC-MS/MS, using ESI in either positive or negative mode or atmospheric pressure chemical ionization (APCI) in positive mode. The analytical method provided a simple and sensitive method for the detection of a wide range of pharmaceuticals with recoveries in deionized water above 80% for all of the compounds. The authors also studied the effect of sample preservatives on the recovery of the pharmaceuticals.[21] They compared formaldehyde and sulfuric acid, obtaining the best results for the latter, which prevented the degradation of the target compounds and did not adversely affect their recoveries. Matrix effects were also examined in this work showing that all compounds detected with (+)ESI and (−)ESI, except hydrocodone, showed a considerable degree of ionization suppression. Hydrocodone, though, showed signal enhancement. In another work the same group reported a methodology using an isotope dilution technique for every analyte to compensate for matrix effects in the ESI source, SPE losses, and instrument variability.[22] The method was tested with three matrices (wastewater, surface water, and drinking water), and the results indicated that the method was very robust using isotope dilution for each target compound. The work described a method that analyzed 15 pharmaceuticals and 4 metabolites using SPE (HLB) coupled with LC-MS/MS with ESI source. Matrix spike recoveries for all compounds were between 88 and 106% for wastewater influent, 85 and 108% for wastewater effluent, 72 and 105% for surface water impacted by wastewater, 96 and 113% for surface water, and 91 and 116% for drinking water. The method detection limits were between 0.25 and 1 ng/L.

A study[23] evaluated different strategies to reduce matrix effects in LC-MS/MS with an ESI interface. First, the peak area of the target compounds in solvent were compared with the target compounds spiked in matrix extracts obtaining signal suppressions in the range of 40 to 90%. Next, internal calibration curves with internal labeled standard in solvent and in spiked matrix extracts were prepared. The overlapping of both curves confirmed that signal losses experienced by the analytes were corrected by the internal standards. Finally, the effectiveness of diluting sample extracts was studied. For this purpose, the signals obtained after sequential dilution of a WWTP effluent and influent extract were compared with the ones obtained for the corresponding concentrations of the standards in the solvent. The authors considered that matrix effects were eliminated with dilutions 1:2 and 1:4, and this approach was selected for this work. For the extraction of the target analytes, one-step SPE testing Oasis HLB, Isolute ENV+ and Isolute C_{18} with and without sample acidification was optimized. Oasis HLB, with sorbent based on a hydrophilic-lipophilic polymer, provided high recoveries for all target compounds at neutral pH. Recoveries were higher than 60% for both surface and wastewaters, with the exception of several compounds: ranitidine (50%), sotalol (50%), famotidine (50%), and mevastatin (34%).

Miao et al.[24] reported a method using SPE (C_{18}) and LC-(-)-ESI-MS/MS for the simultaneous detection of nine acidic pharmaceutical drugs (bezafibrate, clofibric acid, diclofenac, fenoprofen, gemfibrozil, ibuprofen, indomethacin, ketoprofen, and naproxen) in WWTP effluents. The recoveries ranged from 59% (indomethacin) to 92% (fenoprofen) in the WWTP effluent. The specificity of the method was checked spiking samples with analytes at concentration of 0.05 µg/L. Two interfering peaks resulting from endogenous components in the WWTP effluent were detected in the MRM channels for fenoprofen and indomethacin. Coextractives in the WWTP yielded fragmentation patterns similar to fenoprofen and indomethacin. However, the separation efficiency provided by high performance liquid chromatography (HPLC) was sufficient to resolve these interfering compounds from the analytes showing the importance of using LC to improve the selectivity of the analysis.

A multiresidue analytical method using SPE and LC-MS/MS for 28 pharmaceuticals including antimicrobial drugs, two diuretics (furosemide and hydrochlorothiazide), cardiovascular (atenolol and enalapril), antiulcer, psychiatric drugs, an antiinflammatory, a β_2-sympathomimetic (salbutamol), a lipid regulator, some estrogens (17β-estradiol, 17α-ethinylestradiol, estrone), two antitumorals (cyclophosphamide and methotrexate), and two metabolites (clofibric acid and demethyl diazepam) was developed.[25] To optimize the extraction method, several stationary phases and different PH samples were tested. The cartridges were Oasis MCX at pH 1.5/2.0 and 3.0 for all the compounds and at pH 7.0/7.5 for omeprazole; LiChrolute EN at pH 3.0, 5.0, 7.0, and 9.0 for all the compounds; Bakerbond C_{18} at pH 8.0 and 9.5 for the extraction of amoxicillin; and Oasis HLB at pH 7.0 for omeprazole and pH 8.5/9.0 for amoxycillin. They selected Oasis MCX for water samples at pH 1.5/2.0 and LiChrolute EN for water samples at pH 7. Recoveries of the pharmaceuticals were mostly greater than 70% and instrumental and method limits of detection in the order of ng/L.

Vieno et al.[26] developed a method that allowed the quantification of the four β-blockers—acebutolol, atenolol, metoprolol and sotalol, carbamazepine—and the three fluoroquinolones antibiotics—ciprofloxacin, ofloxacin, and norfloxacin—in groundwater, surface waters, and raw and treated sewages. The authors[26] studied the effect of the washing and of the eluting solvent and pH on the extraction step using a single pretreatment (SPE, Oasis HLB). Prior to the elution step, the adsorbent was washed with 2 mL of 5% of methanol in 2% aqueous NH_4OH showing improvements in the MS detection in terms of decreased ion suppression and thus improved detectability of the compounds. Methanol was the solvent of choice yielding the highest recoveries as compared with those obtained with acetonitrile or acetone. The study of the influence of the pH of the water in the extraction methodology was performed at three pH values: 4.0, 7.5, and 10.0. For most of the compounds, the pH did not have a pronounced effect on the recovery, with the exception of atenolol and sotalol, which were poorly recovered at low pH (<10% at pH 4.0). The samples were analyzed with LC-MS/MS using ESI in positive mode showing ion suppression in the ESI source.[26] To evaluate the matrix effects, the authors infused continuously a standard solution into the mass spectrometer and then injected either solvent or a real sample extract onto the LC column. Moreover, SPE extracts of groundwater, surface water, and wastewater influent and effluent were spiked with pharmaceuticals, and spiked samples were analyzed in LC-MS/MS with ESI interface. No ion suppression was noticed for any of the analytes in groundwater extracts; some signal suppression (<8%) was noticed for sotalol, acebutolol, and metoprolol for the surface-water extract; and more severe signal suppression (40%) was observed in the wastewater influents and effluents. The authors[26] reported relative recoveries higher than those reported previously by other authors[24] showing an improvement of the general methodology for all the target analytes, except for ciprofloxacin and norfloxacin.

The determination of selected drugs and their metabolites with a multiresidue methodology was also reported by Zuehlke et al.[27] Carbamazepine, dimethylaminophenazone, phenazone, propyphenazone, 1-acetyl-1-methyl-2-dimethyloxanoyl-2-phenylhydrazide, 1-acetyl-1-methyl-2-phenylhydrazide, two human metabolites of metamizole (formylaminoantipyridine and aminoantipyridine), and two micobiological metabolites (1,5-dimethyl-1,2-dehydro-3-pyrazolone, 4-(2-methylethyl)-1,5-dimethyl-1,2-dehydro-3-pyrazole) were studied. To allow for efficient SPE of the two microbiological metabolites from water on a conventional C_{18} sorbent, the authors[27] prepared the water samples by a simple *in situ* derivatization with acetic anhydride in basic media in order to decrease the polarity and to increase the molecular weight of these substances by acetylation. Only the two analytes 1,5-dimethyl-1,2-dehydro-3-pyrazolone and 4-(2-methylethyl)-1,5-dimethyl-1,2-dehydro-3-pyrazole were derivatized while the other compounds were quantitatively extracted without chemical transformation. The analytes were then separated by LC-APCI-MS/MS and quantified by comparison with the internal standard, dihydrocarbamazepine.[27] Although ESI led to higher peak intensities than APCI, the latter interface was chosen because it provided a matrix-independent ionization resulting in recoveries of ~100% (Table 2.1).

Although the use of GC-MS generally requires the derivatization of polar drugs, Boyd et al.[28] used this approach to analyze acetaminophen, fluoxetine, ibuprofen, naproxen, and clofibric acid, a human metabolite of clofibrate and etofibrate. The target compounds were isolated from wastewater, surface water, and untreated drinking water samples by SPE using a polar SDB-XC Empore disk. Derivatization with *N,O*-bis(trimethylsilyl)-trifluoroacetamide in the presence of trimethylchlorosilane was used to enhance thermal stability of clofibric acid, which thermally degraded in the GC injection port, and to reduce the polarity of specific target analytes (clofibric acid, ibuprofen, and naproxen) in order to facilitate their GC-MS analysis. The limits of detection were between 0.6 (clofibric acid) and 25.8 ng/L (fluoxetine). Although the recoveries for most of the compounds were greater than 47% (Table 2.1), acetaminophen was repeatedly not detected possibly due to the weak retention of this compound on the extraction disk.

Next, analytical methods using multiple extraction methods, different liquid chromatography eluents, or the combination of two detection techniques are reviewed.[10,29,30] Sacher et al.[29] reported the determination of 60 pharmaceuticals including analgesics, antirheumatics, β-blockers, broncholitics, lipid regulators including two metabolites, antiepileptics, vasodilators, tranquilizers, antitumoral drugs, iodated X-ray contrast media and antimicrobials in groundwater with different SPE procedures, and the combination of GC-MS (after derivatization of the acidic compounds) and LC-ESI-MS/MS. Different stationary phases, pH (3, 5, and

TABLE 2.1
Methods for the Analysis of Drugs in Aqueous Environmental Samples

Analytes	Matrix	Extraction Procedure (for SPE: Sorbent, Sample pH; Elution Solvent(s)	Separation and Detection Method	Recovery [%]	Limit of Detection and Quantification	Ref.
Multiresidue method for neutral drugs and 2 metabolites	GW[1], SW[2], WW[3]	SPE[4]: Isolute C$_{18}$, pH 7–7.5; MeOH	LC-(+)ESI-MS/MS	GW>80; SW and WW: 9–97%	MQL[5]; GW:10 ng/L; WW: 25–250 ng/L	[20]
Multiresidue method for neutral and acidic drugs	SW, WW	SPE: HLB, pH 2; MeOH	LC-(+/−)ESI-MS/MS	Deionized water: 80	IDL[6]:12–32 ng/L	[21]
Multiresidue method for 15 drugs and 4 human metabolites	DW7, SW, WW	SPE: HLB; MeOH/MTBE (1:9)	LC-(+/−)ESI-MS/MS	90–110	MDL: 0.25–1 ng/L	[22]
Multiresidue method for neutral and acidic drugs and 5 metabolites	SW, WW	SPE: HLB, pH 7; MeOH [Optimization of stationary phases and pH]	LC-(+/−)ESI-MS/MS	SW: 50–116; WW: 60–102	MDL[8]; SW:1–30 ng/L; WW:3–160 ng/L	[23]
Multiresidue method for neutral and basic drugs	GW, SW, WW	SPE: HLB, pH 10; MeOH [Optimization of washing and eluting solvent, and of pH]	LC-(+)ESI-MS/MS	GW: 50–119; SW: 22–113; WW: 64–115	IQL[9]: 0.46–10.6 µg/L; MQL; GW:1–10 ng/L, SW: 1–24 ng/L, WW: 1.4–29 ng/L	[26]
Multiresidue method for neutral and acidic drugs and 5 metabolites	GW, SW, WW	SPE: C$_{18}$; MeOH	LC-(+)APCI-MS/MS Interface optimization (ESI and APCI)	GW, SW and WW:87–117 except for dimethylaminophenazone	MQL: 10–20 ng/L	[27]
Multiresidue method for acidic drugs	WW	SPE: LiChrolute 100 RP-18, pH 2; MeOH	LC-(−)ESI-MS/MS	59–92	MDL: 5–20 ng/L	[24]

Description	Matrix	SPE/Extraction	Technique	Recovery (%)	Limits	Ref.
Multiresidue method for 28 drugs and 2 metabolites	WW	SPE: 1 Oasis MCX, pH 1.5/2.0; MeOH, MeOH (+2% NH₃) and MeOH (+0.2% NaOH) 2. LiChrolute EN, pH 7; MeOH, EtOAc [Optimization of stationary phases and pH]	LC-(+/−)ESI-MS/MS	>70	IQL: 600–39400 ng/L MQL:0.1–5.2 ng/L	[25]
Multiresidue method for 4 drugs and 1 human metabolite	DW, SW, WW	SPE: SDB-XC, pH 2; MeOH, CH₂Cl₂/MeOH	GC-MS	47–88	IDL:0.6–25.8 ng/L	[28]
Combined method for 60 drugs and their metabolites	SW	SPE: RP C₁₈ pH 3, PPL, Bond-Elut pH 7, LiChrolute EN pH 3, Isolut ENV+ pH 5; MeOH, acetonitrile, water, triethylamine	GC-MS and LC-(+)ESI-MS/MS	36–151	IDL: 1.8–13 ng/L	[29]
Combined method for 11 drugs and 2 metabolites	SW, WW	SPE: Strata X, pH 3; MeOH [Stationary phase optimization]	LC-(+/−)ESI-IT-MS	>60 Except for lofepramine and mefenaminic acid	MQL: 10–50 ng/L	[30]
Combined method for 11 drugs and 2 metabolites	SW, WW	SPE: Strata X, pH 3; MeOH, MeOH (+2% HOAc), MeOH (+2% NH₃) [Stationary phase optimization]	LC-(+/−)ESI-IT-MS	>60 Except for chloroquine and chlosantel	MDL: 1–20 ng/L IQL: 20–105 pg	[32]
Combined method for 13 drugs	SW, GW, DW	SPE: Oasis-MCX, pH 3; MeOH/NH₃ (19:1)	LC-(+/−)ESI-MS/MS and LC-QqToF-MS	60–75 except for fenofibrate (36)	MQL: 5–25 ng/L	[33]

(Continued)

TABLE 2.1
(Continued)

Analytes	Matrix	Extraction Procedure (for SPE: Sorbent, Sample pH; Elution Solvent(s))	Separation and Detection Method	Recovery [%]	Limit of Detection and Quantification	Ref.
Combined method for >50 drugs and their metabolites	SW	SPE: 1:tandem HLB and MCX; MeOH (+5% NH$_3$) 2:HLB: MeOH	LC-(+)ESI-MS	>80	-10	[10]
Combined method for >10 drugs and 3 human metabolites	SW, Seawater	SPE: Oasis HLB; hexane, EtOAc and MeOH	LC-(+)ESI-MS/MS and GC-MS	70–100	MQL:0.07–0.69 ng/L	[34]
Combined method for 5 neutral ad acidic drugs and 1 metabolite	SW, WW	SPE: Oasis HLB; EtOAc/acetone (1:1)[Optimization of washing and eluting solvent]	GC-MS	71–118	MDL 1–10 ng/L	[35]
Methodology by therapeutic class						
Acidic drugs, acetylsalicylic acid and 4 metabolites	SW, DW, WW	SPE: C18, pH 2; MeOH	GC-MS and GC-IT-MS/MS	58–90	MQL: WW: 50–250 (GC-MS) SW:5–20 (GC-MS) DW: 1–10 (GC-IT-MS/MS)	[36]
8 acidic drugs	SW	SPE: C18, pH 2; MeOH	GC-IT-MS/MS	≤90	MQL:1 ng/L	[38]
5 acidic drugs	GW, SW, WW	SPE: Oasis MCX, pH 2; acetone	LC-(−)ESI-MS/MS	GW: 82–103 SW: 75–112 WW: 57–100	MQL:1–25 ng/L	[39]

Analytes	Matrix	SPE/Extraction	Analytical method	Recovery (%)	Limits	Ref.
31 antimicrobials	WW	SPE: (1) Oasis HLB, pH 6 (2) Oasis HLB, pH 3 and Na$_2$EDTA; MeOH	LC-(+)ESI-MS/MS	72–99	MDL: 1–8 ng/L	[44]
18 antimicrobials	GW, SW	SPE: LiChrolute EN + LiChrolute C$_{18}$, pH 3; MeOH Freeze-drying	LC-(+/–)ESI-MS/MS	SPE: 58–120 Freeze-drying: 54–102	MQL$_{SPE}$: 2–5 ng/L MQL$_{Freeze-drying}$: 20–50 ng/L	[46]
13 antimicrobials	GW SW, WW	SPE: Oasis HLB, pH<3; acetonitrile (+1% NH$_3$) [Optimization of pH, sorbent and elution solvent]	LC-(+)-ESI-IT-MS	GW: 51–120 SW: 74–127 WW: 82–126	MDL: 0.027–0.19 µg/L MQL: 0.10–0.65 µg/L	[43]
13 antimicrobials	SW	SPE: Oasis HLB, pH 2–3; MeOH	LC-(+)-ESI-IT-MS and LC-DAD	96–102	MDL: 0.05 µg/L MDL$_{US EPA}$: 0.03–0.5 µg/L MQL: 0.1 µg/L	[47]
13 antimicrobials and 1 human metabolite	WW	SPE: Oasis HLB, pH 4; MeOH/EtOAc (1:1) and MeOH (+1% NH$_3$)	LC-(+)ESI-MS/MS	89–108 (trimethoprim: 47)	MQL: 0.3–77 ng/L	[42]
8 neutral drugs	DW, SW, WW	SPE: C$_{18}$, pH 7.5; MeOH	GC-MS and LC/(+)ESI-MS/MS	54–102	MQL GC-MS: 20–250 ng/L LC/(+)ESI-MS-MS:10ng/L–	[48]
4 blood lipid regulators	GW, SW, WW	SPE: C$_{18}$, pH 7	LC-(+/–)ESI-MS/MS	SW: 71–86 WW: 61–91	IDL: 0.7–15.4 pg MDL: 0.1–15.4 ng/L	[49]
Carbamazepine and 5 metabolites	SW, WW	SPE: Oasis HLB, pH 7; MeOH [Optimization of stationary phases]	LC-(+)ESI-MS/MS	SW: 96–103 WW: 84–104	IDL: 0.8–4.8 pg	[51]

(Continued)

TABLE 2.1
(Continued)

Analytes	Matrix	Extraction Procedure (for SPE: Sorbent, Sample pH; Elution Solvent(s))	Separation and Detection Method	Recovery [%]	Limit of Detection and Quantification	Ref.
2 antitumoral drugs	SW, WW	SPE: Macroporous polystyrene divinylbenzene; MeOH	LC-(+)ESI-MS/MS	75–102	MDL: GW: 0.02–0.1 ng/L SW: 0.02–0.1 ng/L GW: 0.3–2 ng/L	[54]
3 antitumoral drugs	WW (Hospital effluents)	SPE: C$_8$; pH 7.5; MeOH/ CHCl$_3$ (1:2)	LC-Fluorescence	85–87	MDL: 0.05–0.06 µg/L MQL: 0.26–0.29 µg/L	[55]
β-Blockers and β$_2$-symphatomimetics	SW, GW, WW	SPE: C$_{18}$-endcapped, pH 7.5; MeOH	GC-MS and LC-(+)ESI-MS/MS	>70	MQL: SW and GW: 5–10 ng/L WW: 50 ng/L	[48]
7 estrogens	SW, WW	SPE: acetone	GC-IT-MS/MS	SW: 41–90 WW: 56–82	MQL:SW: 0.5–1 ng/ LWW: 1–2 ng/L	[57]
6 estrogens, 3 conjugated estrogens, and 3 progestrogens	Water	—	GC-MS, LC-MS and LC-MS/MS [Optimization of the interface and ionization mode]	—	IDL: C/MS: 1–20 ng/mL LC-ESI-MS: 0.1–20 ng/L LC-ESI-MS/MS: 0.1–10 ng/L	[58]
1 ICM[11]	WW	SPE: ENV+, pH 2.8; MeOH	LC-(+)ESI-MS/MS	75	MDL: 6.7 ng/L MQL: 20 ng/L	[12]

Compound	Matrix	Extraction/method	Technique	Recovery (%)	Limit	Ref.
6 ICM	DW, SW, WW	SPE: ENV+, pH 2.8; MeOH [Optimization of stationary phases]	LC-(+)ESI-MS/MS	SW: 90–116 WW: 57–113 (ioxithalamic acid: 35)	MQL:10–50 ng/L	[62]
4 ICM and 2 possible human metabolites of ICM	SW, WW	SPE: ENV+ and Envi-Carb, pH 3.5; MeOH and acetonitrile/water (1:1) back flush	LC-(+)ESI/MS-MS	55–100	MDL: 50 ng/L	[63, 64]
4 ICM	GW	SPE: ENV, pH 3; MeOH	LC-(+)ESI-MS/MS	9–46	MDL: 2.3–4.8 ng/L	[29]
7 ICM	SW, DW	Direct injection	IC-ICP-MS	97–99	IDL: 0.02–0.04 µg/L	[65]
Cocaine and its metabolite (abuse drug)	SW, WW	SPE: Oasis HLB, pH 2; MeOH and MeOH (+2% NH3)	LC-(+)ESI-MS/MS and LC-ESI-IT-MS	>90	MDL: Cocaine :0.12 ng/L Metabolite: 0.06 ng/L	[16]
1 barbiturate	GW	—	LC-DAD	GW: >95	—	[69]
6 barbiturates	GW, SW, WW	SPE: Oasis HLB, pH 7; acetone and EtOAc	GC-MS	GW: 67–104 SW: 64–105 WW: 52–105	MDL: SW: 1–5 ng/L WW: 10–20 ng/L	[72]

[1] GW: Groundwater
[2] SW: Surface water
[3] WW: Wastewater
[4] SPE: Solid phase extraction
[5] MQL: Method quantification limits
[6] IDL: Instrumental limit of detection
[7] DW: Drinking water
[8] MDL: Method detection limit
[9] IQL: Instrumental quantification limit
[10] Not reported
[11] ICM: Iodinated Contrast Media for X-ray

7), and eluting solvents were used for the determination of the target compounds. Recoveries varied between 75 and 125% in both tap water and surface water, and limits of detection were very low for some of the target compounds (Table 2.1).

A combined method was developed for the determination of 11 pharmaceuticals (dextropropoxyphene, diclofenac, erythromycin, ibuprofen, lofepramine, mefenamic acid, paracetamol, propanolol, sulfamethoxazole, tamoxifen, trimethoprim) and 2 metabolites (N4-acetyl-sulfamethoxazole and clofibric acid). The method relied on a one-step-SPE, an HPLC separation using four different solvent gradients and detection by IT-MS in consecutive reaction monitoring (CRM) mode.[30] A number of stationary phases were evaluated for the extraction of the target compounds (Isolute ENV+, Oasis HLB, Oasis MCX, Isolute C_8, Isolute C_{18}, Varian Bond Elut C_{18}, and Phenomenex Strata X). The latter two sorbents were identified as being the most effective, and Strata X was shown the better phase for extracting the majority of the selected compounds. Recoveries typically higher than 60%, except for lofepramine (not recovered) and mefenamic acid (24%), were found.[30] For some pharmaceuticals, ionization suppression due to solvent gradient is critical and must be optimized accordingly for individual analytes. Areas of ion suppression by the matrices were identified by injecting a blank sample matrix (sewage effluent and freshwater) into a stream of analyte causing an elevated baseline.[31] Only the suppression of N4-acetyl-sulfamethoxazole by the effluent matrix was a cause of concern. Another method using IT-MS in CRM mode for the determination of an innovative list of 10 pharmaceuticals (chloropromazine, chloroquine, closantel, fluphenazine, miconazole midazolam, niflumic acid, prochlorperazine, trifluoperazine, and trifluperidol) listed on the Oslo and Paris Commission for the Protection of the Marine Environment of the North East Atlantic (OSPAR) as well as for fluoxetine in water was developed.[32] The limited occurrence of these compounds was thus not surprising, as some of them are used in fairly small quantities in the country studied. Three extraction materials, Oasis HLB, the mixed mode Oasis HLB cation-exchange cartridges MCX, and Phenomenex Strata X, were tested showing recoveries greater than 60% for the third extraction material for almost all the compounds except for closantel and cloroquine. Method detection limits were in agreement with those reported by Hilton et al.[30] using also IT-MS in CRM mode.

Stolker et al.[33] reported a combined methodology using LC-(+/-)ESI-MS/MS and quadrupole-time of flight mass spectrometry (LC-QqToF-MS) for the analysis of 13 pharmaceuticals, including 4 analgesics (acetylsalicylic acid, diclofenac, ibuprofen, and paracetamol), 3 antimicrobials (sulfamethoxazole, erythromycin, and chloramphenicol), 5 blood-lipid regulators and β-blockers (fenofibrate, bezafibrate, clofibric acid, bisoprolol, and metoprolol), and the antiepileptic drug carbamazepine. The samples were extracted in HLB-MCX SPE column, and the recoveries of the method were between 60 and 75% for all the compounds except fenofibrate, whose recovery was too low—36%; probably because of its relatively nonpolar character, the selected extraction conditions were not optimum for this compound.[33] Other authors[24] reported higher recoveries for fenofibrate—more than 90% using LiChrolute 100 RP-18 as a stationary phase to extract this compound from waters samples. Acetylsalicylic acid presented a recovery of 195%. This could be explained by the phenomenon of ion enhancement for this early eluting compound. LC-QqToF-MS was used only for confirmatory purposes.

In another study,[34] the combination of LC-ESI-MS/MS and GC-MS after derivatization with methylchloromethanoate for the determination of selected pharmaceuticals, among them analgesics with emphasis on ibuprofen and its metabolites (hydroxy-ibuprofen and carboxy-ibuprofen), β-blockers, antidepressants in wastewater and seawater, was reported. The extraction procedure was performed in 6-mL glass cartridges with the same packing material as in Oasis HLB cartridges. Limits of quantification for the entire method were in the range of 0.07 to 0.69 ng/L for GC-MS, and the recoveries were between 70 and 100%. An interesting result of this work[34] was the quantification of ibuprofen and its two metabolites in the two types of water showing characteristic patterns, with hydroxy-ibuprofen being the major component in sewage, whereas carboxy-ibuprofen was dominant in seawater samples. The determination of neutral (carbamazepine) and acidic pharmaceuticals (ibuprofen, naproxen, ketoprofen, diclofenac, and clofibric acid) in surface water and wastewater was also performed with SPE using Oasis HLB. Samples were analyzed by GC-MS after derivatization with diazomethane.[35] The authors analyzed the extract from SPE twice, first directly after the SPE method and then after derivatization. Recoveries for ketoprofen, diclofenac, and carbamazepine were low when methanol was used as eluting solvent. Therefore, solvent mixtures of ethyl acetate-methanol or ethyl acetate-acetone were evaluated. A mixture of ethyl acetate-acetone (50:50) provided the best recoveries for all compounds. Regarding the optimization of the washing solvent mixture, the authors[35] found that up to 20% of methanol in the washing solvent did not affect analyte recoveries, even for the highly polar compound, clofibric acid. However, significant analyte loss occurred for all target compounds when the methanol content was increased to 50%. In order to be more cautious, they used methanol/water (10:90) as a washing solvent for all subsequent experiments. Relative recoveries (corrected with internal standard) were between 71 and 118%. The method detection limits were between 1 and 10 ng/L (i.e., two orders of magnitude higher relative to those reported by Weigel et al.[34]).

2.3   DETERMINATION OF DRUGS ACCORDING TO THEIR CLASS

In this section a comprehensive review of methods developed for specific therapeutic classes is provided. A large number of publications dealing with the determination of drugs in water using advanced mass spectrometric techniques have been published. The extraction and preconcentration techniques involve SPE in which many different sorbent types, eluting schemes, and solvents with or without ion pairing reagents, buffers, and modifiers were used. Discussion here will be limited to aqueous samples because Chapter 3 will be devoted to issues related to extraction and analysis of solid-bound pharmaceuticals.

2.3.1   ANALGESICS AND ANTIINFLAMMATORY DRUGS

Analgesic and antiinflammatory drugs are ubiquitous in wastewater effluents of municipal WWTPs[36] and as a result are found in surface waters. This group of compounds is among the major pharmaceutical pollutants in recipient waters at concentrations of up to μg/L levels.[14] For example, bezafibrate has been found in WWTP effluent and surface water sample at concentrations as high as 4.6 and 3.6 µg/L, respectively.[36]

Most of the members of this group are acidic in nature because they contain carboxylic moieties and one or two phenolic hydroxyl groups showing pK_a values between 3.6 and 4.9. At neutral pH they exist mainly in their ionized form; therefore, the sample pH has to be adjusted to a pH between 2 to 3 in order to protonate the carboxylic and hydroxyl groups in order to achieve high and reproducible recoveries. First works reported the use of GC-MS or GC-MS/MS for the determination of these compounds in water matrices. Ternes et al.[37] described a methodology for the determination of some analgesics, antiinflammatory drugs, and lipid regulators and two metabolites of ibuprofen (hydroxy- and carboxy-ibuprofen), together with compounds such as salicylic acid, the main metabolite of acetylsalicylic acid in sewage, river, and drinking water. The method consisted of SPE using RP-C_{18} followed by methylation of the carboxylic groups with diazomethane, acetylation of phenolic hydroxyl groups with acetanydride/triethylamine, and determination by GC-MS and GC-IT-MS/MS. The MQL down to 10 ng/L were achieved in wastewater effluents as well in river water by GC-MS and down to 1 ng/L using GC-IT-MS/MS. Other authors also used GC-IT-MS/MS for the determination of eight acidic pharmaceuticals in water by SPE on RP-C_{18}. In-port methylation in the GC using trimethylsulfoniumhydroxide improved the detection limits such that concentrations in the ng/L range could be achieved.[38] Recently LC-MS/MS have become the common methodology for the separation and detection of the analgesic, antiinflammatory, and blood lipid regulator drugs. Acidic drugs, most of which are derivatives of phenyl acetic acid, often have been detected under negative ionization mode conditions, and deprotonated molecules [M-H]$^-$ were chosen as a precursor ions.[24] Typically they showed, to varying degrees, the characteristic tendency to lose CO_2, leading to a benzyl anion that is stabilized by conjugation with the aromatic ring and a limited number of other ions. For neutral compounds like fenoprofen, acetaminophen, propylphenazone, and phenylbutazone the analysis has been carried out in positive mode, and all precursor ions were the result of [M+H]$^+$of the molecule.

An analytical methodology for the determination of five acidic pharmaceuticals—ibuprofen, naproxen, ketoprofen, diclofenac, and bezafibrate—in water with Oasis MCX and LC-ESI-MS/MS in negative mode was developed.[39] Absolute and relative recoveries (relative to the recovery of the surrogate standard) were reported for groundwater, surface water, and wastewater. The relative recoveries (Table 2.1) were significantly higher than the absolute recoveries. The analytical procedure gave good recoveries for ibuprofen, naproxen, ketoprofen, and diclofenac. Nevertheless, in the WWTP effluent, the relative recovery of ibuprofen was only 57% and 67% for bezafibrate. Low method limits of detection were reported for ibuprofen, diclofenac, and bezafibrate. They were 1 ng/L in ground and surface waters and 5 ng/L in WWTP samples, and 5 and 25 ng/L for naproxen and ketoprofen, respectively (Table 2.1).

2.3.2 ANTIMICROBIALS

Antimicrobials are widely used in human and veterinary medicine to prevent or treat bacterial infections. In addition, veterinary applications include use of antimicrobials as feed additives at subtherapeutic doses to improve feed efficiency and promote growth.[40] Antimicrobials are of a particular concern because their wide application has led to the selection of resistant bacteria that can threaten the effectiveness of antimicrobials for the treatment of human infections. Antibiotic-resistant bacteria reach the environment through animal and human waste, which could transfer the resistance genes to other bacterial species.[41]

Some studies have reported the occurrence of antimicrobials in sewage sludge of WWTPs[42] and surface waters[43] commonly using LC-MS/MS. For example, sulfonamides and tetracyclines, the two most frequently analyzed families of antimicrobials, are detected commonly by LC-(+)ESI-MS/MS. For tetracyclines, the product ions produced from the protonated molecule $[M+H]^+$ at low collision energies were the losses of H_2O and NH_3 (from tetracyclines containing a tertiary OH moiety at C-6) to finally give abundant $[M-H_2O-NH_3+H]^+$. For the sulfonamides group the dominant process from the protonated sulfonamide was the cleavage of the sulphur-nitrogen bond yielding the stable sulphanilamide moiety detected at m/z 156. Macrolides antibiotics are basic and lipophilic molecules that contain a lacton ring and sugars. These compounds underwent mass fragmentation losing two characteristic sugars (desosamine and cladinose) and water.

Currently, the methodologies developed for antimicrobial determination include a list of compounds representative of different classes of drugs because all are expected to have environmental effects. A methodology for 31 antimicrobials from the macrolide, quinolone, sulfonamide, and tetracycline classes using SPE and LC-(+)-MS/MS was developed.[44] Quantitative recoveries for all compounds, even for tetracyclines, were obtained (Table 2.1). The authors used for the extraction of tetracyclines Na_2EDTA as a chelating agent to decrease the tendency for those compounds to bind to cations into the matrix. To improve the resolution and peak shape of the tetracyclines in the chromatographic column, some authors[45] added oxalic acid to the mobile phase. In this study[44] the authors used oxalic acid and ESI operated to 380°C because nonvolatile reagent may accumulate in the ESI source, and at elevated probe temperatures oxalic acid decomposes to carbon dioxide and water. In the absence of stable isotope-labeled surrogate standards for quantitation, they prepared a series of standard solutions by spiking the analytes into filtered effluent samples and extraction by SPE and analysis by LC-ESI-MS/MS. Analytical data from the spiked samples were used to construct standard calibration curves for quantifying the analytes in the unspiked samples. These calibration curves compensated for both variations in the SPE recoveries and matrix effects that can either suppress or enhance signals with LC-ESI-MS/MS instrumentation.

Another multianalyte method for the determination of 18 antimicrobials in water using SPE (mixed LiChrolute EN and LiChrolute C_{18} materials) or freeze drying (100 mL of water) and LC-(-)ESI-MS/MS for all the compounds except for chloramphenicol in positive mode was described.[46] The analytes belonged to different groups of antimicrobials such as penicillins, tetracyclines, sulfonamides, and macrolide antibiotics. Except for dehydrated-erythromycin, trimethoprim, and tetracycline, recoveries from freeze drying were greater than 80%, and for SPE method recoveries were slightly lower than for the previous methodology. Tetracyclines were not recovered in the SPE methodology because Na_2EDTA was not added to the water.

However, Na$_2$EDTA was added in the freeze drying methodology. Method quantification limits using SPE were one order of magnitude lower due to the 1-L sample volume used for the determination of antimicrobials in water. Batt and Aga[43] developed an analytical method for the simultaneous determination of 13 antimicrobials belonging to 5 classes (fluoroquinolones, lincosamides, macrolides, sulfonamides, and tetracyclines) in wastewater, surface water, and groundwater. The authors optimized SPE methodology and LC-(+)-ESI-IT-MS as a detection method. The comparison between different pH value type of cartridges (C$_{18}$ and Oasis HLB), and different eluting solvents was performed. The optimum condition proved to be the use of samples adjusted to pH3, with Na$_2$EDTA added, and extraction using Oasis HLB. They used Na$_2$EDTA to efficiently extract the macrolides and tetracyclines. Although the pH adjustment did not affect the extraction efficiency of the majority of the compounds, the recovery for fluoroquinolones was reduced below 35% at no pH adjustment. Fluoroquinolones have exhibited acceptable recoveries in both basic and acidic pH; however, tetracyclines in acidic pH were better recovered. The authors applied an ion trap data-dependent scanning method, which simultaneously collected full scan and MS/MS data for unequivocal identification of target analytes. Other authors[47] also reported the determination of 13 antimicrobials, sulfonamides, and tetracyclines, in surface waters using SPE (Oasis HLB) and LC-(+)ESI-IT-MS but in CRM mode. For the enrichment of the water samples, pH between 2 and 3 was used because tetracyclines are not stable at pH < 2. They checked the effect of the column diameter, flow rate, and temperature (15°C, 25°C, 35°C) on the peak shape in the chromatographic separation, and 15°C was the temperature of choice because this lower temperature resulted in a better peak shape symmetry. To determine the matrix effects, the authors checked the performance of the internal standard simatone in the ESI interface during 7 months. No significant variability in the peak area during this time was observed, concluding that this standard was not affect by the matrix suppression. The authors checked the ion suppression of tetracyclines and sulfonamides, and the surface-water matrix effects were significant when measuring tetracyclines but not sulfonamides. Although the limits of detection and quantitation depend on the volume of sample extracted, the complexity of water matrices, and injection volume of extract, similar limits of quantitation were obtained in these studies.[43,47] To correct matrix suppression and losses in the SPE method, a method for 13 antimicrobials and the metabolite N4-acetyl-sulfamethoxazole in wastewater using five labeled internal standards was reported.[42] The method combined SPE (Oasis HLB) and LC-(+)ESI-MS/MS showing recoveries above 80% (Table 2.1), with the exception of trimethoprim, where they ranged between 30 and 47%, probably because of the use of nonideal surrogate standard (^{13}C$_6$ Simazine).

2.3.3 ANTIEPILEPTICS, BLOOD LIPID REGULATORS, AND PSYCHIATRIC DRUGS

Some of these classes of compounds are neutral pharmaceuticals without any acidic functional groups and therefore can be enriched at neutral pH on reverse phase materials; they can generally be analyzed by GC-MS without derivatization. However, LC-MS/MS is the method of choice because it has been shown to have better limits of detection and better selectivity. Ternes et al.[48] compared the determination by GC-MS and LC-ESI-MS/MS of eight neutral drugs—carbamazepine, clofibrate, dimethyl-aminophenazone, diazepam, etofibrate, fenofibrate, phenazone, and pentoxifylline. For three of these drugs—carbamezapine, phenazone, and pentoxifylline—the detection limits were improved to 10 ng/L, independent from the water matrix.

For the analysis of fibrates and statins, LC-MS/MS with ESI interface is preferred. With this technique the sensitivity is approximately tenfold higher than in an APCI source.[33] For the analysis of some statins, a class of compounds belonging to the blood lipid regulators like the fibrates, negative ionization mode is usually the method of choice. Miao and Metcalfe[49] reported the analysis of some statins (atorvastatin, lovastatin, pravastatin, and simvastatin) in waters with SPE and LC-ESI-MS/MS in both negative and positive mode. LC-(+)ESI-MS/MS with methyl-ammonium acetate as an additive in the mobile was more sensitive than negative mode for all compounds. Protonated atorvastatin and methylammonium-adducted lovastatin, pravastatin, and simvastatin were selected as precursor ions, and product ions were detected by MRM. The instrumental detection limits of atorvastatin, lovastatin, pravastatin, and simvastatin were 0.7, 0.7, 8.2, and 0.9 pg, respectively, and the method detection limits were between 0.1 and 15.4 ng/L.

Carbamazepine belongs to the group of antiepileptic drugs where it represents the basic therapeutic agent for treatment of epilepsy. This compound is one of the most frequently detected pharmaceuticals in wastewater and river water. This drug is generally analyzed with ESI interface in positive mode due to the higher sensitivity found compared with APCI. The protonated molecule undergoes fragmentation to the loss of 43 Da, which corresponds to the carbamoyl group.

Carbamazepine undergoes extensive hepatic metabolism by cytochrome P450 system. Thirty-three metabolites of carbamazepine have been identified from human and rat urine.[50] A quantitative method for simultaneous determination of carbamazepine and 5 of its 33 metabolites (10,11-dihydro-10,11-epoxycarbamaze-pine, 10,11-dihydro-10,11-dihydroxycarbamazepine, 2-hydroxycarbamazepine, 3-hydroxycarbamazepine, 10,11-dihydro-10-hydroxycarbamazepine) was reported.[51] The developed method encompassed an SPE procedure optimizing the stationary phase (Oasis HLB, Supelclean-18 and LC-18) and extracting the water samples at pH 7 followed by separation and detection with LC-(+)ESI-MS/MS. The Oasis HLB was finally chosen for SPE because of its superior extraction efficiencies showing recoveries for all the analytes including carbamazepine and its metabolites exceeding 80% in both water matrices. Cross-talk among some MRM channels was studied. The metabolite 10,11-dihydro-10,11-epoxycarbamazepine could be observed in channels m/z 253 $\rightarrow$ 210 and m/z 253 $\rightarrow$ 180, which were used to monitor 2-hydroxycarba-mazepine and 3-hydroxycarbamazepine. Therefore, chromatographic separation of the analytes was critical and was optimized on a C8 column using a tertiary solvent system.[51] All analytes and internal standard were resolved chromatographically with total run of 11 min. Matrix effects were also studied with four kinds of matrices, HPLC water, surface water, and influent and effluent from a WWTP. Ion suppression was highest in the influent. To correct ion suppression, 10,11-dihydrocarbamazepine was used as internal standard.

2.3.4   ANTITUMORAL DRUGS

Antitumoral drugs have carcinogenic, mutagenic, teratogenic, and fetotoxic properties. Cyclophosphamide and ifosfamide, both isomeric alkylating N-lost derivatives, are among the most frequently used antitumorals. Traces of cyclophosphamide and ifosfamide were detected in hospital effluents[52] as well as WWTP influents and effluents.[53] Apparently, ifosfamide reaches surface waters because no biodegradation occurs in WWTPs. One study determined the expected concentration of ifosfamide in German surface waters at 8 ng/L based on data modeling.[53] Buerge et al.[54] developed a highly sensitive analytical method based on SPE (macroporous polystyrene-divinylbenzene) and LC-(+)ESI-MS/MS for the determination of the antitumoral drugs ifosfamide and cyclophosphamide in wastewater and surface waters. Recoveries ranged from 74 to 94% for cyclophosphamide and from 75 to 102% for ifosfamide. Method detection limits ranged from 0.002 to 0.1 ng/L in groundwater and between 0.2 and 2 ng/L for wastewaters (Table 2.1). Mahnik et al.[55] reported a method to determine anthracyclines in hospital effluents by SPE (C_8) combined with LC-fluorescence detection. To extract the anthracyclines from water, bovine serum was added in order to have linear standard curves because these compounds adsorb on surfaces. The authors obtained quantitative recoveries >80% and limits of quantitation in the low μg/L.

2.3.5   CARDIOVASCULAR DRUGS (β-BLOCKERS) AND β₂-SYMPATHOMIMETICS

These compounds contain a secondary aminoethanol structure as well as several hydroxy groups. Due to their polarity these compounds are usually determined with LC-MS/MS and as ionization mode ESI in positive mode due to their basic character. The protonated molecule is the selected precursor ion, and the most intense diagnostic ion is m/z 116 corresponding to [(N-isopropyl-N-2-hydroxypropylamine)].

Ternes et al[48] reported the determination of several β-blockers and β₂-sympathomimetics in waters comparing two methods of separation and detection: GC-MS and LC-MS/MS. For GC-MS, the sample preparation included SPE (C_{18}-endcapped), a two-step derivatization by silylation of the hydroxyl groups and trifluoroacetylation of the secondary amino moieties. For LC-MS/MS, only the extraction of the water with SPE was necessary. The recoveries exceeded 70% for both methodologies; only atenolol, sotalol, and celiprolol were not detected by GC-MS. The method limits of quantification were comparable for the both techniques, being 5 to 10 ng/L in drinking water and surface water and 50 ng/L in wastewater.[48] The authors recommended the use of LC-(+)ESI-MS/MS for the analysis of these polar molecules in the environment, because the derivatization of the hydroxyl groups required for GC-MS analysis was incomplete.

2.3.6   ESTROGENS

Recently, a multitude of chemicals have shown to act as endocrine disrupters disturbing the hormonal systems of aquatic organisms. These compounds can be classified into naturally occurring and xenobiotic compounds.[56] Natural substances include sex hormones (estrogens, progesterone, and testosterone) and phytoestrogens, while xenobiotic endocrine disruptors include synthetic hormones, such as the contraceptive 17α-ethinylestradiol and man-made chemicals and their by-products (e.g., pesticides and flame retardants). Natural hormones and contraceptives are endocrine disruptors with effective concentrations at low ng/L levels.[57] Estrogens encompass a group of compounds with steroid structures containing phenolic and sometimes aliphatic hydroxy groups. Both natural and synthetic can be analyzed simultaneously because of their physicochemical properties.

A sensitive method using SPE (combination of RP-C18 and LiChrolute EN) and GC-IT-MS/MS for the quantification of seven estrogens in sewage samples and river water was developed.[57] Recoveries of the analytes in groundwater after SPE, cleanup, and derivatization generally exceed 75%. Method limit of quantification in different waters were between 0.5 ad 1 ng/L (Table 2.1). The confirmation provided with GC-IT-MS/MS was essential, because 17α-ethynilestradiol and an unknown compound exhibited exactly the same retention time.

Another paper compared different mass spectrometric approaches (derivatized sample with N,O-bistrimethylsilyl-trifluoroacetamide and detected with GC/MS as well as LC/MS and LC-MS/MS without derivatization) for the analysis of estrogens (both free and conjugated) and progestogens.[58] For LC-MS and LC-MS/MS, different instruments, ionization techniques (ESI and APCI), and ionization modes (positive and negative) were employed. Although LC-ESI-MS showed instrumental detection limits comparable with those obtained with LC-ESI-MS/MS (0.1-10 ng/mL), LC-ESI-MS/MS was the method of choice based on the selectivity of this method that provides the feature to avoid false-positive determinations.

2.3.7 X-Ray Contrast Agents

Iodinated X-ray contrast media (ICM), such as iopromide and diatrizoate, are widely used in human medicine for imaging of organs or blood vessels during diagnostic tests. They are metabolically stable in the human body and are excreted almost completely within a day. As such they are frequently detected in WWTP effluents and surface waters due to their persistence and high usage.[59] Monitoring studies of iopromide and diatrizoate in municipal WWTPs showed no significant removal of these compounds throughout the plant.[60,61]

The various analytical methodologies commonly used for the determination of ICM in aqueous matrices are summarized in Table 2.1. All but one method for the environmental analysis of ICM described in the literature encompass an enrichment of the aqueous sample by means of SPE. Hirsch et al.[62] optimized a protocol for the determination of six ICM (diatrizoate, iomeprol, iopamidol, iopromide, iothalamic acid, and ioxithalamic acid) in aqueous matrices with SPE and LC-(+)ESI-MS/MS. Several SPE sorbents (LiChrolute RP-C_{18}, LiChrolute EN, combination of the cartridges RP-C_{18} and EN, and Isolute ENV+) were tested. SPE using Isolute ENV+ material proved to be the method of choice since higher recoveries and lower quantitation limits were achieved (Table 2.1). Regarding the matrix suppression effects in the ESI interface, the authors studied the recoveries of the ICM in different water samples. Thanks to the addition of a surrogate standard, the relative recoveries from surface-water samples were generally quantitative. However, analyzing the WWTP effluents, some matrix influences for ioxithalamic acid were present, reducing its recovery to 5%. Adopting this approach, several authors reported similar method performance.[12]

Putschew et al.[63,64] proposed another analytical method based on a sequential SPE for the isolation of five ICM (diatrizoate, iopromide, iotrolan, iotrolan, iotroxin acid) and their possible metabolites (ipha and phipha) using LiChrolute EN and Enviro-Carb as extraction materials and LC-(+)ESI/MS-MS as the analyzing method. The recoveries for all compounds were higher than 70°%; only the ionic compounds were recovered at levels of between 55 and 61%. The method detection limits were 50 ng/L; lower detection limits could be achieved if trifluoroacetic acid was used in the mobile phase, but other acid could have affected negatively the separation of the ICM. In contrast, in another study[29] the recoveries obtained for the four ICM—iopamidol, iopromide, iomeprol, and diatrizoate—were low (<50%) when only LiChrolute EN sorbent was used. This can be attributed to the extremely high polarity and water solubility of these compounds.

The same group, in another study[65] using ion chromatography with inductively coupled plasma mass spectrometry (IC-ICP-MS) without previous sample preconcentration, found that limits of detection below 0.2 µg L^{-1} could be achieved. Reproducibility was below 6% for the six ICM studied. Comparing the sensitivity and specificity of the two methodologies, direct injection and detection IC-ICP-MS[65] and SPE and LC-MS/MS,[29] reported by the same group, LC-MS/MS offered a significantly higher sensitivity (MDL below 10 ng/L) and specificity. However, the IC-ICP-MS method offered the possibility of detecting other iodine-containing compounds besides the target analytes.

For the determination of ICM in environmental samples, ESI usually operated in the positive ion mode has been the preferred method for the sensitive detection of these polar analytes with molecular weights of up to 1600 Da. For monomeric structures the protonated precursor ion usually produces the loss of H_2O and HI.[66] The application of (−)-ESI mode was particularly attractive for those ICM bearing a free carboxylic acid, though nonionic ICM have also been reported to produce [M−H]$^-$ ions.[67] Negative ionization was also successfully applied to the compound-class specific detection of ICM all of which carrying aromatic iodine. Operating the ion source at a high cone voltage led to in-source fragmentation of the deprotonated parent molecules resulting in the formation of the diagnostic iodide anion that was monitored at m/z 127 during a selected ion monitoring acquisition.[67] Although this approach suffered selectivity as compared with the MRM mode traditionally used in triple quadrupole instruments, monitoring a single ion during the entire chromatographic run added some sensitivity to the technique.

2.3.8   DRUGS OF ABUSE

Although many substances are included in this group (heroin, tetrahydrocannabinol, cocaine, phencyclidine, LSD, psilocybin, and mescaline), only one study about the occurrence of cocaine in the aquatic environment has been published. Zuccato et al.[16] developed a method to analyze cocaine and its main human metabolite, benzoylecgonine, in surface and wastewaters using an SPE (Oasis HLB) method and LC-(+)ESI-MS/MS and LC-ESI-IT-MS. This was presented as a "nonintrusive" approach to determine abuse drug usage in the community. Recoveries were >90% for both compounds, and method limits of detection were 0.06 and 0.12 ng/L for benzoylecgonine and cocaine, respectively.

2.3.9   OTHER DRUGS

Barbiturates are derivatives of barbituric acid and act as central nervous system depressants; therefore, they produce a wide spectrum of effects—from mild sedation to anesthesia. Some are also used as anticonvulsants. Today, barbiturates are infrequently used as anticonvulsants and for the induction of anesthesia.[68] Benzodiazepines were mainly used as replacements, and since the introduction of diazepam (the first benzodiazepine prescribed for clinical use) in 1963, barbiturates have been gradually phased out. Nowadays, due to low usage, few reports of barbiturates in the environment are reported. However, two studies recently pointed out the need to investigate these compounds in the environment. Holm et al.[69] first reported on leachates carrying pharmaceuticals from a landfill. High concentrations (mg/L) of numerous sulfonamides and barbiturates (5,5-diallylbarituric acid) analyzed with LC-Diode array detector (DAD) from domestic waste and from a pharmaceutical manufacturer were found in leachates close to the landfill. Two studies also reported the occurrence of barbiturates in the environment, pentobarbital and 5,5-diallylbarituric acid in groundwater[70] and phenobarbital and the metabolite of the antiepileptic primidone in the effluent of a WWTP.[71] In the WWTP these two drugs were found at concentrations of 30 and 1000 µg/L, respectively.

Recently, Peschka et al.[72] reported the study of the occurrence of some barbiturates, including butalbital, secobarbital, pentobarbital, hexobarbital, aprobarbital, and phenobarbital, in surface and groundwaters. A method using an SPE (Oasis HLB) and GC was developed, showing method limits of detection down to 1 ng/L. Good recoveries of selected barbiturates were obtained from spiked surface water samples, with values between 64 and 105%, and groundwater and wastewater effluents with, in general, slightly lower values ranging from 67 to 104% and 52 to 105%, respectively (Table 2.1). The drugs were found in surface and groundwaters, indicating a strong recalcitrance of these compounds, which had been used at high levels in 1960.

2.4   CONCLUSION

Drugs and their metabolites are present in the environment at low levels, and the use of advanced analytical methods that afford low limits of detection and high selectivity has allowed for the detection of them at low ng/L levels. Moreover, thanks to these methodologies, the number of analytes detected has increased considerably during the last decade. The proper choice of an extraction and cleanup methodology for the determination of these emerging contaminants is an important step in the development of an analytical method, because the success of the analytical determination depends on the type of stationary phase and the washing and eluting solvents used for that purpose, as has been described in this chapter. The determination of human metabolites of drugs is challenging, because these biotransformation products are usually more polar than the parent compound and they might not be retained by conventional sorbents. Some authors reported the addition of derivatizing agents to the water in order to allow for efficient extraction of these polar metabolites with conventional extraction procedures. An additional challenge is the determination of some excreted metabolites. They are formed by conjugation with glucuronic acid or other polar moieties that are expected to be cleaved by microorganisms into the unchanged pharmaceuticals in the environment.

LC-MS/MS is the most frequently employed separation and detection technique in drug analysis, due to its high sensitivity and because it allows for unequivocal identification of the analytes. Although the use of this technique in ESI mode exhibits several advantages, ion suppression is an important process to be taken into account in the quantification of the analytes in view of the reported studies showing drastic matrix effects when ESI source was used.

Matrix effects (signal suppression or enhancement) are believed to result from the competition of the analyte ions and the matrix components for access to the droplet surface to the gas-phase emission. A feasible solution to address this issue is to use standard addition, but this is time consuming and cost intensive. Another approach relies on the use of an APCI source, which is much less subject to matrix-dependent ionization interferences. Although for many polar compounds ESI usually leads to higher peak intensities as compared with APCI, ESI signals for these compounds can be affected by the matrix of the sample. To minimize matrix effects an optimization of the parameters of the extraction and cleanup of the sample, including stationary phases, and washing and eluting solvent, can be a feasible approach for that purpose. Dilution of extracts has also been reported as an economic methodology for reducing matrix effects. Finally, the use of isotope dilution to compensate matrix effects also has been discussed in this chapter.

Multiresidue methods offer advantages in terms of providing a more comprehensive picture of the occurrence and fate of the contaminants in the environment examined. In addition, the simultaneous determination of a large number of analytes by a single method represents a less time-consuming and hence more economic approach as compared with applying several drug-class specific protocols.

ACKNOWLEDGMENTS

The work described in this article was supported by the EU Project (EMCO-INCO-CT-2004-509188) and by the Spanish Ministerio de Educación y Ciencia Project EVITA (CTM2004-06255-CO3-01). This work reflects only the author's views, and the European Community is not liable for any use that may be made of the information contained therein. SP acknowledges a postdoctoral contract from I3P Program (Itinerario Integrado de Inserción Profesional), cofinanced by CSIC and European Social Funds.

REFERENCES

1. Daughton, C. and Ternes, T.A. 1999. Pharmaceuticals and personal care products in the environment: agents of subtle change? *Environ. Health Perspect.* 107:907.
2. Daughton, C. 2004. Non-regulated water contaminants: emerging research. *Environ. Impact Assess. Rev.* 24:711.

3. Jones, O.A., Lester J.N., and Voulvoulis, N. 2005. Pharmaceuticals: a threat to drinking water? *Trends Biotechnol.* 23:163.

4. Mellon, M., Benbrook, C., and Benbrook, K.L. *Hogging it: estimates of antimicrobial abuse in livestock.* A Report of Union of Concerned Scientists, Cambridge, MA. 2001.

5. Schulman, L.J., Sargent, E.V., Naumann, B.D., Faria, E.C., Dolan, D.G., and Wargo, J.P. A human health risk assessment of pharmaceuticals in the aquatic environment. 2002. *Hum. Ecol. Risk. Assess.* 8:657.

6. Erickson, B.E. Analyzing the ignored environmental contaminants. 2002. *Environ. Sci. Technol.* 36:141A.

7. Hopfgartner, G., Varesio, E., Tschäppät, V., Grivet, C., Bourgogne, E., and Leuthold, L.A. 2004. Triple quadrupole linear ion trap mass spectrometer for the analysis of small molecules and macromolecules. *J. Mass Spectrom.* 39:845.

8. Hirsch, R., Ternes, T., Haberer, K., and Kratz, K.L. 1999. Occurrence of antibiotics in the aquatic environment. *Sci. Total Environ.* 225:109.

9. Petrovic, M., Hernando, M.D., Diaz-Cruz, M.S., and Barceló, D. 2005. Liquid chromatography-tandem mass spectrometry for the analysis of pharmaceuticals residues in environmental samples: a review. *J. Chromatogr. A* 1067:1.

10. Kolpin, D.W., Furlong, E.T., Meyer, M.T., Thurman, E.M., Zaugg E.D., Barber, L.B., and Buxton, H.T. 2002. Pharmaceuticals, hormones and other organic wastewater contaminants in U.S. streams, 1999–2000: a national reconnaissance. *Environ. Sci. Technol.* 36:1202.

11. Pérez, S., Eichhorn, P., and Aga, D.S. 2005. Evaluating the biodegradability of sulfamethazine, sulfamethoxazole, sulfathiazole and trimethoprim at different stages of sewage treatment. *Environ. Toxicol. Chem.* 24:1361.

12. Carballa, M., Omil, F., Lema, J.M., Llompart, M., García-Jares, C., Rodríguez, I., Gómez, M., and Ternes, T. 2004. Behavior of pharmaceuticals, cosmetics and hormones in a sewage treatment plant. *Water Res.*, 38:2918.

13. Glassmeyer, S.T., Furlong, E.T., Kolpin, D.W., Cahill, J. D., Zaugg, S. D., Werner, S. L., Meyer, M.T., and Kryak, D.D. 2005. Transport of chemical and microbial compounds from known wastewater discharges: potential for use as indicators of human fecal contamination. *Environ. Sci. Technol.* 39:5157.

14. Heberer, T. 2002. Occurrence, fate, and removal of pharmaceutical residues in the aquatic environment: a review of recent research data. *Toxicol. Lett.* 131:5.

15. Calamari, D., Zuccato, E., Castiglioni, S., Bagnati, R., and Fanelli, R. 2003. Strategic survey of therapeutic drugs in the rivers Po and Lambro in northern Italy. *Environ. Sci. Technol.* 37:1241.

16. Zuccato, E., Chiabrando, C., Castiglioni, S., Calamari, D., Bagnati, R., Schiarea, S., and Fanelli, R. 2005. Cocaine in surface waters: a new evidence-based tool to monitor community drug abuse. *Environ. Health: A Glob. Acc. Sci. Source* 4:14.

17. Quintana, J.B., Weiss, S., and Reemtsma, T. 2005. Pathways and metabolites of microbial degradation of selected acidic pharmaceutical and their occurrence in municipal wastewater treated by a membrane bioreactor. *Water Res.* 39:2654.

18. Kalsch, W. 1999. Biodegradation of the iodinated X-ray contrast media diatrizoate and iopromide. *Sci. Total Environ.* 255:143.

19. Pérez, S. and Barceló, D. 2007. Application of advanced mass spectrometric techniques in the analysis and identification of human and microbial metabolites of pharmaceuticals in the aquatic environment. *Trends Anal. Chem.* 26:494.

20. Ternes, T.A., Bonerz, M., and Schmidt, T. 2001. Determination of neutral pharmaceuticals in wastewater and rivers by liquid-chromatography-electrospray tandem mass spectrometry. *J. Chromatogr.* 938:175.

21. Vanderford, B.J., Pearson, R.A., Rexig, D.J., and Snyder, S. 2003. Analysis of endocrine disruptors, pharmaceuticals, and personal care products in water using liquid chromatography/tandem mass spectrometry. *Anal. Chem.* 75:6265.

22. Vanderford, B.J., and Snyder S. 2006. Analysis of pharmaceuticals in water by isotope dilution liquid chromatography/tandem mass spectrometry, *Environ. Sci. Technol.* 40:7312.

23. Gros, M., Petrovic, M., and Barceló, D. 2006. Development of a multi-residue analytical methodology based on liquid-tandem mass spectrometry (LC-MS/MS) for screening and trace level determination of pharmaceuticals in surface and wastewaters. *Talanta* 70:678.

24. Miao, X.S., Koenig, B.G., and Metcalfe, C.D. 2002. Analysis of acidic drugs in the effluents of sewage treatment plants using liquid chromatography—Electrospray ionization tandem mass spectrometry. *J. Chromatogr. A* 952:139.

25. Castiglioni, S., Bagnati, R., Calamari, D., Fanelli, R., and Zuccato, E., 2005. A multiresidue analytical method using solid-phase extraction and high-pressure liquid chromatography tandem mass spectrometry to measure pharmaceuticals of different therapeutic classes in urban wastes. *J. Chromatogr.* 1092:206.

26. Vieno, N.M., Tuhanen, T., and Kronberg, L. 2006. Analysis of neutral and basic pharmaceuticals in sewage treatment plants and in recipient rivers using solid-phase extraction and liquid chromatography-tandem mass spectrometry detection. *J. Chromatogr. A* 1134:101.

27. Zuehlke, S., Duennbier, U., and Heberer, T. 2004. Determination of polar drug residues in sewage and surface water applying liquid chromatography-tandem mass spectrometry, *Anal. Chem.* 76:6548.

28. Boyd, G.R., Reemstma, H., Grimm, D.A., and Mitra, S. 2003. Pharmaceuticals and personal care products (PPCPs) in surface and treated waters of Louisiana, USA, and Ontario Canada. *Sci. Total Environ.* 311:135.

29. Sacher, F., Lange, F.T., Brauch, H.J., and Blankenhorn, I. 2001. Pharmaceuticals in groundwaters: analytical methods and results of a monitoring program in Baden-Württemberg, Germany. *J. Chromatogr. A* 938:199.

30. Hilton, M.J. and Thomas, K.V. 2003. Determination of selected human pharmaceutical compounds in effluent and surface water samples by high-performance liquid chromatography-electrospray tandem mass spectrometry. *J. Chromatogr. A.* 1015:129.

31. Nelson, M.D. and Dolan, J.W. 2002. Ion suppression in LC-MS-MS—A case study. LC GC North America (January) http://www.lcgceurope.com/lcgceurope/data/articlestandard/lcgceurope/062002/9103/article.pdf.

32. Roberts, P.H. and Bersuder, P. 2006. Analysis of OSPAR priority pharmaceuticals using high-performance liquid chromatography-electrospray ionization tandem mass spectrometry. *J. Chromatogr. A* 1134:143.

33. Stolker, A.A.M., Niesing, W., Hogendoorn, E.A., Versteegh, J.F.M., Fuchs, R., and Brinkman, U.A.Th. 2004. Liquid chromatography with triple-quadrupole or quadrupole-time of flight mass spectrometry for screening and confirmation of residues of pharmaceuticals in water. *Anal. Bioanal. Chem.* 378:955.

34. Weigel, S., Berger, U., Jensen, E., Kallenborn, R., Thoresen, H., and Hühnerfuss, H. 2004. Determination of selected pharmaceuticals and caffeine in sewage and seawater from Tromsø/Norway with emphasis on ibuprofen and its metabolites. *Chemosphere* 56:583.

35. Öllers, S., Singer, H., Fäsler, P., and Müller, S.R. 2001. Simultaneous quantification of neutral and acidic pharmaceuticals and pesticides at the low-ng/L level in surface and wastewater. *J. Chromatogr. A* 911:225.

36. Ternes, T.A. 1998. Occurrence of drugs in German sewage treatment plants and rivers. *Water Res.* 32:3245.

37. Ternes, T.A., Stumpf, M., Schuppert, B., and Haberer, K. 1998. http://scholar.google.com/url?sa=U&q=http://cat.inist.fr/%3FaModele%3DafficheN%26cpsidt%3D2192137, *Vom Wasser* 90:295.

38. Sacher, F., Lochow, E., Bethmann, D., and Brach, H.J. 1998. Occurrence of drugs in surface waters. *Vom Wasser* 90:233.
39. Lindqvist, N., Tuhkanen, T., and Kronberg, L. 2005. Occurrence of acidic pharmaceuticals in raw and treated sewages and in receiving waters. *Water Res.* 39:2219.
40. Jorgensen, S.E. and Halling-Sørensen, B. 2000. Drugs in the environment. *Chemosphere* 40:691.
41. Reinthaler, F.F., Posch, J., Feierl, G., Wüst, G., Haas, D., Ruckenbauer, G., Manscher, F., and Marth, E. 2003. Antibiotic resistance of *E. coli* in sewage and sludge. *Water Res.* 37:1685.
42. Göbel, A., McArdell, C.S., Suter, M.J.F., and Giger, W. 2004. Trace determination of macrolide and sulfonamide antimicrobials, a human sulfonamide metabolite, and trimethoprim in wastewater using liquid chromatography coupled to electrospray tandem mass spectrometry. *Anal. Chem.* 76:4756.
43. Batt, A.L. and Aga, D.S. 2005. Simultaneous analysis of multiple classes of antibiotics by ion trap LC/MS/MS for assessing surface water and groundwater contamination. *Anal. Chem.* 77:2940.
44. Miao, X.S., Bishay, F., Chen, M., and Metcalfe, C.D. 2004. Occurrence of antimicrobials in the final effluents of wastewater treatment plants in Canada. *Environ. Sci.. Technol.* 38:3533.
45. Kennedy, D.G., McCracken, R.J., Cannavan, A., and Hewitt, S.A. 1998. Use of liquid chromatography-mass spectrometry in the analysis of residues of antibiotics in meat and milk. *J. Chromatogr. A* 812:77.
46. Hirsch, R., Ternes, T.A., Merz, A., Haberer, K., and Wilken, R.D. 2000. A sensitive method for the determination of iodine containing diagnostic agents in aqueous matrices using LC-electrospray-tandem-MS detection. *Fres. J. Anal. Chem.* 366:835.
47. Yang, S., Cha, J., and Carlson, K. 2004. Quantitative determination of trace concentrations of tetracycline and sulfonamide antibiotics in surface water using solid-phase extraction and liquid chromatography/ion trap tandem mass spectrometry. *Rapid Commun. Mass Spectrom.* 18:2131.
48. Ternes, T.A., Hirsch, R., Mueller, J., and Haberer, K. 1998. Methods for the determination of neutral drugs as well as betablockers and b_2 sympathomimetics in aqueous matrices using GC/MS and LC/MS/MS. *Fres. J. Anal. Chem.* 362:329.
49. Miao, X.S. and Metcalfe, C.D. 2003. Determination of cholesterol-lowering statin drugs in aqueous samples using liquid chromatography–electrospray ionization tandem mass spectrometry. *J. Chromatogr. A* 998:133.
50. Valentine, C.R., Valentine, J.L., Seng, J., Leakey, J., and Casciano, D. 1996. The use of transgenic cell lines for evaluating toxic metabolites of carbamazepine. *Cell. Biol. Toxicol.* 12:155.
51. Miao, X.S. and Metcalfe, C.D. 2003. Determination of carbamazepine and its metabolites in aqueous samples using liquid chromatography–electrospray ionization tandem mass spectrometry. *Anal. Chem.* 75:3731.
52. Steger-Hartmann, T., Kümmerer, K., and Schecker, J. 1996. Trace analysis of the antineoplastics ifosfamide and cyclophosphamide in sewage water by two-step solid-phase extraction and gas chromatography–mass spectrometry. *J. Chromatogr. A*, 726:179.
53. Kümmerer, K., Steger-Hartmann, T., and Meyer, M. 1997. Biodegradability of the antitumor agent ifosfamide and its occurrence in hospital effluents and communal sewage. *Water Res.* 31:2705.
54. Buerge, I.J., Buser, H.R., Poiger, T., and Müller, M.D. 2006. Occurrence and fate of the cytostatic drugs cyclophosphamide and ifosfamide in wastewater and surface waters. *Environ. Sci. Technol.* 40:7242.
55. Mahnik, S.N., Rizovski, B., Fuerhacker, M., and Mader, R.M. 2006. Development of an analytical method for the determination of anthracyclines in hospital effluents. *Chemosphere* 65:1419.

56. López de Alda, M.J., Diaz-Cruz, S., Petrovic, M., and Barceló, D. 2003. Liquid chromatography-(tandem) mass spectrometry of selected emerging pollutants (steroid sex hormones, drugs and alkylphenoloc surfactants) in the aquatic environment. *J. Chromatogr. A* 1000:503.

57. Ternes, T.A., Stumpf, M., Müller, J., Haberer, K., Wilken, R.-D., and Servos, M. 1999. Behavior and occurrence of estrogens in municipal sewage treatment plants-1. Investigations in Germany, Canada, and Brazil. *Sci. Total Environ.* 225:81.

58. Diaz-Cruz, S., López de Alda, M.J., López, R., and Barceló, D. 2003. Determination of estrogens and progstogens by mass spectrometric techniques (GC/MS, LC/MS and LC/MS/MS). *J. Mass Spectrom.* 38:917.

59. Ternes, T.A., Bonerz, M., Herrmann, N., Loffler, D., Keller, E., Bago Lacida, B., and Alder, A.C. 2005. Determination of pharmaceuticals, iodinated contrast media and musk fragrances in sludge by LC/tandem MS and GC/MS. *J. Chromatogr. A* 1067:213.

60. Putschew, A., Wischnack, S., and Jekel, M. 2000. Occurrence of triiodinated X-ray contrast agents in the aquatic environment. *Sci. Total Environ.* 255:129.

61. Ternes, T.A. and Hirsch, R.A. 2000. Occurrence and behaviour of X-ray contrast media in sewage facilities and the aquatic environment. *Environ. Sci. Technol.* 34:2741.

62. Hirsch, R., Ternes, T.A., Lindart, A., Haberer, K., and Wilken, R.D. 2000. A sensitive method for the determination of iodine containing diagnostic agents in aqueous matrices using LC-electrospray-tandem-MS detection. *Fres. J. Anal. Chem.* 366:835.

63. Putschew, A., Wischnack, S., and Jekel, M. 2000. Occurrence of triiodinated X-ray contrast agents in the aquatic environment. *Sci. Total Environ.* 255:129.

64. Putschew, A., Schittko, S., and Jekel, M. 2001. Quantification of triiodinated benzene derivatives and X-ray contrast media in water samples by liquid chromatography–electrospray tandem mass spectrometry. *J. Chromatogr. A* 930:127.

65. Sacher, F., Raue, B., and Brauch, H.J. 2005. Analysis of iodinated X-ray contrast agents in water samples by ion chromatography and inductively-coupled plasma mass spectrometry. *J. Chromatogr. A* 1085:117.

66. Pérez, S., Eichhorn, P., Celiz, M.D., and Aga, D.S. 2006. Structural characterization of metabolites of the X-ray contrast agent iopromide in activated sludge using ion trap mass spectrometry. *Anal. Chem.* 78:1866.

67. Putchew, A. and Jekel, M. 2003. Induced in-source fragmentation for the selective detection of organic bound iodine by liquid chromatography/electrospray mass spectrometry. *Rapid Commun. Mass Spectrom.* 17:2279.

68. Cozanitis, D.A. 2004. One hundred years of barbiturates and their saint. *J.R. Soc. Med.* 97:594.

69. Holm, J.V., Rügge, K., Bjerg, P.L., and Christensen, T.H. 1995. Occurrence and distribution of pharmaceutical organic compounds in the groundwater downgradient of a landfill (Grindsted, Denmark). *Environ. Sci. Technol.* 29:1415.

70. Eckel, B.P., Ross, B., and Esensee, R.K. 1993. Pentobarbital found in groundwater. *Ground Water* 31:801.

71. Heberer, T. 2002. Tracking persistent pharmaceutical residues from municipal sewage to drinking water. *J. Hydrol.* 266:175.

72. Peschka, M., Eubeler, J.P., and Knepper, T.P. 2006. Occurrence and fate of barbiturates in the aquatic environment. *Environ. Sci. Technol.* 40:7200.

3 Sample Preparation and Analysis of Solid-Bound Pharmaceuticals

Christine Klein, Seamus O'Connor, Jonas Locke, and Diana Aga

Contents

3.1 INTRODUCTION

Pharmaceuticals and personal-care products used by humans are often excreted or washed down drains to wastewater treatment plants, where they can be bound to particulate in sludge or discharged into local waters and eventually bind to sediments. Similarly, antibiotics and hormones that are used in farm animal operations can become bound to manure, soil that is amended with this manure, and also on air particulate originating from those farms. When pharmaceuticals are bound to particles, it is less likely that they will undergo biotransformation. However, these compounds can desorb and become more bioavailable should conditions change, making this process favorable. Therefore, it is important for researchers to report the fraction of pharmaceuticals that are bound to solids in environmental studies instead of just concluding that the pharmaceutical has been biodegraded or otherwise removed from the system.

The extent to which these compounds become bound to solids can be characterized by a compound's Kd (a solid-water partitioning coefficient) value. See Chapter 6 for more information on sorption processes. The environmental solids these pharmaceuticals can bind to are diverse in composition. Because the degree of binding depends highly on the nature of the sorbent-sorbate interactions, the available methods for extraction of pharmaceuticals from solid matrices vary widely. Hence, the applicability of extraction and analytical methods needs to be evaluated for different pharmaceutical compounds, and conditions need to be optimized for various types of solids.

Once an extraction method is found to be suitable for removing sorbed pharmaceuticals from environmental solids, it is often necessary to perform some degree of cleanup on these samples before detection and quantification. Environmental matrices such as manure, soil, sludge, and sediment all contain natural organic matter (NOM), which is generally described as a poorly defined mixture of organic substances with variable properties in terms of acidity, molecular weight, and molecular structure. Many times extraction methods will coextract portions of the environmental matrix, which can interfere with analysis in a variety of ways. Through extraction and cleanup, one is able to remove many matrix interferences; however, the more extensive these procedures are, the greater the possibility becomes for analyte loss. Additionally, because the levels of pharmaceuticals are low, these samples will often need to be preconcentrated in order to be detectable even by the most modern instrumentation. Unfortunately, this also leads to the concentration of interferences, which are generally much more abundant in a sample than the analytes themselves.

The extent to which samples containing solid-bound pharmaceuticals are manipulated, and the best overall analytical method used, depends on several factors. As when preparing most environmental samples, an analyst must balance the advantages and disadvantages of extraction, cleanup, and concentration techniques. Compromises are often made, and these decisions are ultimately driven by factors such as the required detection limit for the purpose of the study, analytical instrumentation available, the amount and type of extracted material that contaminates a sample, and the concentrations of analytes present in the samples. This chapter will review various sample preparation strategies employed in the analysis of pharmaceuticals in soil, manure, and sludge and discuss the advantages and limitations of each technique.

3.2   MATRICES OF SOLID-BOUND PHARMACEUTICALS

Solid-bound pharmaceuticals have been found in matrices ranging from household and farm dust to the sediments that receive treated wastewater. Assessing the chemical composition of a sample and knowledge of properties such as polarity and binding sites can help an analyst determine the best extraction method to desorb their analyte from the matrix and best cleanup method to remove interfering matrix components. Additional sample preparation steps such as sample drying, either by air or freeze-drying methods, or mechanical separation by sieving and grinding must be taken into consideration because of their labor intensiveness and effect on the time it takes to prepare a sample. In general, the less natural organic matter content in a matrix, the easier it is to extract organic analytes. For example, when comparing extraction efficiencies from different soil types, higher recoveries for tetracyclines (TCs) are observed in sandier soil than in soil with more organic matter.[1]

The compositions of typical matrices encountered when conducting environmental analysis are described in this section. Air particulate has been found to contain antibacterial agents such as triclosan[2] in household dust and tetracyclines, sulfamethazine, tylosine, and chloramphenicol in animal confinement buildings.[3] The dust from the interior of an animal confinement building was analyzed and found to contain approximately 85% organic material, composed of protein (from skin), animal feed, endotoxins, fungi, and bacteria.[4] This composition will vary, depending on the source.

Sludges have been found to be contaminated with a variety of compounds ranging from personal-care products and pharmaceuticals, which are washed and flushed down drains, to hormones and antibiotics, which are used in animal production. Wastewater sludge contains many compounds and is primarily organic in nature. The composition of this organic portion of sludge is made up of sugars, proteins, fatty acids, cellulose, and plant macromolecules with phenolic and aliphatic structures but is still not completely characterized.[5] It also has microorganisms and exocellular material and residues originating from wastewater (for example, paper plant residues, oils, fats, and fecal material).[6] Sludges from different sources can have enormous compositional variation, which necessitates validation of the extraction efficiency for each sludge source.[7]

Sediment is comprised of the particulate in surface waters that settles to the bottom of the water column or remains suspended and transported in waterways. Sediment particles have both organic and inorganic fractions, which can contain humic material; metal oxides such as iron and manganese; and also trace metals, silicates, sulphides, and minerals.[8] The particle size of sediments is often indicative of its components and will determine the types of compounds that are sorbed to it; therefore, it often needs to be sieved during its preparation before it is extracted for contaminants.

Soil is made from eroded earth that is mixed with decayed plant and animal tissues. It contains mostly organic carbon, inorganic clays, and sand. Pharmaceuticals are found in soil when it is amended with sludges and manure to give it more nutrient content in the form of carbon and nitrogen. Like sediment, soil needs to be sieved before analysis.

3.3   SAMPLE EXTRACTION TECHNIQUES

Pharmaceuticals that are bound to solids must be removed, or extracted, from these solids prior to analysis. However, a standard method for extraction does not exist, and an extraction procedure must be optimized for the conditions that an analyst encounters. One of the most widely used methods for assessing the efficiency of an extraction procedure is by determining the percent recovery and extraction yield of a method through spiking experiments. In this type of experiment an analyst will take a sample of the matrix that they want to extract a compound from, add a known amount of that compound, and extract it using the selected method to determine how much is removed. Issues that must be considered when performing this type of experiment are (1) solvent selection; (2) contact time; (3) spiking level; and (4) possible effects on binding/transformation by the microbial community present in the sample. It also is important to validate methods when comparing extraction of different types of solids because efficiencies can differ, leading to gross underestimation or overestimation of analyte concentration. Simply adding an internal standard does not excuse the validation of an extraction procedure, because the contact time with the solids may not be appropriate for assessing extraction efficiency.

Solvent selection is probably the most important step in developing an extraction method for solid-bound pharmaceuticals. The analyte should have high solubility in the extraction solvent in order to desorb the analytes efficiently. Many pharmaceutical compounds, such as antibiotics and hormones, have low water solubility and are relatively hydrophobic, making it necessary to use organic solvents for extraction. But even when pharmaceuticals have high water solubility, Kd may be high due to interactions other than hydrophobic.[9] Pressurized liquid extraction (PLE) and supercritical fluid extraction (SFE) techniques can lead to higher extraction efficiencies relative to traditional solid–liquid extraction. Solvent modifiers such as acids or bases are sometimes added to extraction solvents to increase the solubility of analytes in the extraction solvents and to improve extraction efficiencies. Some compounds such as tetracyclines are known to form complexes with di- and trivalent cations in the clay minerals or to hydroxyl groups at the surface of soil particles.[10–12] Hence, complexing agents such as ethylene diamine tetraacetic acid (EDTA) are often added to the extraction buffer to improve percent extraction recovery.[13]

The contact time of the analyte with the solid matrix prior to extraction is an important parameter to consider when validating and optimizing extraction procedures. It has been shown for 17α-estradiol and sulfonamide antibiotics that the longer the contact time between the soil and the analytes, the lower the percent extraction recoveries obtained.[14,15] In addition, when short contact time between the solid and the analyte was allowed prior to PLE extraction, temperature had little effect on the extraction efficiencies of the spiked soils. However, when 17 days of contact time was allowed, an increase in extraction temperature significantly improved percent recoveries.[15]

Spiking a solid matrix at environmentally relevant concentrations is also important when determining extraction efficiency. It may be tempting for an analyst to spike at higher levels because this can alleviate problems associated with detection. However, this can mislead one into believing that the extraction efficiency is higher or more reproducible than it actually is at the lower concentrations typically observed in the natural environment. For instance, in an experiment conducted in our laboratory to find optimized extraction conditions for tetracyclines from soil, the percent recovery at low concentrations was significantly higher than at the spiked concentrations. When soil was spiked (n = 3) at concentrations of 100 ng/g, the recoveries of tetracyclines ranged from 89 to 92%, with standard deviations at or below 10% (see Table 3.1). However, when the spiking levels approached environmentally relevant concentrations (below 20 ng/g), exaggerated recoveries (>100%) and very high

TABLE 3.1

Tetracycline Recoveries in Soil Using Accelerated Solvent Extraction (ASE) and Solid Phase Extraction (SPE) Cleanup (n = 3)

Spiking Concentration Tetracycline/Soil	Tetracycline Compound	Percent Recovery	Standard Deviation
100 ng/g	TC	91%	3%
100 ng/g	OTC	89%	10%
100 ng/g	CTC	92%	6%
50 ng/g	TC	140%	10%
50 ng/g	OTC	151%	5%
50 ng/g	CTC	155%	10%
25 ng/g	TC	140%	47%
25 ng/g	OTC	138%	47%
25 ng/g	CTC	171%	50%

Note: As spiking level approaches environmentally relevant concentrations below 25 ng/g, recovery becomes less reproducible.

TC = Tetracycline
OTC = Oxytetracycline
CTC = Chlortetracycline

standard deviations (>50%) were observed, suggesting significant matrix interference that is highly variable.

Microbial communities present in the matrix could potentially alter the sorption of pharmaceutical compounds in soil via biodegradation or biotransformation. Therefore, it is important to ensure proper sample storage and to account for possible biodegradation when evaluating extraction methods. For instance, the extraction recoveries for ibuprofen were improved from 25 to 94%, and for trimethoprim from 68 to 86%, when the solid samples were first autoclaved before fortification with the analytes, demonstrating the influence of live microbial community on the amount of recovered pharmaceuticals. It is possible that during the contact time of 14 hours used in the study, the microorganisms have either incorporated the pharmaceuticals into the organic matter content of the soil or have degraded the pharmaceuticals into other compounds that were not monitored by the method. The adsorption isotherms for these compounds in sediment remained unaltered by the autoclaving process, despite the potential effects of autoclaving on the sediment, organic matter, and cation-exchange capacity.[16]

Tetracyclines and hormones (such as estrogens) that are introduced into the environment present unique challenges that are not encountered in other biological or food samples. For instance, the strong interaction of tetracyclines with natural organic matter and with clay components in soil can lead to poor extraction efficiencies and large variability in percent recoveries.[17] While tetracyclines are fairly polar (Kow 0.8), the zwitterionic character of these compounds causes them to complex with ions present in soil or sludge. Estrogens, on the other hand, are fairly nonpolar (Kow 4.2) and thus have high sorption to solid matrices with high organic matter content.

3.3.1 SOLID–LIQUID EXTRACTION

Solid–liquid extraction is the most basic extraction method for solid samples. It involves mechanical agitation of a mixture consisting of the sample to be extracted and an excess of solvent. The solvent and sample are extracted for a period of time, typically longer than 10 minutes. In a typical solid–liquid extraction, the sample is then centrifuged, and the supernatant is removed and extraction is repeated several times. The supernatants are combined, at which time the sample can be analyzed, or more typically the extract volume is reduced to facilitate analysis. This can be achieved through several different methods and is often dependent on the equipment available and the analyte being examined. Solid–liquid extraction is simple and cost effective, since the equipment needed is minimal. However, one major drawback is the relatively large amount of solvent used for this technique compared with the other techniques discussed below.

Solid–liquid extraction has been used to examine a variety of pharmaceuticals in environmental solids. Steroid estrogens, such as estradiol, estrone, estriol, and ethinylestradiol, have been extracted from sediment and sewage sludge with percent recovery ranging from 61 to 71%.[18] In another study, the effect of solvent composition was examined for extracting tetracyclines from sediment, and it was found that higher citric acid concentration (0.1%) in the solvent improved extraction efficiencies up to 105%. However, as the concentration of the chelating agent Na_2EDTA increased, the extraction efficiency decreased.[19] This conflicts with other reports that suggest that the addition of EDTA into the extraction solvent improves extraction efficiency by releasing metal-complexed tetracyclines.[20]

Solid–liquid extraction has also been used to determine the concentrations of the antibiotics tylosin, sulfamethazine, chloramphenicol, and tetracyclines from the dust found in an animal confinement shelter; however, the recoveries were not reported.[3] Information on recoveries would have been interesting because the concentration in dust is typically very low. Although solid–liquid extraction is frequently used in the analysis of a wide range of pharmaceuticals, its use is limited because compounds that are strongly sorbed to a solid matrix may need more rigorous extraction conditions, such as those provided by the other techniques discussed below.

3.3.2 SONICATION-ASSISTED EXTRACTION

In sonication assisted extraction (SAE) samples are mixed with an extraction solvent and placed into a sonication bath. The sample mixture is subjected to acoustic vibrations with frequencies above 20 kHz. These ultrasonic waves travel through the sample, leading to expansion and compression cycles in the solvent. The expansion cycles cause a negative pressure in the liquid, and if the amplitude of these waves is strong enough, cavities or bubbles in the solvent can be observed. Upon the collapse of these bubbles, localized temperatures and pressures can exceed 5000 K and 1000 atm, respectively, creating shockwaves which in turn increase the desorption of analytes from the matrix surface.[21] Additionally, the collapse of these bubbles in the presence of suspended particles can lead to asymmetric collapses that form high-speed microjets toward the solid's surface, leading to erosion and cleavages. The increase in surface area also improves extraction of analytes.[22] This technology has been used for the extraction of pharmaceuticals and natural hormones from matrices such as manure and soil amended with manure,[23,24] and sludge,[7] but primarily in river sediments.[16,25–28]

In a study that aimed to determine antibiotics in pig slurry, sonication was used with a solvent system composed of methanol, McIlvane buffer, and EDTA. The recoveries for oxytetracycline and sulfachloropyridazine ranged from 77 to 102% and 58 to 89%, respectively, using a concentration range of 1 to 20 mg/L. A similar extraction method (with methanol added in the solvent) was used to extract soil that was spiked at concentrations ranging from 0.2 to 0.5 ug/g. The recoveries were reported as 27 to 75%, 68 to 85%, and 47 to 105% for oxytetracycline, sulfachloropyridazine, and tylosin, respectively, in four different types of soils. In general, lower recoveries were observed in soils with higher clay and organic carbon content, especially for oxytetracycline and tylosin, which have higher Kd values.[24] However, the spiking levels used in these studies are orders of magnitude above environmentally relevant concentrations, and therefore these methods may not be applicable to real environmental samples. Additionally, no mention is made regarding the contact time used during the recovery studies; this ignores the effect of aging on the extractability of pharmaceuticals from soil. Similarly, estrogen analysis was attempted in freeze-dried solids from hog lagoon samples using sonication and a methanol/acetone mixture as an extraction solvent.[23] However, extraction recoveries were not reported; hence, no assessment can be made on how successful the sonication method is for estrogen extraction from solids.

3.3.3 Pressurized Liquid Extraction (PLE)

PLE, also known as accelerated solvent extraction (ASE), involves the use of pressurized extraction vessels at elevated temperatures to achieve efficient extraction of analytes. The sample is placed in an extraction cell with an inert solid dispersant. The solid dispersant, such as sand or diatomaceous earth, serves a twofold purpose. First, it fills the empty cell volume to minimize excess solvent consumption, and second, it increases the surface area of the sample that is exposed to the extraction solvent. The filled extraction cell is then pressurized with extraction solvent and placed in an oven to increase the temperature. Higher temperatures increase solubility of analytes in the solvent and decrease viscosity of the solvent, leading to better penetration in to interstitial spaces present in the sample. Additionally, raising the temperature increases extraction kinetics such as desorption. Elevated pressures also contribute somewhat to increasing solvent contact with the sample, but mainly they serve to keep the solvents liquid at the increased temperature. PLE offers the advantage in many cases of automation, leading to increased sample throughput. Furthermore, PLE reduces the solvent requirement to extract samples, saving on analysis cost and minimizing organic solvent waste.

In studies that compare the effects of solvent, temperature, pressure, and time using PLE, solvent selection generally has the most significant effect, followed by temperature.[15,29] Another parameter that is often optimized in PLE is the number of extraction cycles and the need to prewet the soil.[29] However, it appears that for some compounds, such as fluoroquinolones, the best extraction recovery is obtained with no prewetting of the solid sample.

Thermal degradation studies must be conducted when using elevated temperatures in PLE. Typically, they are conducted by spiking the analytes on quartz sand and extracting the sand by PLE at various temperature settings. It is important to assess the effect of high temperature on the stability of the analytes because some pharmaceutical compounds are thermally labile. Another important consideration is the amount of sample used for extraction. The amount of sample must be minimized (5 to 10 g) to avoid unnecessary extraction of large amounts of organic matter and other matrix components.[15] A coextracted matrix not only interferes with the analyte detection, but could also clog the extraction vessels, as has been observed during the extraction of 25 g soil.[1] Furthermore, additives and buffers that are used in other extraction methods can precipitate and clog the lines of the PLE apparatus.[1] PLE has been used to extract tetracyclines,[1] macrolides, ionophores, sulfonamides, fluoroquinolones,[29,30] and estrogens[31,32] from soil, sludge, and sediments.

3.3.4 MICROWAVE-ASSISTED SOLVENT EXTRACTION

Another emerging technique in environmental analysis is microwave-assisted extraction (MAE). This technique involves the use of microwaves to heat a sample in a closed vessel so that temperatures above the normal boiling point of the extraction solvent can be utilized. The increased temperature improves analyte solubility, extraction kinetics, and solvent contact (wetting) with the matrix, similar to PLE. The main advantages of MAE include: decreased extraction time, increased sample throughput by means of automated and simultaneous extraction of several samples, and decreased solvent consumption relative to soxhlet extraction. The disadvantages to using MAE for soil extraction include: thermal decomposition of analytes, nonselective extraction of matrix components, and limited solvent choices. Only microwave-active solvents such as water and methanol can be used; nonpolar organic solvents such as hexane, cyclohexane, and methylene chloride are not useable in MAE.

MAE has been successfully used for the extraction of estrogens from sediments[33] and quinolone antibiotics from soils and sediments.[34,35] It should be noted, however, that during extraction, estradiol was oxidized to estrone in sediment containing low organic matter.[33] It was suspected that the manganese oxides present in the sediment have catalyzed this reaction when exposed to microwaves. The study by Morales-Munoz[35] showed the importance of solvent pH with respect to the analyte and pKa/ionizable functional groups. When an analyte is protonated or deprotonated, its solubility in water (used as the extracting solvent) is increased. The study by Prat[34] illustrates the advantages of using MAE over conventional extraction techniques. For instance, the use of MAE in a 15-minute extraction of fluoroquinolone resulted in approximately 80% recovery, while 1-hour of mechanical shaking resulted in less than 40% recovery. Further, MAE allowed recoveries of greater than 90% using three extraction cycles for fluoroquinolone from soil.[35]

3.3.5 Supercritical Fluid Extraction

SFE exploits the properties of a fluid when elevated temperatures and pressures are applied above their critical point. Because a supercritical fluid exhibits thermal and physical properties of both a liquid and a gas, the surface tension is nonexistent, causing the diffusivity to increase. This affords supercritical fluids the ability to readily penetrate porous and fibrous solids, including environmental matrices. An advantage of using SFE is that it is often conducted using carbon dioxide; therefore, large amounts of organic wastes are not generated.

The use of SFE affords the user a higher degree of selectivity, with minimum amount of coextracted matrix obtained relative to the other extraction techniques. For example, SFE was reported to produce the cleanest soil extracts compared to other methods, such as PLE.[36] SFE has been used to extract nonsteroidal antiinflammatory drugs (NSAIDs) from river sediment,[28] as well as 4-nonylphenol and bisphenol A from sludge.[37,38] The recoveries of NSAIDs such as naproxen and ketoprofin from river sediment using SFE were comparable to the recoveries obtained using PLE (78 to 79%) and MAE (81 to 82%). On the other hand, the recoveries of 4-nonylphenol and bisphenol A were disappointing and were lower than the recoveries observed using PLE. However, the only parameter that was altered in SFE was the solvent composition; it might be possible to obtain higher recoveries if other SFE extraction parameters are optimized.

3.3.6 Matrix Solid Phase Dispersion

Matrix solid phase dispersion (MSPD) is another technique that can be used for extraction of analytes from solid samples. Briefly, this technique uses a solid phase sorbent, usually octadecylsilyl silica packing material similar to those used in reversed phase high-performance liquid chromatography (HPLC) and solid phase extraction (SPE). This solid phase is conditioned with the appropriate solvents and then mixed with the solid sample using a mortar and pestle. The mixture is then transferred to a column, and the analytes are eluted with organic solvents. This extraction procedure offers the advantage of providing direct contact between the solid extracting materials and the analytes in the solid sample. The large surface area of the derivatized silica particles used in MSPD facilitates efficient transfer of analytes from soil to the extracting solid phase.

MSPD has been applied in the analysis of tetracycline antibiotics in several food-related matrices, such as milk and catfish tissue, but there has been no report on its application in the analysis of tetracyclines in soil.[39–41] Using an octadecylsilyl deriva-tized silica solid phase, with oxalic acid and EDTA as modifiers, recovery from catfish tissue was 81% with a limit of detection (LOD) of 50 µg/kg.[40] The method was slightly more variable when using milk samples, with recoveries ranging from 64 to 94%.[41] The extraction of steroids such as estradiol, testosterone, and proges-terone, were compared in poultry, porcine, and beef meats. Using MSPD, extraction efficiencies greater than 90% for all three compounds were observed.[42] MSPD has also been used to extract triclosan and parabens from household dust, with good recoveries, ranging from 80 ± 5 – 114 ± 9% over spiking concentrations of 50ng/g to 300 ng/g.[2]

MSPD technique is not readily adaptable to large numbers of sample, however, because the mixing of sample and solid phase is done manually. In addition, the detection limits reported are slightly higher than what is needed for tetracyclines in environmental soil residue analysis and are in the high ppt range for steroids, which may not be low enough for environmental samples.

3.4 SAMPLE CLEANUP TECHNIQUES

Natural organic matters, such as humic and fulvic acids, present in environmental samples are coextracted with the analytes and often complicate analytical detection. The problems associated with sample analysis due to coextracted compounds are collectively termed "matrix effects." Hence, the amount of coextracted natural organic matter must be minimized to achieve a successful analysis. The most widely used technique to concentrate environmental samples and to reduce matrix effects is SPE. Gel permeation chromatography has also been used to separate proteinaceous and high molecular weight materials from smaller molecular weight target analytes. Recently, molecularly imprinted polymers (MIPs) have been used as selective sorbents for SPE. The extent of sample cleanup needed will depend on the susceptibility of the analytical instrument to matrix effects. Therefore, the implementation of a cleanup method must be carefully considered for each analyte, sample type, and instrumentation.

3.4.1 SOLID PHASE EXTRACTION

SPE is used to separate compounds in a sample based on their polarities and solubilities in specific solvents and is based on principles similar to those of liquid chromatography. SPE is typically conducted by passing a large-volume aqueous sample through a cartridge packed with an appropriate sorbent. Ideally, the analytes will sorb to the packing material, while most interfering compounds are unretained and pass through the cartridge. The sorbed analytes are then eluted with a relatively small volume of solvent and collected. The eluate may need to be evaporated to a smaller volume and may be solvent exchanged to an appropriate solvent for further analysis. On the other hand, the SPE procedure may be designed so that the unwanted matrix can be captured in the SPE sorbent while the analytes pass through the cartridge and are collected for further concentration. Either way, the main purpose of SPE is the removal of matrix components such as salts and some organic matter, while concentrating analytes. SPE has replaced many conventional liquid–liquid extraction techniques due to advantages gained by minimizing solvent consumption, increased selectivity through choices in both stationary phase and elution solvent, and ability to automate extraction. The stationary phase is available in reversed phase, normal phase, and ion exchange mode, and in cartridge, disk, pipette tip, and microtiter plate format.

The SPE process can be best described as happening in four steps. The first is a conditioning step, where the sorbent is "wetted." Solvents are used to rinse the sorbent and to allow contact of the analytes with the solid phase. The next step involves the "loading" of the sample onto the cartridge. The sample is passed through the cartridge by pressure differential. In the third step the cartridge is washed to clean any residual matrix left behind on the sorbent. Finally, a strongly eluting solvent is passed through the cartridge and the eluate is collected. SPE can be implemented for various classes on compounds and can be used across classes by using multimodal and mixed phase extractions; "piggybacking" cartridges on one another and performing tandem SPE (see Figure 3.1). For example, a strong anion exchange (SAX), SPE cartridge in tandem with a reversed-phase SPE resin (such as the hydrophilic–lipophilic balanced, HLB, cartridge) is often used for the removal of NOM from soil extracts. In the analysis of tetracycline antibiotics, the use of tandem SPE reduces the amount of NOM considerably. Figure 3.2 semiquantitatively shows the amount of NOM removal when using both C-18 and HLB cartridges in tandem with SAX. The ultraviolet (UV) trace for each type of SPE setup was performed on a soil extract, and the corresponding SPE setup is shown in Figure 3.2. This indicates that the use of tandem SPE for soil samples removes most of the coextracted NOM from an extract resulting in more accurate quantification.

3.4.2 MOLECULARLY IMPRINTED POLYMERS

One promising technique that has been recently applied to address problems in preparation of environmental samples is molecular imprinting.[43] This technique involves using the analyte as a template molecule and creating specific interaction sites within a polymeric solid. The selectivity of the sites depends on the interactions between the template and the monomer used to develop the imprint (see Figure 3.3). A comprehensive review on the analytical applications of MIPs has been conducted by Andersson.[44]

MIPs have been synthesized using TCs as templates.[45,46] These materials showed good selectivity toward the tetracycline template relative to the control. Xiong et al. developed an imprinted polymer that was used to recover spiked tetracycline from fish tissue.[46] The polymer was used as an SPE sorbent and achieved recoveries between 97 and 106% from the matrix. Caro and colleagues recently developed an imprinted polymer for the selective removal of tetracyclines from pig kidney tissue that was selective for several tetracycline analogues.[45] Suedee and colleagues synthesized an MIP that was class selective for tetracyclines, which was subsequently used in an affinity membrane for the removal of tetracyclines from water.[47] This latter study demonstrated that the use of MIPs with broad selectivity to isolate tetracycline degradation products is feasible. However, research is still needed to investigate whether or not these materials could be used for the selective preconcentration of tetracyclines in highly complex matrices, such as soil and sludge extracts.

MIPs have also been synthesized to remove estrogens and estrogenic compounds from environmental matrices such as natural waters.[48,49] Estradiol was successfully

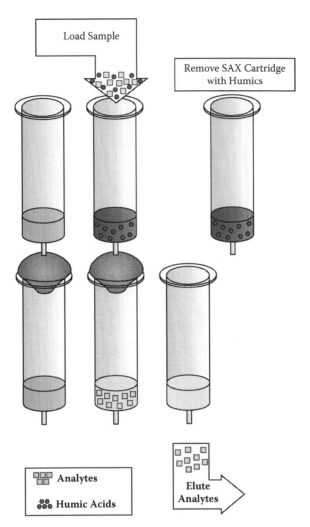

FIGURE 3.1 A sample containing analytes and humic acids is passed through tandem SAX and HLB SPE cartridges. The humics are retained by the SAX cartridge and the tetracyclines pass through to the HLB cartridge, where they are retained. The SAX cartridge is then removed, and the tetracyclines are eluted from the HLB cartridge, minimizing the concentration of humic acids in the sample.

removed from different waters using an MIP with recoveries of 103 to 104% for drinking water, pond water, and well water.[48] The binding of estrogens to an MIP developed to aid in the analysis of estrogens in natural waters was lower than the binding of estrogens in deionized waters; however, selective binding up to 76 nmol/mg of MIP was observed in natural waters, and the MIP proved to be reusable for at least five uses without any loss of performance.[49] Along the lines of an MIP, a selective technique using antibodies in immunoaffinity chromatography has also been used to isolate estrogens from environmental samples.[50] However, antibodies are susceptible to fouling, may be denatured by organic solvents, and can be expensive

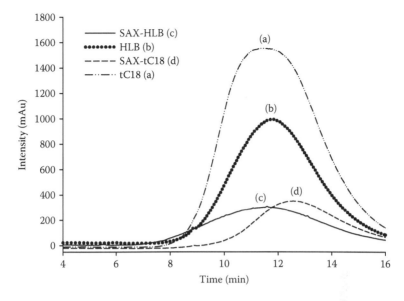

FIGURE 3.2 UV chromatograms of a standard 1.0% humic acids after cleanup with (a) tC18 SPE, (b) HLB SPE, (c) tandem SAX and HLB SPE, and (d) tandem SAX and tC18 SPE.

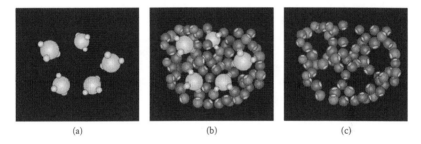

FIGURE 3.3 A solution containing analyte molecules (a), is surrounded by monomers that bind to the analyte and are then polymerized to form a solid (b). The analyte is then removed, and a solid remains with sites that will bind specifically to the analyte (c).

to make. Hence, the use of immunoaffinity columns for sample cleanup prior to the analysis of solid-bound pharmaceutical contaminants has been very limited.

3.4.3 Size Exclusion Chromatography

Size exclusion chromatography (SEC) exploits size differences between the analyte and the interfering matrix components to achieve cleaner samples. In SEC a sample is passed through a column with packing material (typically a gel) containing fixed pore sizes. The larger molecules are excluded from the pore volume and pass through the column. As the size of the molecules decreases, approaching the size of pores or hydrodynamic volume in the packing material, the compound penetrates the pores more deeply and is eluted chromatographically with respect to size. Smaller molecules freely elute in and out of the pores and elute at the total permeation volume. This makes gel permeation chromatography an excellent technique for removing proteins and lipids from samples. It has been used as a cleanup method for steroid estrogens in sediment and sewage sludge.[18,32] Humic and fulvic acids are often contaminants of environmental samples. Humics, categorized as having a molecular weight range greater than 1000 Daltons (Da), can be removed from smaller pharmaceuticals using gel permeation chromatography (GPC); however, these samples still contain fulvic acids, which have a molecular weight less than 1000 Da.

3.5 SPECIAL CONSIDERATIONS IN SAMPLE ANALYSIS

3.5.1 LIQUID CHROMATOGRAPHY

Liquid chromatographic (LC) methods are the most widely used techniques for the analysis of pharmaceuticals in the environment because most of these compounds are polar and not amenable to gas chromatographic analysis. Often when analytes from environmental matrices are extracted and concentrated using common methods, they are dissolved in organic solvent. However, most modern liquid chromatography methods are reversed phase, and samples often must be dissolved in mostly aqueous solvent before they can be injected into the instrument. Analytes and matrix components that are soluble in organic solvents may have limited solubility in aqueous solutions, and therefore precipitate during the process of solvent exchange. This can cause a significant loss of analyte in the solution, particularly if the analyte sorbs strongly to the precipitate formed. Analyte losses of up to 30% have been observed during the solvent exchange step in the analysis of estrogens extracted from sediments.[7]

The most widely used detector for liquid chromatography is UV/V because it is less expensive than mass spectrometry (MS). However, the analysis of environmentally derived samples with these types of detectors can be problematic because matrix components also absorb light at the same wavelength as the analytes and interfere with the detection. To overcome these challenges, analysts have used LC with fluorescence detection. This technique uses a derivatization step to convert the analyte to a fluorescent conjugate and improve selectivity. However, an 8-hour reaction time was needed for maximum intensity and stability of the derivatized compound.[24] Despite the time required for derivatization, successful detection of sulfachloropyridazine, oxytetracycline, and tylosine was achieved in the ppb range in soil and pig slurry samples.

To date, MS detection of pharmaceutical contaminants has become the method of choice. Analysis of many pharmaceuticals by LC/MS often employs positive electrospray ionization ((+)-ESI), although negative electrospray ionization, atmospheric pressure chemical ionization (APCI), and fast atom bombardment ionization have also been used[52–56] utilizing ion-trap and quadrupole mass spectrometers.[57,58] However, ESI is susceptible to ionization suppression or enhancement due to matrix interferences by humic acids (HA) coextracted from soil, sediment, and sludge, or from mobile phase additives that are used in improving separation.[59,60] In addition, ionization enhancement has also been shown to occur when using an APCI interface for mass spectrometry.[61] Ionization suppression is strongly sample dependent and leads to inaccurate and nonreproducible quantitation of analytes in complex matrices. Matrix effects can even be magnified, particularly when using alkaline solvents for extraction[62] because a significant portion of NOM is coextracted from matrices and subsequently preconcentrated with the analytes.[63]

In the analysis of ethinylestradiol from a freeze-dried sludge extract, the ionization suppression is remarkable. A 5-g sample of freeze-dried sludge was spiked with 500 ng of ethinylestradiol and extracted using methanol and acetone with ASE. The extract was subjected to SPE cleanup using C-18 cartridges and analyzed using LC/MS. The sample was compared with a standard containing 500 ng of ethinylestradiol. However, it did not appear that ethinylestradiol was extracted from the sample. To ensure that the retention time had not shifted, an additional 500 ng and then another 1000 ng of ethinylestradiol was added to the sample. The ionization suppression is shown in Figure 3.4. This demonstrates the need for an effective cleanup step when using LC/MS for analysis.

3.5.2 GAS CHROMATOGRAPHY

Gas chromatography/mass spectrometry (GC/MS) has been used in the analysis of pharmaceutical contaminants in the environment. However, most GC/MS applications for pharmaceutical compounds require derivatization because most of these

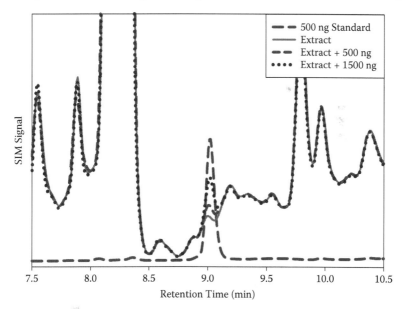

FIGURE 3.4 Chromatogram of a standard containing 500 ng of EE2 [m/z (–) 295] is compared with samples containing the extract of freeze-dried sludge solids spiked with 500 ng of EE2. Initial analysis of the extracted sample is not able to determine if extraction is working or if matrix effects have cause peak shifting. An additional 500 ng and then 1000 ng of EE2 is added. Ionization suppression due to matrix effects still gives a smaller signal in a sample containing 3x the EE2 as in a standard. (SIM: selected monitoring; EE2: Ethynilestradiol.)

analytes are fairly polar and are not amenable to direct GC/MS analysis. Many pharmaceutical compounds have been derivatized for GC/MS analysis, such as NSAIDs,[28] triclosan,[2] and estrogens.[7,18,25,63] Like liquid chromatographic methods, cleanup for these samples also employs SPE and GPC; however, other methods such as HPLC fractionation before derivatization are also used.[7,63] While GC/MS analysis with derivatization can be tedious, it can offer additional selectivity and improve detection limits relative to LC/MS analysis.

3.6 CONCLUSION

Many studies of pharmaceuticals in the environment examine the concentrations of these compounds in water and exclude the analysis of solids due to the difficulties encountered during solid analysis. However, many pharmaceuticals released into the environment are unaccounted for and may be bound to solids and, in turn, can become bioavailable. Limitations of solid analysis occur because the available methods do not suit the extraction of hydrophobic or strongly sorbed compounds without coextraction of undesirable matrix interferences. Further development of better methods is needed in order to fully understand and assess the extent to which sorbed compounds may affect exposed biota in the environment.

REFERENCES

1. Jacobsen, A.M.H.-S., Bent; Flemming, Ingerslev; Hansen, Steen Honore, 2004. Simultaneous extraction of tetracycline, macrolide and sulfonamide antibiotics from agricultural soils using pressurized liquid extraction, followed by solid phase extraction and liquid-tandem mass spectrometry. *Journal of Chromatography A* 1038: 157–170.
2. Canosa, P., Rodriguez, I., Rubi, E., and Cela, R., 2007. Determination of parabens and triclosan in indoor dust using matrix solid-phase dispersion and gas chromatography with tandem mass spectrometry. *Analytical Chemistry* 79(4): 1675–1681.
3. Hamscher, G.P., Theresia; Sczesny, Silke; Nau, Heinz; Hartung, Jorg, 2003. Antibiotics in dust originating from a pig-fattening farm: a new source of health hazard for farmers. *Environmental Health Perspectives* 111(13): 1590–1594.
4. Hartung, J., 1997. Dust exposure of livestock (In German). *Zentralbl Arbeitsmed* 47: 65–72.
5. Dignac, M.-F., Ginestet, P., Rybacki, D., Bruchet, A., Urbain, V., and Scribe, P., 2000. Fate of wastewater organic pollution during activated sludge treatment: nature of residual organic matter. *Water Research* 37(17): 4185–4194.
6. Degrémont, ed., *Water Treatment Handbook,* 1991. Cachan: Lavoisier Publishing.
7. Ternes, T.A., Andersen, H., Gilberg, D., and Bonerz, M., 2002. Determination of estrogens in sludge and sediments by liquid extraction and GC/MS/MS. *Analytical Chemistry* 74(14): 3498–3504.
8. El Bilali, L., Rasnussen, P.E., Hall, G.E.M., and Fortin, D., 2002. Role of sediment composition in trace metal distribution in lake sediments. *Applied Geochemistry* 17: 1171–1181.
9. Tolls, J., 2001. Sorption of veterinary pharmaceuticals in soils: a review. *Environmental Science and Technology* 35(17): 3397–3406.
10. Albert, A. and Rees, C.W., 1956. Avidity of the tetracyclines for the cations of metals. *Nature* (London, United Kingdom) 177: 433–434.

11. Sithole, B.B. and Guy, R.D., 1987. Models for tetracycline in aquatic environments. II. Interaction with humic substances. *Water, Air, and Soil Pollution* 32(3–4): 315–321.

12. Sithole, B.B. and Guy, R.D., 1987. Models for tetracycline in aquatic environments. I. Interaction with bentonite clay systems. *Water, Air, and Soil Pollution* 32(3–4): 303–314.

13. O'Connor, S. and Aga, D.S., 2007. Analysis of tetracycline antibiotics in soil: advances in extraction, clean-up and quantification. *Trends in Analytical Chemistry*, in press.

14. Lee, L.S., Strock, T.J., Sarmah, A.K., and Rao, P.S.C., 2003. Sorption and dissipation of testosterone, estrogens, and their primary transformation products in soils and sediment. *Environmental Science and Technology* 37(18): 4098–4105.

15. Stoob, K., Singer, H.P., Stettler, S., Hartmann, N., Mueller, S.R., and Stamm, C.H., 2006. Exhaustive extraction of sulfonamide antibiotics from aged agricultural soils using pressurized liquid extraction. *Journal of Chromatography A* 1128(1–2): 1–9.

16. Loeffler, D. and Ternes, T.A., 2003. Determination of acidic pharmaceuticals, antibiotics and ivermectin in river sediment using liquid chromatography-tandem mass spectrometry. *Journal of Chromatography A* 1021(1–2): 133–144.

17. Kulshrestha, P., Giese, R.F., Jr., and Aga, D.S., 2004. Investigating the molecular interactions of oxytetracycline in clay and organic matter: insights on factors affecting its mobility in soil. *Environmental Science and Technology* 38: 4097–4105.

18. Gomes, R.L., Avcioglu, E., Scrimshaw, M.D., and Lester, J.N., 2004. Steroid-estrogen determination in sediment and sewage sludge: a critique of sample preparation and chromatographic/mass spectrometry considerations, incorporating a case study in method development. *TrAC, Trends in Analytical Chemistry* 23(10–11): 737–744.

19. Kim, S.-C. and Carlson, K., 2007. Quantification of human and veterinary antibiotics in water and sediment using SPE/LC/MS/MS. *Analytical and Bioanalytical Chemistry* 387(4): 1301–1315.

20. Lindsey, M.E., Meyer, M., and Thurman, E.M., 2001. Analysis of trace levels of sulfonamide and tetracycline antimicrobials in groundwater and surface water using solid-phase extraction and liquid chromatography/mass spectrometry. *Analytical Chemistry* 73(19): 4640–4646.

21. Hyoetylaeinen, T. and Riekkola, M.-L., 2004. Approaches for on-line coupling of extraction and chromatography. *Analytical and Bioanalytical Chemistry* 378(8): 1962–1981.

22. Santos Jr., D., Krug, F.J., Pereira, M.D.G., and Korn, M., 2006. Currents on ultrasound-assisted extraction for sample preparation and spectroscopic analytes determination. *Applied Spectroscopy Reviews* 41(3): 305–321.

23. Hutchins, S.R., White, M.V., Hudson, F.M., and Fine, D.D., 2007. Analysis of lagoon samples from different concentrated animal feeding operations for estrogens and estrogen conjugates. *Environmental Science & Technology* 41(3): 738–744.

24. Blackwell, P.A., Holten Lutzhoft, H.-C., Ma, H.-P., Halling-Sorensen, B., Boxall, A.B.A., and Kay, P., 2004. Ultrasonic extraction of veterinary antibiotics from soils and pig slurry with SPE clean-up and LC-UV and fluorescence detection. *Talanta* 64(4): 1058–1064.

25. Braga, O., Smythe, G.A., Schaefer, A.I., and Feitz, A.J., 2005. Steroid estrogens in ocean sediments. *Chemosphere* 61(6): 827–833.

26. Hajkova, K., Pulkrabova, J., Schurek, J., Hajslova, J., Poustka, J., Napravnikova, M., and Kocourek, V., 2007. Novel approaches to the analysis of steroid estrogens in river sediments. *Analytical and Bioanalytical Chemistry* 387(4): 1351–1363.

27. Lopez de Alda, M.J. and Barcelo, D., 2001. Use of solid-phase extraction in various of its modalities for sample preparation in the determination of estrogens and progestogens in sediment and water. *Journal of Chromatography A* 938(1–2): 145–153.

28. Antonic, J. and Heath, E., 2007. Determination of NSAIDs in river sediment samples. *Analytical and Bioanalytical Chemistry* (387): 1337–1342.

29. Golet, E.M., Strehler, A., Alder, A.C., and Giger, W., 2002. Determination of fluoroquinolone antibacterial agents in sewage sludge and sludge-treated soil using accelerated solvent extraction followed by solid-phase extraction. *Analytical Chemistry* 74(21): 5455–5462.

30. Schlusener, M.P., Spiteller, M., and Bester, K., 2003. Determination of antibiotics from soil by pressurized liquid extraction and liquid chromatography-tandem mass spectrometry. *Journal of Chromatography A* 1003: 21–28.

31. Chun, S., Lee, J., Geyer, R., and White, D., 2005. Comparison of three extraction methods for 17b-estrdiol in sand, bentonite and organic rich silt loam. *Journal of Environmental Science and Health, Part B* 40: 731–740.

32. Houtman, C.J., van Houten, Y.K., Leonards, P.E.G., Brouwer, A., Lamoree, M.H., and Legler, J., 2006. Biological validation of a sample preparation method for ER-CALUX bioanalysis of estrogenic activity in sediment using mixtures of xeno-estrogens. *Environmental Science & Technology* 40(7): 2455–2461.

33. Labadie, P. and Hill, E.M., 2007. Analysis of estrogens in river sediments by liquid chromatography-electrospray ionisation mass spectrometry. *Journal of Chromatography* A 1141(2): 174–181.

34. Prat, M.D., Ramil, D., Compano, R., Hernandez-Arteseros, J.A., and Granados, M., 2006. Determination of flumequine and oxolinic acid in sediments and soils by microwave-assisted extraction and liquid chromatography-fluorescence. *Analytica Chimica Acta* 567(2): 229–235.

35. Morales-Munoz, S.L., Luque-Garcia, J.L., and Luque de Castro, M.D., 2004. Continuous microwave-assisted extraction coupled with derivatization and flouimetric monitoring for the determination of flouroquinolone antibacterial agents from soil samples. *Journal of Chromatography A* 1059: 25–31.

36. Hawthorne, S.B., Grabanski, C.B., Martin, E., Miller, D.J., 2000. Comparison of Soxhlet extraction, pressurized liquid extraction, supercritical fluid extraction and subcritical water extraction for environmental solids: recovery, selectivity and effects on sample matrix. *Journal of Chromatography A* 892: 421–433.

37. Lin, J.-G., Arunkumar, R., and Liu, C.-H., 1999. Efficiency of supercritical fluid extraction for determining 4-nonylphenol in municipal sewage sludge. *Journal of Chromatography A* 840(1): 71–79.

38. Meesters Roland, J.W. and Schroder Horst, F., 2002. Simultaneous determination of 4-nonylphenol and bisphenol A in sewage sludge. Anal Chem FIELD Full Journal Title: *Analytical Chemistry* 74(14): 3566–3574.

39. Brandsteterova, E., Kubalec, P., Bovanova, L., Simko, P., Bednarikova, A., and Machackova, L., 1997. SPE and MSPD as pre-separation techniques for HPLC of tetracyclines in meat, milk, and cheese. Zeitschrift fuer Lebensmittel-Untersuchung und Forschung A: *Food Research and Technology* 205(4): 311–315.

40. Long, A.R., Hsieh, L.C., Malbrough, M.S., Short, C.R., and Barker, S.A., 1990. Matrix solid phase dispersion isolation and liquid chromatographic determination of oxytetracycline in catfish (Ictalurus punctatus) muscle tissue. *Journal—Association of Official Analytical Chemists* 73(6): 864–867.

41. Long, A.R., Hsieh, L.C., Malbrough, M.S., Short, C.R., and Barker, S.A., 1990. Matrix solid-phase dispersion (MSPD) isolation and liquid chromatographic determination of oxytetracycline, tetracycline, and chlortetracycline in milk. *Journal—Association of Official Analytical Chemists* 73(3): 379–384.

42. Gentili, A., Sergi, M., Perret, D., Marchese, S., Curini, R., and Lisandrin, S., 2006. High- and low-resolution mass spectrometry coupled to liquid chromatography as confirmatory methods of anabolic residues in crude meat and infant foods. *Rapid Communications in Mass Spectrometry* 20(12): 1845–1854.

43. Cai, W. and Gupta, R.B., 2004. Molecularly-imprinted polymers selective for tetracycline binding. *Separation and Purification Technology* 35(3): 215–221.

44. Andersson, L.I., 2000. Molecular imprinting: developments and applications in the analytical chemistry field. *Journal of Chromatography, B: Biomedical Sciences and Applications* 745(1): 3–13.

45. Caro, E., Marce, R.M., Cormack, P.A.G., Sherrington, D.C., and Borrull, F., 2005. Synthesis and application of an oxytetracycline imprinted polymer for the solid-phase extraction of tetracycline antibiotics. *Analytica Chimica Acta* 552(1–2): 81–86.

46. Xiong, Y., Zhou, H., Zhang, Z., He, D., and He, C., 2006. Molecularly imprinted on-line solid-phase extraction combined with flow-injection chemiluminescence for the determination of tetracycline. *Analyst* (Cambridge, United Kingdom) 131(7): 829–834.

47. Suedee, R., Srichana, T., Chuchome, T., and Kongmark, U., 2004. Use of molecularly imprinted polymers from a mixture of tetracycline and its degradation products to produce affinity membranes for the removal of tetracycline from water. *Journal of Chromatography. B: Analytical Technologies in the Biomedical and Life Sciences* 811(2): 191–200.

48. Bravo, J.C., Fernandez, P., and Durand, J.S., 2005. Flow injection fluorimetric determination of b-estradiol using a molecularly imprinted polymer. *Analyst* (Cambridge, United Kingdom) 130(10): 1404–1409.

49. Meng, Z., Chen, W., and Mulchandani, A., 2005. Removal of estrogenic pollutants from contaminated water using molecularly imprinted polymers. *Environmental Science and Technology* 39(22): 8958–8962.

50. Reddy, S. and Brownawell, B.J., 2005. Analysis of estrogens in sediment from a sewage-impacted urban estuary using high-performance liquid chromatography/time-of-flight mass spectrometry. *Environmental Toxicology and Chemistry* 24(5): 1041–1047.

51. Zhu, J., Snow, D.D., Cassada, D.A., Monson, S.J., and Spaldin, R.F., 2001. Analysis of oxytetracycline, tetracycline, and chlortetracycline in water using solid-phase extraction and liquid chromatography-tandem mass spectrometry. *Journal of Chromatography A* 928: 177–186.

52. Blanchflower, W.J., McCracken, R.J., Haggan, A.S., and Kennedy, D.G., 1997. Confirmatory assay for the determination of tetracycline, oxytetracycline, chlortetracylcine and its isomers in muscle and kidney using liquid chromatography-mass spectrometry. *Journal of Chromatography B* 692: 351–360.

53. Oka, H., Ikai, Y., Hayakawa, J., Harada, K.-I., Asukabe, H., and Suzuki, M., 1994. Improvement of chemical analysis of antibiotics. 22. Identification of residual tetracyclines in honey by frit FAB/LC/MS using a volatile mobile phase. *Journal of Agriculture and Food Chemistry* 42: 2215–2219.

54. Nakazawa, H., Ino, S., Kato, K., Watanabe, T., Ito, Y., and Oka, H., 1999. Simultaneous determination of residual tetracyclines in foods by high-performance liquid chromatography with atmospheric pressure chemical ionization tandem mass spectrometry. *Journal of Chromatography B* 732: 55–64.

55. Delepee, R., Maume, D., Bizec, B.L., and Pouliquen, H., 2000. Preliminary assays to elucidate the structure of oxytetracycline's degradation products in sediments. Determination of natural tetracyclines by high-performance liquid chromatography–fast atom bombardment mass spectrometry. *Journal of Chromatography B* 748: 369–381.

56. Kamel, A.M. and Fouda, H.G., 2002. Mass spectral characterization of tetracyclines by electrospray ionization, H/D exchange, and multiple stage mass spectrometry. *Journal of American Society for Mass Spectrometry* 13: 543–557.

57. Vartanian, V.H., Boolsby, B., and Brodbelt, J.S., 1998. Identification of tetracycline antibiotics by electrospray ionization in a quadropole ion trap. *Journal of American Society for Mass Spectrometry* 9: 1089–1098.

58. Gustavsson, S.A., Samskog, J., Markides, K.E., and Langstrom, B., 2001. Studies of signal suppression in liquid chromatography–electrospray ionization mass spectrometry using volatile ion-pairing reagents. *Journal of Chromatography A* 937: 41–47.

59. Mallet, C.R., Lu, Z., and Mazzeo, J.R., 2004. A study of ion suppression effects in electrospray ionization from mobile phase additives and solid-phase extracts. *Rapid Communications in Mass Spectrometry* 18: 49–58.

60. Liang, H.R., Foltz, R.L., Meng, M., and Bennett, P., 2003. Ionization enhancement in atmospheric pressure chemical ionization and suppression in electrospray ionization between target drugs and stable-isotope-labeled internal standards in quantitative liquid chromatography/tandem mass spectrometry. *Rapid Communications in Mass Spectrometry* 17: 2815–2821.

61. Dams, R., Huestis, M.A., Lambert, W.E., and Murphy, C.M., 2003. Matrix effect in bio-analysis of illicit drugs with LC-MS/MS: influence of ionization type, sample preparation, and biofluid. *Journal of the American Society of Mass Spectrometry* 14: 1290–1294.

62. Schnitzer, M., 1978. Humic substances: chemistry and reactions. *Developments in Soil Science* 8(Soil Org. Matter): 1–64.

63. Williams, R.J., Johnson, A.C., Smith, J.J.L., and Kanda, R., 2003. Steroid estrogens profiles along river stretches arising from sewage treatment works discharges. *Environmental Science and Technology* 37(9): 1744–1750.

4 Gadolinium Containing Contrast Agents for Magnetic Resonance Imaging (MRI)
Investigations on the Environmental Fate and Effects

Claudia Neubert, Reinhard Länge, and Thomas Steger-Hartmann

Contents

4.1 INTRODUCTION

Mainly due to progress in analytical instrumentation, there has been an increased awareness of the presence of pharmaceutical compounds as environmental contaminants in recent years.[1,2] Although concentrations of pharmaceuticals in the aquatic environment are usually only in the parts-per-billion or parts-per-trillion levels, there is growing concern over their release because of their biological activity, which is not limited to human targets.

As a result of that concern, specific ecological risk assessment procedures have been refined, which led to the introduction of guidelines in some of the major human pharmaceutical markets (Europe, United States). Essentially these procedures consist of an estimation of the environmental concentration, on the one hand, and the experimental determination of a no-effect concentration (NOEC) of the pharmaceutical on the other hand.[3,4] Because the aquatic environment represents the primary recipient of pharmaceuticals that are being discharged from wastewater treatment plant effluents, risk assessment has focussed on the aquatic ecosystem. The European risk assessment guideline[3] proposes a tiered system in which exposure estimation and risk screening are included, as well as the determination of physicochemical properties of new human pharmaceuticals and diagnostic agents.

To assess the potential effects of contaminants on the aquatic environment, a battery of selected organisms, each representing a specific level of the aquatic ecosystem (see Figure 4.1), is investigated. Furthermore, in order to assess persistence and thus temporal development of exposure, tests on biodegradation are conducted. Screening tests for biodegradation allow a first qualitative assessment of the potential of sewage treatment plants or natural surface waters to degrade the compound of interest.

Among the first pharmaceutical compounds that were analytically detected in the aquatic environment[5] and subsequently assessed for their ecotoxicological risk were iodinated X-ray contrast agents.[6] Fewer data are currently available for the second class of contrast agents used in MRI, even though those compounds have been detected in ground water as early as in 1996.[7]

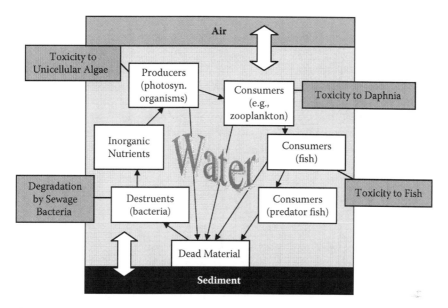

FIGURE 4.1   Interactions in an aquatic ecosystem and derived test systems (gray) on different trophic levels.

This chapter reports the results of ecotoxicological studies and biodegradability tests of several gadolinium-containing contrast enhancing agents for MRI and provides an environmental risk assessment based on the information obtained. MRI is an essential tool in the noninvasive diagnostics of various diseases, such as tumors, to improve lesion identification and characterization. In order to improve the sensitivity and specificity of diagnoses, several contrast enhancing agents have been developed in the last few decades by various pharmaceutical manufacturers and are marketed worldwide.[8]

Gadolinium (Gd), a lanthanide, is the most widely used metal in MRI contrast agents. Its ion has paramagnetic properties (seven unpaired electrons) and a very long electronic relaxation time. Due to the toxicity of free Gd, which is caused by an interaction with calcium channels,[9,10] and a precipitation tendency above pH 6 with subsequent trapping in the liver,[11–13] clinical use is only possible in a complexed form. Commonly used chelating agents are polyamino-polycarboxylic ligands such as diethylenetriaminepentaacetic (DTPA). The complexes formed by the different chelates can be grouped, according to their size and structure, into:

- macrocyclic chelates such as gadobutrol (Gadovist®) and
- linear chelates such as dimeglumine gadopentetate (Gd–DTPA) (Magnevist®) or Gadodiamide, Gd–diethylenetriamine pentaacetate bismethylamide (Gd–DTPA–BMA) (Omniscan®)

Due to the exceptional stability of these highly hydrophilic chelates and the lack of human metabolism, the contrast media are quantitatively excreted unchanged after administration within hours, and are subsequently emitted into the aquatic environment. Several studies have shown notable increases in Gd concentrations in surface or groundwaters receiving sewage effluents, an observation which has been termed "Gd anomaly."[14–16]

The Gd anomaly results from the use of MRI contrast agents for which the most significant entry route is the effluent from wastewater treatment works.[16] Relatively little information on the aquatic toxicity of Gd or Gd-chelates has been published. Therefore, in a first step, the aquatic toxicity of these compounds was investigated in short-term tests on standard aquatic species at high concentrations. Furthermore, the biological stability under the incubation with activated sludge bacteria was studied in screening tests.

4.2 METHODS

All described tests were performed according to internationally standardized guidelines and in accordance with the good laboratory practice (GLP) principles. Dimeglumine gadopentetate, gadobutrol, and gadoxetic acid disodium were manufactured by Bayer Schering Pharma AG, Germany, gadofosveset trisodium by Mallinckrodt Inc., United States. Table 4.1 shows the structures and selected physicochemical properties of the tested compounds.

4.2.1 BIODEGRADABILITY OF DIMEGLUMINE GADOPENTETATE, GADOBUTROL, GADOXETIC ACID, DISODIUM, AND GADOFOSVESET TRISODIUM

Test systems for ready biodegradability were originally established for household detergents and are required by the European Reserach Area (EU ERA) guideline to assess the degradation of a human pharmaceutical. The test compounds dimeglumine gadopentetate, gadoxetic acid disodium, and gadofosveset trisodium were investigated according to the test guideline of the Organization for Economic Cooperation and Development (OECD), 301E.[17] Briefly, the compounds were incubated in aqueous solutions including nutrients with microorganisms from a municipal sewage treatment plant for 62 days (test compound: dimeglumine gadopentetate, in duplicate) and 28 to 29 days (test compound: gadoxetic acid disodium, gadofosveset trisodium, in triplicate).

The test concentration for the substances was adjusted to 20 mg dissolved organic carbon (DOC) per liter corresponding to 56.7 mg dimeglumine gadopentetate, 52.5 mg gadoxetic acid disodium, and 51.65 mg gadofosveset trisodium.

Additionally, a reference substance (sodium acetate) was tested at the same DOC concentration in order to verify the viability and activity of the degrading microorganisms. Furthermore, one flask containing both the test substance and the reference substance was tested as a toxicity control. Three additional vessels without any test or reference substances were used as blank (control).

The biological degradation of the test and reference substances was evaluated by the decrease of DOC in the solutions. Total organic carbon (TOC) and DOC were measured by a TOC analyzer. Additionally, for this specific case, the concentration

TABLE 4.1

Structure, International Union of Pure and Applied Chemistry (IUPAC) Names, Molecular Weight, and Water Solubility of the Tested Compounds

IUPAC Names, Molecular Weight, and Water Solubility	Structure
Compound: Magnevist Active agent: Dimeglumine gadopentetate	IUPAC name: Diethylenetriamine-pentaacetic acid, Gadolinium Complex, dimeglumine salt
Molecular weight: 938	Water solubility: ≤469 g/L
Compound: Gadovist Active agent: Gadobutrol	IUPAC name: 10-[(1SR, 2RS) – 2,3 – Dihydroxy – 1 – hydroxymethylpropyl] – 1, 4, 7, 10 – tetraazacyclododecane – 1, 4, 7 – triacetic acid, Gadolinium – Complex
Molecular weight: 604.7	Water solubility: $1081 \pm$ g/L
Compound: Primovist® Active agent: Gadoxetic acid disodium	IUPAC name: (4S) – 4 – (4 – Ethoxybenzyl) – 3, 6, 9 – tris (carboxy-latomethyl) – 3, 6, 9 – triazaundecanedioic acid, Gadolinium – Complex, Disodium salt)
Molecular weight: 725.7	Water solubility: 1057 g/L
Compound: Vasovist® Active agent: Gadofosve-set trisodium	IUPAC name: Trisodium {N – (2 – {bis[(carboxy-kappa O) methyl]amino-kappa N}ethyl) – N - [(R) – 2 – {bis[(carboxy-kappa O) methyl] amino – kappa N} – 3 – {[(4,4 – diphenylcyclohex yloxy)phosphinato – kappa O]oxy} propyl]glycinato(6 -) – kappa N, kappa O} gadolinate(3 -)
Molecular weight: 957.9	Water solubility: ≤247 g/L

of dimeglumine gadopentetate was analyzed by high-performance liquid chromatography/ultraviolet (HPLC/UV). Specific concentration analysis of gadoxetic acid disodium and gadofosveset trisodium was not performed because it is not required by the OECD guideline.

Gadobutrol was tested for microbial degradation in agreement with the test guideline 3.11 of the Environmental Assessment Technical Assistance Handbook,[18] which slightly differs from the OECD 301 E procedure in using an inoculum from a municipal sewage treatment plant mixed with a filtered suspension of garden soil. The inoculum was preadapted in an aqueous solution including nutrients with the test substance gadobutrol or the reference substance (glucose monohydrate) for 14 days. Afterward, the test substance (10 mg/L), reference substance (10 mg/L), and the blank solution were incubated with the preadapted microorganisms for 27 days.

The biological degradation of the test and reference substances was evaluated by the measurement of the carbon dioxide (CO_2) produced during the test period. CO_2 was absorbed by $Ba(OH)_2$. CO_2 production was determined by titration of the $Ba(OH)_2$ solution as described in the guideline.

4.2.2 ACUTE TOXICITY TEST OF DIMEGLUMINE GADOPENTETATE, GADOBUTROL, GADOXETIC ACID DISODIUM, AND GADOFOSVESET TRISODIUM WITH FISH

Fish represent the nonmammalian consumer of an aquatic ecosystem (Figure 4.1). In order to assess the toxicity of the test compound to representative species of this trophic level, the acute toxicity of gadobutrol and gadoxetic acid disodium was determined with rainbow trout (*Oncorhynchus mykiss*) on the basis of the guideline Freshwater Fish Acute Toxicity, Environmental Assessment Technical Assistance Handbook, Technical Assistance Document 4.11[19] with a test duration of 96 hours. The acute toxicity of dimeglumine gadopentetate and gadofosveset trisodium to the zebrafish (*Danio rerio*) was conducted in accordance with the test guideline OECD 203[20] and the EC Guideline Part 2—Testing Methods, Part C. 1.[21]

Ten fish were used for each concentration of the test compound and for the control group. The fish were exposed for a period of 96 hours to the dilution water and to various concentrations of the substances (0.1, 1.0, 10.0, 100.0, and 1000.0 mg/L in case of gadobutrol and gadoxetic acid disodium, 100 mg/L in case of dimeglumine gadopentetate, and 1000 mg/L in case of gadofosveset trisodium).

Mortalities and visual abnormalities, as well as pH value, oxygen concentration, and temperature, were recorded at approximately 3, 6, 24, 48, 72, and 96 hours. Samples for the concentration analysis by inductively coupled plasma/mass spectrometry (ICP/MS) (inductively coupled plasma/atomic emission spectrometry [ICP/AES] in the case of gadofosveset trisodium) were taken in regular intervals. The analytical method determined the Gd concentration on the basis of which the test substance concentration was calculated.

4.2.3 Acute Immobilization Test of Dimeglumine Gadopentetate, Gadobutrol, Gadoxetic Acid Disodium, and Gadofosveset Trisodium with *Daphnia magna*

The crustacean *Daphnia magna* represents the primary feeder of an aquatic ecosystem (Figure 4.1). In order to assess the toxicity of the test compound to representative species of this trophic level, the test compound gadobutrol was investigated in agreement with the test guideline: Daphnia Acute Toxicity, Environmental Assessment Technical Assistance Handbook, Technical Assistance Document 4.08,[22] whereas the test compounds dimeglumine gadopentetate, gadoxetic acid disodium, and gadofosveset trisodium were investigated according to the guideline of the OECD 202 and the EC guideline part C.2.[23,24] Different guidelines were used for these tests because they were performed for the use in different regulatory regions.

The test was performed with five juvenile daphnia in each vessel and four replicates for each concentration. The crustaceans were exposed for a period of 48 hours under static conditions. Immobilization was recorded at 24 and 48 hours. The pH value, oxygen concentration, and temperature were measured at 0 and 48 hours.

The test solutions had nominal concentrations of 0.1, 1.0, 10.0, 100.0, and 1000.0 mg/L (test compound: gadobutrol); 100 mg/L (test compounds: dimeglumine gadopentetate and gadoxetic acid disodium); and 90 mg/L (test compound: gadofosveset trisodium).

Samples for the concentration analysis of gadobutrol and gadoxetic acid disodium by ICP/MS were taken daily. The method included a detection of Gd, and the final concentrations for gadobutrol and gadoxetic acid disodium were calculated accordingly.

For dimeglumine gadopentetate and gadofosveset trisodium only nominal values were available. Since these compounds are very well soluble in water (≤469 g/L for dimeglumine gadopentetate and ≤247 g/L for gadofosveset trisodium, respectively) and are very stable, the actual concentration was assumed to be in agreement with the nominal.

4.2.4 Growth Inhibition Test of Dimeglumine Gadopentetate, Gadobutrol, Gadoxetic Acid Disodium, and Gadofosveset Trisodium on Green Algae

Green algae are the main primary producers in freshwater ecosystems. Unicellular green algae are established in ecotoxicity testing, since they represent the main part of the floral biomass. The studies were conducted with an algae population of *Chlorella vulgaris* (test compound: gadobutrol) and *Desmodesmus subspicatus* (test compounds: dimeglumine gadopentetate, gadoxetic acid disodium, and gadofosveset trisodium) in agreement with the OECD guideline 201 and the EC guideline part C.3.[25,26]

The test substances were incubated in an aqueous solution including nutrients for the duration of approximately 72 hours. The nutrient solution was made up of mainly nitrate, phosphates, and some trace elements. Due to the long-time course of the experiments and to the changing guideline requirements, the tested concentrations were not identical for the different contrast agents. The nominal test concentrations were 0, 1.25, 2, 4, 10, 20, and 100 mg/L for the test compound dimeglumine gadopentetate; 0, 40, 88, 194, 426, 937, and 2062 mg/L for the test compound gadobutrol; 0, 2, 4, 10, 20, 40, and 80 mg/L for the test compound gadofosveset trisodium; and 63, 125, 250, 500, and 1000 mg/L for the test compound gadoxetic acid disodium. In an additional test with gadoxetic acid disodium, solutions with nominal loadings of 1000, 5000, and 10,000 mg/L were prepared.

The algae were exposed to each concentration in triplicate. Six vessels were prepared for the control. The algae were incubated in an incubator shaker under continuous light. As a parameter for the growth of the algae population, the cell concentrations of the test and control solutions were counted with an electronic particle counter ("Coulter Counter") at approximately 24, 48, and 72 hours. The pH value was measured at the beginning and at the end of the test.

For the study with dimeglumine gadopentetate and gadofosveset trisodium, an incubating apparatus (Abimed Algen Test XT) was used. In this case the cell number was determined via measurement of chlorophyll fluorescence. The increase of biomass and the growth rate was calculated on the basis of the cell counts. The calculated biomass and growth rate of each concentration were compared to those of the controls, and the inhibition was calculated. Concentration analysis was not performed.

4.2.5 GROWTH INHIBITION TEST OF DIMEGLUMINE GADOPENTETATE ON DIFFERENT MICROORGANISMS

Microorganisms play a role as degraders in the aquatic environment, thus lowering the exposure with introduced contaminants. Furthermore, some of the microorganisms (bluegreen algae) also represent the trophic level of producers.

The growth inhibition test of dimeglumine gadopentetate was conducted in agreement with the standard DIN 38 412 L8.[27] It was incubated in an aqueous solution including nutrients, with a bacterial population containing *Pseudomonas putida* for the test duration of approximately 16 hours.

The test concentrations were 0.1, 1.0, 10.0, 100.0, and 1000.0 mg/L and a control. All test concentrations were incubated in duplicate. As a parameter for the test growth of the bacterial population, the turbidity of the test and control solutions was analyzed photometrically at a wavelength of 436 nm. A concentration analysis was not performed.

The effect of gadobutrol on different microbes was studied in a growth inhibition test in agreement with the test guideline Microbial Growth Inhibition, Environmental Assessment Technical Assistance Handbook, Technical Assistance Document 4.02.[28] Different bacterial, fungal, and algal microbes (*Pseudomonas putida, Azotobacter beijerinckii, Aspergillus niger, Caetomium globosom,* and *Nostoc ellipsosporum)* were exposed to graduated concentrations of gadobutrol. The microbes were incubated on agar plates containing nutrients and the test substance over periods of 20 hours (*Pseudomonas putida*), 48 hours (*Azotobacter beijerinckii*), 3 days (*Aspergillus niger, Caetomium globosom*), and 10 days (*Nostoc ellipsosporum*) under appropriate conditions. The concentrations of the test substance were 0.1, 1.0, 10.0, 100.0, and 1000.0 mg/L. The growth of the microbes was assessed at the end of the respective incubation period. Growth was defined as appearance of colonies.

Concentration analysis was not performed.

4.3 RESULTS

4.3.1 BIODEGRADABILITY OF DIMEGLUMINE GADOPENTETATE, GADOBUTROL, GADOXETIC ACID DISODIUM, AND GADOFOSVESET TRISODIUM

Figure 4.2 summarizes the results of the degradation tests at the end of the incubation period. Microbial degradation was only observed in the test with dimeglumine gadopentetate, which was likely due to the degradation of meglumine (see Section 4.4).

The individual degradation curves of dimeglumine gadopentetate and sodium acetate are depicted in Figure 4.3. Degradation of the test compound started between day 15 and day 21, and degradation values of approximately 40% were reached after 43 days.

Figure 4.4 shows the concentrations of dimeglumine gadopentetate [mg/L] measured by HPLC/UV. They varied between 53.6 and 62.1 mg/L. The analysis for free Gd was negative, indicating that no Gd was released from the chelate. The results of the degradation of gadoxetic acid disodium, gadobutrol, and gadofosveset trisodium showed that none of these compounds was readily biodegradable and none of the compounds was toxic to the degrading bacteria.

4.3.2 ACUTE TOXICITY OF DIMEGLUMINE GADOPENTETATE, GADOBUTROL, GADOXETIC ACID DISODIUM, AND GADOFOSVESET TRISODIUM TO FISH

The measured substance concentrations were approximately 90 to 120% of the nominal values. The time course of the results demonstrates that the substance solutions were stable during the whole exposure period. The results of the measured

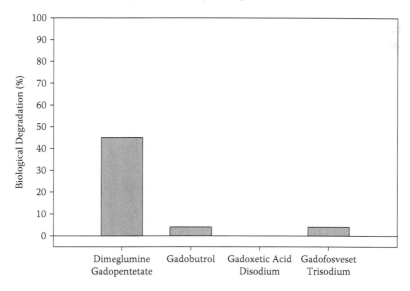

FIGURE 4.2 Biological degradation of dimeglumine gadopentetate, gadobutrol, gadoxetic acid disodium, and gadofosveset trisodium at the end of the degradation tests [%].

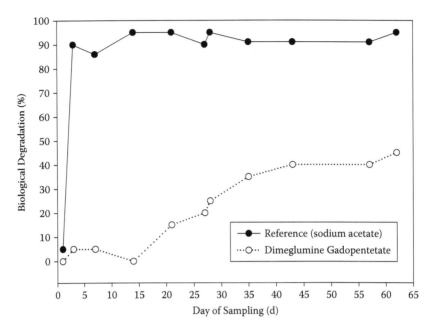

FIGURE 4.3 Biological degradation of dimeglumine gadopentetate and the reference compound sodium acetate [%] in the modified OECD screening test.

concentrations of the test compounds of the studies on the acute toxicity to fish and waterflea are summarized in Table 4.2.

No substance-related mortality or abnormal behavior was observed in the tests during the whole exposure time. On the basis of the given results the LC_{50}/96 hours for gadobutrol, gadoxetic acid disodium, and gadofosveset trisodium was >1000 mg/L, for dimeglumine gadopentetate >100 mg/L.

4.3.3 ACUTE IMMOBILIZATION TEST OF DIMEGLUMINE GADOPENTETATE, GADOBUTROL, GADOXETIC ACID DISODIUM, AND GADOFOSVESET TRISODIUM WITH *DAPHNIA MAGNA*

Immobilized daphnia were not observed in either the test or in the control solutions of dimeglumine gadopentetate, gadobutrol, and gadofosveset trisodium. In the test with gadoxetic acid disodium, one daphnia was immobilized in the control. Table 4.2 summarizes the results of the measured concentrations of the test compounds of the studies.

4.3.4 GROWTH INHIBITION TEST OF GADOBUTROL, DIMEGLUMINE GADOPENTETATE, GADOXETIC ACID DISODIUM, AND GADOFOSVESET TRISODIUM ON GREEN ALGAE

Figure 4.5 gives the inhibition [%] of the growth of *Chlorella vulgaris* after 72 hours exposure to gadobutrol on the basis of the biomass (integral) and the growth rate. In order to illustrate the data on which the inhibition [%] is calculated, cell numbers

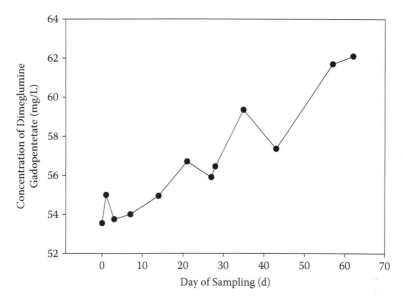

FIGURE 4.4 Concentrations of dimeglumine gadopentetate [mg/L] in the modified OECD screening test.

TABLE 4.2
Measured Concentrations in Acute Toxicity Tests on Fish and Waterflea

Test Compound	Measured Concentrations in the Acute Toxicity Tests on Fish [% of the Nominal Concentrations]	Measured Concentrations in the Acute Toxicity Tests on Waterflea [% of the Nominal Concentrations]
Dimeglumine gadopentetate	106.01 (mean)	—
Gadobutrol	90–120*	91–100+
Gadoxetic acid disodium	90–100	90
Gadofosveset trisodium	97.71	—

* An exceptionally low concentration at the nominal value of 1.0 mg/L (72 hours) was excluded from further calculations.

+ The analysis of the control solution yielded a detectable concentration of gadobutrol after 24 hours (mean value (MV) = 0.671 mg/L, standard deviation (SD) = 0.0009).

and standard deviation (SD) of *Chlorella vulgaris* are added. A clear inhibition of the algae growth was determined at the highest concentration of 2062 mg/L (76% for biomass, 32% for the growth rate). In all other concentrations there was a slightly higher growth compared with the controls. EC_{50} values for growth inhibition could not be calculated since only the highest concentrations of 2062 mg/L showed a clear effect. The EC_{50} value (biomass integral) can be estimated to lie in a range between 937 and 2062 mg/L.

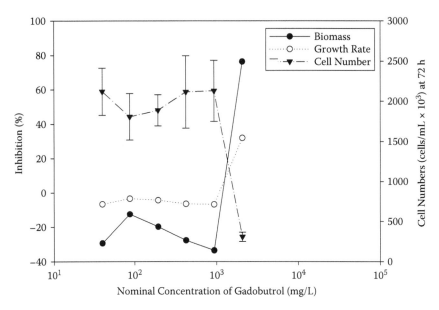

FIGURE 4.5 Inhibition of the growth rate and the biomass of *Chlorella vulgaris* [%] and cell numbers (cells/mL x 10^3 ± SD) of *Chlorella vulgaris* after 72-hour exposure to gadobutrol.

Figure 4.6 shows the percentage inhibition of the growth rate and the biomass of gadoxetic acid disodium after 72 hours exposure time, including cell numbers.

No adverse effects were observed in the growth inhibition test of dimeglumine gadopentetate up to a concentration of 100 mg/L and in the growth inhibition test of gadofosveset trisodium up to a concentration of 80 mg/L. The NOEC and EC_{50}-values are summarized in Table 4.3.

4.3.5 GROWTH INHIBITION TEST OF DIMEGLUMINE GADOPENTETATE AND GADOBUTROL ON DIFFERENT MICROORGANISMS

No inhibitory effect of dimeglumine gadopentetate on the growth of *Pseudomonas putida* was observed. EC_{10}—and EC_{50}—values were therefore higher than 1000 mg/L. None of the tested microorganisms were growth inhibited by gadobutrol. The minimum inhibitory concentration (MIC) was therefore higher than 1000 mg/L.

4.4 DISCUSSION

A series of ecotoxicity tests was conducted to assess the environmental risk of selected Gd-containing contrast-enhancing agents for MRI. First, the results are discussed for each test system; second, a risk assessment is performed based on these data.

The compounds were tested over a long period (about 15 years) and for different regulatory regions (Europe, United States). During this time span, the guidelines changed for various reasons and the applied test procedures were modified because

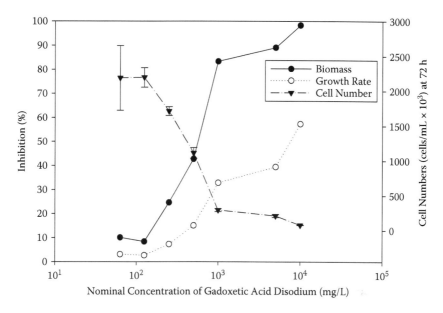

FIGURE 4.6 Inhibition of the growth rate and the biomass of *Desmodesmus subspicatus* [%] and cell numbers (cells/mL x 10^3 ± SD) of *Desmodesmus subspicatus* after 72-hour exposure to gadoxetic acid disodium. (For the concentration of 1000 mg/L the values of a second test [32.9 % Inhibition, 83.4 %, respectively] were chosen in this figure.)

TABLE 4.3
EC_{50} and NOEC of Various Test Compounds in Algae Tests

Test Compound (Species)	NOEC	EC_{50}
Dimeglumine gadopentetate (Desmodesmus subspicatus)	100 mg/L	>100 mg/L
Gadobutrol (Chlorella vulgaris)	937 mg/L	>937 mg/L
Gadoxetic acid disodium (Desmodesmus subspicatus)	125 mg/L	>500 mg/L
Gadofosveset trisodium (Desmodesmus subspicatus)	80 mg/L	> 80 mg/L

of scientific progress and experiences in the laboratory. For these reasons the studies were not conducted according to exactly identical procedures.

4.4.1 Degradation Tests

The test compounds dimeglumine gadopentetate, gadobutrol, gadoxetic acid disodium, and gadofosveset trisodium were not readily biodegradable under the conditions of the tests because a degradation of more than 60 to 70% was not achieved within 10 days. The degradation of the reference substances fulfilled the quality criteria set by the guidelines (i.e., the inoculum was viable and active).

The elimination of organic carbon in the test solution of dimeglumine gadopentetate, as indicated by DOC measurement, cannot be attributed to the degradation of the Gd–DTPA complex, because the chemical analysis by HPLC/UV was specific for this complex and did not show any degradation. The slight increase as shown in Figure 4.4 is assumed to be an inaccuracy of the measurement. It is therefore most likely that the dimeglumine salt, which contains 14 C atoms, was degraded to a large extent toward the end of the experiment. This interpretation is further confirmed by the fact that no free Gd was found in the assay, demonstrating that the lanthanide was not released from the complex during the test. A chemical characterization of the dimeglumine salt or its degradation products was not performed.

No decrease in degradation of the reference compound was observed in any of the toxicity controls, indicating that the compounds have no microcidal properties. Most probably, the tested compounds are not amenable to degradation due to their large size and complex molecular structure, which impedes internalization into the bacteria and subsequent enzymatic attack.[29] Due to the high thermodynamic stability of the Gd-chelates, it is not likely that the complexes release Gd. The high thermodynamic stability constants of the tested compounds indicate equilibrium far on the side of the complex:

Dimeglumine gadopentetate	$\log K = 22.52$[30]
Gadobutrol	$\log K = 21.75$[30]
Gadoxetic acid	$\log K = 23.46$[31] and
Gadofosveset	$\log K = 22.06$[32]

Even if small amounts of the chelates would decomplex, the resulting free ligands would not necessarily be readily degradable. Pitter et al. (2001) found in the Zahn-Wellens test for inherent biodegradability that the biodegradability of ethylene(propylene)di(tri)amine-based complexing agents depends on the character and number of substituents and nitrogen atoms in the molecule. Tetra(penta)substituted derivatives with two or more tertiary nitrogen atoms and carboxymethyl or 2-hydroxyethyl groups in the molecule (ethylenediaminetetraacetic acid [EDTA], DTPA, propylenediaminetetraacetic acid [PDTA], hydroxyethylethylenediaminetriacetic acid [HEDTA]) show a high stability under environmental conditions. On the other hand, disubstituted derivatives with two secondary nitrogen atoms in the molecule (e.g., ethylene diamine diacetic acid [EDDA]) are potentially biodegradable. Readily degradable are analogous compounds with substituents, which can be hydrolyzed (e.g., acetyl derivatives with $-COCH_3$ groups) as N,N'-diacetylethylenediamine (DAED) and N,N,N',N'-tetracetylethylenediamine (TAED).[33] Because all tested compounds contain Gd–DTPA or derivatives thereof, the above results are in line with the low degradability observed in our studies.

4.4.2   Ecotoxicity Tests

The measured substance concentration values were approximately 90 to 120% of the nominal values. The time course of the results demonstrates that the substance solutions were stable during the whole exposure period. Nevertheless, it can be stated that in none of the tests acute toxic effects were observed up to the maximum tested concentrations, with the exception of high concentrations of gadobutrol (2062 mg/L) and gadoxetic acid disodium (>125 mg/L), which had adverse effects on the growth of green algae. However, these concentrations are far beyond environmental relevance. The effects of growth inhibition observed at high concentrations could also be explained by the high osmotic pressure of the tested substances.

In summary, our results show that contrast-enhancing agents containing Gd have no acute toxic effects on the tested aquatic organisms up to concentrations of at least 80 mg/L. Algae seem to be the most sensitive species, although only at high concentrations toxic effects were observed.

Very little information is available in the literature regarding the ecotoxicity of Gd-complexes. A study conducted with *Caenorhabditis elegans* showed that Gd-complexes are only toxic at extremely high concentrations, which are no longer environmentally relevant. The nematode was exposed to Gd–DTPA (dimeglumine gadopentetate), 2[1,4,7,10-tetraaza-4,7-bis(carboxymethyl)-10-(2-hydroxypropyl)cyc lododecyl]acetic acid, gadolinium salt (Gd[HP-DO3A]) and 2-(1,4,7,10-tetraaza-4,7-bis(carboxymethyl)-10-([N-carboxymethyl)-N-(4-cyclohexylphenyl)carbamoyl]me thyl)cyclododecyl)acetic acid, monosodium gadolinium salt (Gd[CPA-DO3A])⁻ at various concentrations. Gd–DTPA²⁻ and Gd(HP-DO3A) produced no lethality up to 200 g/L, and Gd(CPA-DO3A)⁻ produced lethality of 17% and 31% at 24-hour exposures of 100 g/L and 200 g/L.[34]

The toxicity of metals and their chelates is influenced by the uptake into the organisms. The bioconcentration of rare earth elements (REEs) in algae was studied by Sun et al.[35] The authors were able to show that bioconcentration was largely dependent on chemical speciation. Adding organic ligands (EDTA, Nitrilotriacetic acid [NTA], Citrate [Cit]), which can form RE-organic complex species, led to major reduction of the REEs bioconcentration in algae. The order from high to low was REE³⁺>REE–Cit>REE–NTA>REE–EDTA complex, which is in the reverse order of the thermodynamic stability constants. The authors found that the relationship of REEs concentration in algae and their concentration in culture medium can be described by the Freundlich adsorption isotherm equation. They concluded that an adsorption process which is rate-limiting controls the rate of the uptake. The presence of organic ligands which form metal-organic complexes would thus reduce the bioconcentration by competing with the membrane binding sites for the available metal ion.[35]

The distribution and bioavailability of REEs in various species (duckweed, daphnia, shellfish, and goldfish) were studied by Yang et al.[36] They found that the accumulated levels of REEs in duckweed were far higher than those in daphnia, shellfish, and goldfish. The low accumulation in fish was further confirmed for the carp.[37] A significant accumulation of Gd was only found in duckweed, suggesting that plants are more likely affected when exposed to exogenous REEs in the aquatic environment. This is in line with the observation that the most sensitive species in our studies were green algae, which may be attributed to a higher bioconcentration.

4.4.3 ENVIRONMENTAL RELEVANCE

To complete the environmental risk assessment, the concentrations of the contrast-enhancing agents in the aquatic environment had to be estimated. Figure 4.7 illustrates the risk assessment procedure based on the determination of the PEC/PNEC ratio. Because of its highest market volume, dimeglumine gadopentetate is chosen as an example. It is a product already marketed for several years in both the United States and European markets. Therefore, actual market data can be used for the calculation of the PEC.

The European Medicines Evaluation Agency (EMEA) guidance on environmental risk assessment[3] proposes the following equation for the calculation:

$$PEC_{surfacewater} = \frac{Dose_{water} \times F_{pen}}{Wastew_{inhab} \times Dilution}$$

where:

$PEC_{surfacewater}$ — Local surface water concentration
$Wastew_{inhab}$ — Amount of wastewater per inhabitant per day ($200\ L\ inh^{-1}\ d^{-1}$)
$Dilution$ — Dilution factor (10)
$DOSE_{ai}$ — Maximum daily dose of active ingredient (42,000 mg)
F_{pen} — Percentage of market penetration (0.0008 % for Germany)

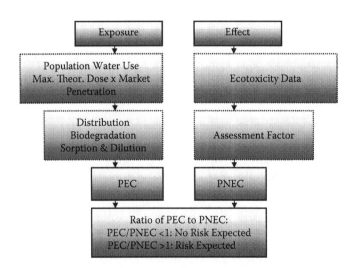

FIGURE 4.7   Risk assessment procedure: predicted environmental concentration/predicted no effect concentration (PEC/PNEC) ratio.

For dimeglumine gadopentetate a value of 0.17 µg/L (equivalent to 0.029 µg/L Gd) in German surface waters is obtained by this calculation. Due to the high number of contrast-enhanced MRI examinations in Germany, this country can be regarded as a worst-case scenario within the industrialized world.

To determine the concentration of a substance below which adverse effects are not expected to occur in the aquatic environment the PNEC is used. It is calculated by applying an assessment factor (AF) to the no observed effect concentration(s) (NOEC) from relevant effects studies. The following formula is used:

$$PNEC = \frac{NOEC}{AF}$$

Since no long-term studies are yet available for Gd-contrast agents, the results of the short-term tests are used here. The use of a factor of 1000 on short-term toxicity data is a conservative and protective factor and is designed to ensure that substances with potential to cause adverse effects are identified in the effects assessments.[4]

In the acute toxicity tests the lowest NOEC was observed for dimeglumine gadopentetate in algae with 100 mg/L. If an assessment factor of 1000 is applied, a PNEC in water of 0.10 mg/L (equivalent to 100 µg/L) is obtained. Thus, the ratio of PEC (0.17 µg/L) to PNEC for dimeglumine gadopentetate in the aquatic compartment is 1.7×10^{-3}, far below the critical PEC/PNEC threshold of 1. The low PEC/PNEC ratio for dimeglumine gadopentetate clearly indicates that the introduction of this diagnostic product into surface water is of little environmental risk.

Gadobutrol, gadoxetic acid disodium, and gadofosvest trisodium have a lower market volume and are therefore assumed to occur in lower concentrations in the aquatic environment.

Considering the estimated environmental concentrations and the results of the ecotoxicological investigations of the tested compounds, they are not assumed to represent a risk for the aquatic environment. The calculated PECs can be viewed in relation to the environmental concentrations of Gd reported in literature. The geogenic background concentrations of Gd in surface waters are generally low. Bau and Dulski[7] reported Gd concentrations in Swedish and Japanese rivers, which drain thinly populated, nonindustrialized areas to vary between 0.001 and 0.012 µg/L. Significantly higher Gd concentrations are found at sites close to sewage effluent discharges, especially if these effluents receive contributions from hospital wastewater, which in turn contains excreted MRI contrast media. For instance, at the wastewater discharge of the large treatment plant Berlin/Ruhleben (Germany), Gd concentrations of 7087 pmol/kg corresponding to 1.114 µg/L were found, while the receiving River Havel contained Gd concentrations in the range of 0.11 to 0.18 µg/L.[7] Möller et al.[14] reported a natural Gd background 0.001 to 0.002 µg/L Gd in the Spree/Havel (Berlin) area and an anthropogenic contribution in the mentioned river waters of 0.03 to 1.07 µg/L, the latter close to a sewage effluent entry point. Accordingly, the calculated PEC of Gd resulting from the use of dimeglumine gadopentetate is reached in the real environment only in areas with densely populated areas close to the point of discharge (i.e., the PEC represents a worst-case scenario).

4.5 SUMMARY AND OUTLOOK

Studies with various contrast-enhancing agents confirmed the expected low toxicity in acute aquatic toxicity tests. The chelates are stable and are not readily biodegradable. Algae were shown to be the most sensitive organisms. The most plausible interpretation of this higher sensitivity is a higher bioconcentration of traces of the free Gd. Further investigations are currently performed into degradation under more environmentally relevant conditions in model wastewater treatment plants or aquatic sediment systems.

To further investigate the toxic potential of free Gd in comparison with the toxicity of the contrast agents, acute aquatic toxicity tests are currently being carried out with $GdCl_3$. Furthermore, long-term tests with the MRI-contrast agents to assess the chronic toxicity are being performed.

REFERENCES

1. Stumpf, M., Ternes, T.A., Haberer, K., Seel, P., and Baumann, W., Nachweis von Arzneimittelrückständen in Kläranlagen und Fließgewässern, *Vom Wasser,* 86, 291, 1996.
2. Halling-Sørensen, B., Nors Nielsen, S., Lanzky, P.F., Ingerslev, F., Holten Lützhøft, H.C., and Jørgensen, S.E., Occurrence, fate and effects of pharmaceutical substances in the environment—A review, *Chemosphere,* 36, 2, 357, 1998.
3. Committee for medicinal products for human use (CHMP), Guideline on the environmental risk assessment of medicinal products for human use, European Medicines Agency, 2006.
4. European Commission, Technical Guidance Document on Risk Assessment Part II, *European Communities,* 2003.
5. Ternes, T.A. and Hirsch, R., Occurrence and behaviour of X-ray contrast media in sewage facilities and the aquatic environment, *Environ. Sci. Technol.,* 34, 2741, 2000.
6. Steger-Hartmann, T., Länge, R., and Schweinfurth, H., Environmental risk assessment for the widely used iodinated X-ray contrast agent iopromide (Ultravist), *Ecotox. Environ. Saf.* 42, 274, 1999, Environmental Research, Section B.
7. Bau, M. and Dulski, P., Anthropogenic origin of positive gadolinium anomalies in river waters, *Earth Plan. Sci. Lett.,* 143, 245, 1996.
8. Debatin, J.F. and Hany, T.F., MR-based assessment of vascular morphology and function, *Eur. Radiol.,* 8, 528, 1998.
9. Krasnow, N., Effects of lanthanum and gadolinium ions on cardiac sarcoplasmic reticulum, *Biochim. Biophys. Acta,* 282, 187, 1972.
10. Bourne, G.W. and Trifaro, J.M., The gadolinium ion: a potent blocker of calcium channels and catecholamine release from cultured chromaffin cells, *Neuroscience,* 7, 1615, 1982.
11. Evans, C.H., *The occurrence and metabolism of lanthanides,* Plenum, New York, 285, 1990.
12. Durbin, P.W., Williams, M.H., Gee, M., Newman, R., and Hamilton, J.G., Metabolism of the lanthanons in the rat, *Proc. Soc. Exp. Biol. Med.,* 91, 78, 1956.
13. Magnusson, G., The behavior of certain lanthanons in rats, *Acta Pharmacol. Toxicol.,* 20, 1, 1963.
14. Möller, P., Dulski, P., Bau, M., Knappe, A., Pekdeger, A., and Sommer-von Jarmerstedt, C., Anthropogenic gadolinium as a conservative tracer in hydrology, *J. of Geochem. Explor.,* 69–70, 409, 2000.
15. Elbaz-Poulichet, F., Seidel, J.L., and Othoniel, C., Occurence of an anthropogenic gadolinium anomaly in river and coastal waters of southern France, *Water Res.,* 36, 1102, 2002.

16. Möller, P., Paces, T., Dulski, P., and Morteani, G., Anthropogenic Gd in surface water, drainage system, and the water supply of the city of Prague, Czech Republic, *Environ. Sci. Technol.*, 36, 2387, 2002.
17. OECD Guidelines for testing chemicals: Ready biodegradability: 301 E Modified OECD Screening Test, *OECD,* 1993.
18. Environmental Assessment Technical Assistance Handbook, 3.11 Aerobic Biodegradation in Water, Food and Drug Administration, Report Accession no. 87-175345/AS, NTIS, Springfield, 1987.
19. Environmental Assessment Technical Assistance Handbook: Technical Assistance Document 4.11: Freshwater Fish Acute Toxicity, Food and Drug Administration, 1987.
20. OECD guidelines for testing of chemicals: 203: Fish, acute toxicity test, *OECD*, 1993.
21. European Commission: Classification, packaging and labelling of dangerous substances in the European Union. Part 2—Testing methods. Part C.1—Acute toxicity for fish, European Commission, 276, 1997.
22. Environmental Assessment Technical Assistance Handbook: Technical Assistance Document 4.08: Daphnia Acute Toxicity, Food and Drug Administration, 1987.
23. OECD guidelines for testing of chemicals: 202, Daphnia sp. Acute immobilization test and reproduction test, Part I and II, OECD, 1993.
24. European Commission: Classification, packaging and labelling of dangerous substances in the European Union. Part 2—Testing methods. Part C.2—Acute toxicity for Daphnia, European Commission, 285, 1997.
25. OECD guidelines for testing of chemicals: 201: Algae, growth inhibition test, OECD, 1993.
26. European Commission: Classification, packaging and labeling of dangerous substances in the European Union. Part 2—Testing methods. Part C.3—Algae inhibition test, European Commission, 285, 1997.
27. DIN 38 412 L8: Deutsches Einheitsverfahren zur Wasser-, Abwasser- und Schlammuntersuchung, Testverfahren mit Wasserorganismen (Gruppe L), Bestimmung der Hemmwirkung von Wasserinhaltsstoffen auf Bakterien (Pseudomonas-Zellvermehrungs-Hemmtest) L8, Beuth, Berlin, DIN Deutsches Institut für Normung e. V., 1988.
28. Environmental Assessment Technical Assistance Handbook: Technical Assistance Document 4.02: Microbial Growth Inhibition, Food and Drug Administration, 1987.
29. Hoffmann, J. and Viedt, H., *Biologische Bodenreinigung—Ein Leitfaden für die Praxis.* Springer-Verlag, Berlin, 1998.
30. Schmitt-Willich, H., Brehm, M., Ewers, C.L., Michl, G., Muller-Fahrnow, A., Petrov, O., Platzek, J., Raduchel, B., and Sulzle, D., Synthesis and physicochemical characterization of a new Gadolinium chelate: the liver-specific magnetic resonance imaging contrast agent Gd-EOB-DTPA, *Inorg. Chem.,* 38, 1134, 1999.
31. Toth, E., Kiraly, R., Platzek, J., Radüchel, B., and Brücher, E., Equilibrium and kinetic studies on complexes of 10-[2,3-dihydroxy-(1-hydroxymethyl)-propyl]-1,4,7,10-tetraazacyclododecane-1,4,7-triacetate, *Inorg. Chim. Acta,* 249, 191, 1996.
32. Caravan, P., Comuzzi, C., Crooks, W., McMurry, T.J., Choppin, G.R., and Woulfe, S.R., Thermodynamic stability and kinetic inertness of MS-325, a new blood pool agent for magnetic resonance imaging, *Inorg. Chem.,* 40, 2170, 2001.
33. Sykora, V., Pitter, P., Bittnerova, I., and Lederer, T., Biodegradability of ethylenediamine-based complexing agents, *Water Res.*, 35, 2010, 2001.
34. Williams, P.L., Anderson, G.L., Johnstone, J.L., Nunn, A.D., Tweedle, M.F., and Wedeking, P., Caenorhabditis elegans as an alternative animal species, *J. of Toxic. Environ. Health, Part A*, 61, 641, 2000.
35. Sun, H., Wang, X., and Wang, L., Bioconcentration of rare earth elements lanthanum, gadolinium and yttrium in algae (*Chlorella vulgarize beijerinck*): Influence of chemical species, *Chemosphere*, 34, 1753, 1997.

36. Yang, X., Yin, D., Sun, H., Wang, X., Dai, L., Chen, Y., and Cao, M., Distribution and bio-availability of rare earth elements in aquatic microcosm, *Chemosphere*, 39, 2443, 1999.
37. Tu, Q., Wang, X.R., Tian, L.Q., and Dai, L.M., Bioaccumulation of the rare earth elements lanthanum, gadolinium and yttrium in carp (Cyprinus carpio), *Environ. Pollut.*, 85, 345, 1994.

Part II

---

*Environmental Fate
and Transformations of
Veterinary Pharmaceuticals*

5 Fate and Transport of Veterinary Medicines in the Soil Environment

Alistair B.A. Boxall

Contents

5.1 INTRODUCTION

Veterinary medicines are widely used to treat disease and protect the health of animals. Dietary enhancing feed additives (growth promoters) are also incorporated into the feed of animals reared for food in order to improve their growth rates. Following administration to a treated animal, medicines are absorbed and in some instances may be metabolized. Release of parent veterinary medicines and their metabolites to the environment can then occur both directly, for example, the use of medicines in fish farms, and indirectly, via the application of animal manure (containing excreted products) to land or via direct excretion of residues onto pasture.[1-4] Over the past 10 years the scientific community has become increasingly interested in the impacts of veterinary medicines on the environment, and there have been significant developments in the regulatory requirements for the environmental assessment of veterinary products. A number of groups of veterinary medicines including sheep dip chemicals, fish farm medicines, anthelmintics, and antibiotics have been well studied in recent years, and a large body of data is now available.[5] This chapter reviews our understanding of the inputs of livestock medicines to the environment and synthesizes the available information on the fate and transport of veterinary pharmaceuticals in manure and soils. Toward the end of this chapter, gaps in the current knowledge are highlighted and recommendations are made for future research.

5.2 INPUTS OF LIVESTOCK MEDICINES TO THE ENVIRONMENT

Large quantities of animal health products are used in agriculture to improve animal care and increase production. Some drugs used in livestock production are poorly absorbed by the gut, and the parent compound or metabolites are known to be excreted in the feces or urine, irrespective of the method of application.[6-11] The main routes of input to the soil environment and subsequent transport routes are illustrated in Figure 5.1. During livestock production, veterinary drugs enter the environment through removal and subsequent disposal of waste material (including manure/slurry and "dirty" waters), via excretion of feces and urine by grazing animals, through spillage during external application, via washoff from farmyard hard surfaces, or by direct exposure/discharge to the environment.

For hormones, antibiotics, and other pharmaceutical agents administered either orally or by injection to animals, the major route of entry of the product into the environment is probably via excretion following use and the subsequent disposal of contaminated manure onto land.[12] Many intensively reared farm animals are housed indoors for long periods at a time. Consequently, large quantities of farmyard manure, slurry, or litter are produced, which are then disposed of at high application rates onto land.[13] Although each class of livestock production has different housing and manure production characteristics, the emission and distribution routes for veterinary medicines are essentially similar. Manure or slurry will typically be stored before it is applied to land. During this storage time it is possible that residues of veterinary medicines will be degraded. A number of studies have therefore explored the

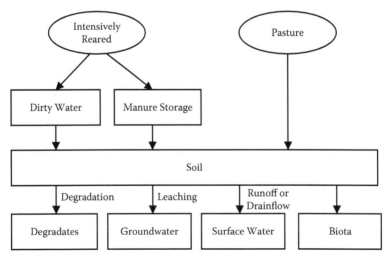

FIGURE 5.1 Input routes and fate and transport pathways for veterinary medicines in the soil environment.

persistence of a range of veterinary substances in different manure/slurry types.[14-17] For example, macrolides and β-lactam antibiotics have been shown to be rapidly degraded in a range of manure types, whereas avermectins and tetracyclines are likely to persist for months. Available data indicate that the dissipation of veterinary medicines in manure or slurry can be very different from the dissipation behavior in soils.[14] One possible explanation is that the mechanism of degradation in manure and slurry stores is anaerobic, whereas degradation in soils is most likely due to aerobic organisms.

Drugs administered to grazing animals may be deposited directly to land or surface water through dung or urine, exposing soil organisms to high local concentrations.[6,13,18-20]. Another significant route for environmental contamination is the release of substances used in topical applications. Various substances are used externally on animals and poultry for the treatment of external or internal parasites and infection. Sheep in particular suffer from a number of external insect parasites for which treatment and protection is sometimes obligatory. The main methods of external treatment include plunge dipping, pour-on formulations, and the use of showers or jetters. With all externally applied veterinary medicines, both diffuse and point source pollution can occur. Sheep dipping activities provide several routes for environmental contamination. In dipping practice, chemicals may enter watercourses through inappropriate disposal of used dip, leakage of used dip from dipping installations, and from excess dip draining from treated animals. Current disposal practices rely heavily on spreading used dip onto land. Wash-off of chemicals from the fleeces of recently treated animals to soil, water, and hard surfaces may occur on the farm, during transport, or at stock markets. Medicines washed off, excreted, or spilt onto farmyard hardsurfaces (e.g., concrete) may be washed off to surface waters during periods of rainfall.

5.3   FATE OF VETERINARY MEDICINES IN SOILS

Once a veterinary medicine is released to the environment, its behavior will be determined by its underlying physical properties (including water solubility, lipophilicity, volatility, and sorption potential). In the following sections information on the fate and transport of veterinary medicines in the soil environment is reviewed.

5.3.1   SORPTION IN SOIL

Data are available on the sorption behavior of antibiotics, sheep dip chemicals, and avermectins in soils (Table 5.1). The degree to which veterinary medicines may adsorb to particulates varies widely. Consequently, the mobility of different veterinary medicinal products also varies widely. Chapter 6 and Chapter 7 in this book discuss sorption and mobility of selected veterinary pharmaceuticals in more detail.

Available data indicate that sulfonamide antibiotics and organophosphate parasiticides will be mobile in the environment, whereas tetracycline, macrolide, and fluoroquinolone antibiotics will exhibit low mobility. The variation in partitioning for a given compound in different soils can be significant and cannot be explained by variations in soil organic carbon. For instance, the maximum reported organic-

TABLE 5.1

Measured Sorption Coefficients (Koc) for a Range of Veterinary Medicines

	Mean	Minimum	Maximum
Avermectin	17650	5300	30000
Chlorfenvinphos	295	—	—
Ciprofloxacin	61000	—	—
Enrofloxacin	392623	16506	768740
Cumaphos	13449	5778	21120
Deltamethrin	8380000	460000	16300000
Diazinon	889	229	1549
Fenbendazole	815.5	631	1000
Metronidazole	47	38	56
Ofloxacin	44143	—	—
Olaquindox	81	46	116
Oxytetracycline	60554	27792	93317
Sulfamethazine	60	—	—
Tetracycline	40000	—	—
Tylosin	4270.5	553	7988
Carbadox	8508.5	184	16833
Sulfamethoxazole	296.1	62.2	530
Sulfadiazine	125	—	—
Sulfapyridine	219	—	—
Sulfachloropyridazine	75.5	69	82
Sulfadimethixine	144	—	—

(Data taken from review of Boxall et al.[40] With permission.)

carbon normalized sorption coefficient for carbadox is approximately two orders of magnitude greater than the lowest reported value. These large differences in sorption behavior are explained by the fact that many veterinary medicines are ionizable with pKa values in the pH range of natural soils. Medicines can therefore occur in the environment as negative, neutral, zwitterionic, and positively charged species.[21,22] Depending on the species, interactions with soil can occur through electrostatic attraction, surface bridging, hydrogen bonding, or hydrophobic interactions.[22] The sorption behavior is also influenced by the properties of the soil, including pH, organic carbon content, metal oxide content, ionic strength, and cation-cation exchange capacity.[22–25] The complexity of the sorbate-sorbent interactions means that modelling approaches developed for predicting the sorption of other groups of chemicals (e.g., pesticides and neutral organics) are inappropriate for use on veterinary medicines (Figure 5.2). Manure and slurry may also alter the behavior and transport of medicines. Recent studies have demonstrated that the addition of these matrices can affect the sorption behavior of veterinary medicines and that they may

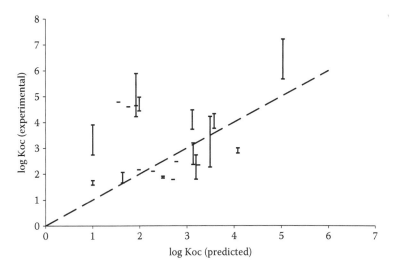

FIGURE 5.2   Relationship between measured and predicted soil sorption coefficients for a range of veterinary medicine classes. Koc predictions were obtained using the Syracuse Research Corporation PCKOC program.

affect persistence.[26,27] These effects have been attributed to changes in pH or the nature of dissolved organic carbon in the soil/manure system.

5.3.2 PERSISTENCE IN SOIL

The main route for degradation of veterinary medicines in soils is via aerobic soil biodegradation. Degradation rates in soil vary, with half-lives ranging from days to years. Degradation of veterinary medicines is affected by environmental conditions such as temperature and pH and the presence of specific degrading bacteria that have developed to degrade groups of medicines.[28,29] As well as varying significantly between chemical classes, degradation rates for veterinary medicines also vary within a chemical class. For instance, of the quinolones, olaquindox can be considered to be only slightly persistent (half-life 6 to 9 days), while danofloxacin is very persistent (half-life 87 to 143 days). In addition, published data for some individual compounds show persistence varies according to soil type and conditions. For example, diazinon was shown to be relatively impersistent (half-life 1.7 days) in a flooded soil that had been previously treated with the compound, but was reported to be very persistent in sandy soils (half-life 88 to 112 days).[5] Of the available data, coumaphos and emamectin benzoate were the most persistent compounds in soil, with half-lives of 300 and 427 days, respectively, while tylosin and dichlorvos were the least persistent with half-lives of 3 to 8 and <1 day, respectively.

For some veterinary medicines, degradation rates in manure can be faster than degradation in soil. For example, under methanogenic conditions the degradation half-life for tylosin A was less than 2 days and was enhanced by increasing concentrations of manure particles in the incubation medium under aerobic conditions.[15] Moreover, when manure is combined with soil, degradation may be enhanced for selected medicines. When manure or slurry is combined with soil, temperature has been shown to significantly affect the rate of degradation of a compound. For example, a half-life of 91 to 217 days was recorded for ivermectin in a soil/feces mixture during winter weather conditions.[30] In contrast, the compound was shown to degrade much more rapidly in a soil/feces mixture during the summer period, with a half-life of 7 to 14 days being measured.[31] The timing of application of manure/slurry to land may therefore be a significant factor in determining the subsequent degradation rate of a compound. Depending on the nature of the chemical, other degradation and depletion mechanisms may occur, including soil photolysis and hydrolysis.[32] The degradation processes may well result in the formation of degradation products.[17] In some instances, these degradation products may be more ecotoxic than the parent compound, more persistent, and more mobile.[2] It is therefore important that the fate of the degradation products in soils is considered when assessing the impact of a veterinary medicine on the environment.

5.4 TRANSPORT OF VETERINARY MEDICINES IN SOIL SYSTEMS

Contaminants applied to soil can be transported to aquatic systems via surface runoff, subsurface flow, and drainflow. The extent of transport via any of these processes is determined by a range of factors, including: the solubility, sorption behavior, and persistence of the contaminant; the physical structure, pH, organic carbon content, and cation exchange capacity of the soil matrix; and climatic conditions such as temperature and rainfall volume and intensity. Most work to date on contaminant transport from agricultural fields has focused on pesticides, nutrients and bacteria, but recently a number of studies have explored the fate and transport of veterinary medicines. Lysimeter, field-plot, and full-scale field studies have investigated the transport of veterinary medicines from the soil surface to field drains, ditches, streams, rivers, and groundwater.[33–41] A range of experimental designs and sampling methodologies has been used. These investigations are described in more detail below.

5.4.1 LEACHING TO GROUNDWATER

The movement of sulfonamide and tetracycline antibiotics in soil profiles was investigated at the field scale using suction probes.[33,42] In these studies sulfonamides were detected in soil pore water at depths of both 0.8 and 1.4 m, but tetracyclines were not (most likely due to their high potential for sorption to soil). Carlson and Mabury[43] reported that chlortetracycline applied to agricultural soil in manure was detected at soil depths of 25 and 35 cm, but monensin remained in the upper soil layers. There are only a few reports of veterinary medicines in groundwater.[42,44] In an extensive monitoring study conducted in Germany,[44] no antibiotics were detected in groundwater in most of the monitoring regions with intensive livestock production. However, residues of sulfonamide antibiotics were detected at four sites. While contamination at two of the sites was attributed to irrigation of agricultural land with domestic sewage, the authors concluded that contamination of groundwater by the veterinary antibiotic sulfamethazine at two of the sites was due to applications of manure.[44]

5.4.2 Runoff

Transport of veterinary medicines via runoff (i.e., overland flow) has been observed for tetracycline antibiotics (i.e., oxytetracycline) and sulfonamide antibiotics (sulfadiazine, sulfamethazine, sulfathiazole, and sulfachloropyridazine).[41,45,46] The transport of these substances is influenced by the sorption behavior of the compounds, the presence of manure in the soil matrix, and the nature of the land to which the manure is applied. Runoff of highly sorptive substances, such as tetracyclines, was observed to be significantly lower than the more mobile sulfonamides.[41] However, even for the relatively water soluble sulfonamides, total mass losses to surface are small (between 0.04% and 0.6% of the mass applied) under actual field conditions.[47] Manure and slurry have been shown to increase the transport of sulfonamides via runoff by 10 to 40 times in comparison to runoff following direct application of these medicines to soils.[36] Possible explanations for this observation include physical "sealing" of the soil surface by the slurry or a change in pH as a result of manure addition that alters the speciation and fate of the medicines.[36] It has been shown that overland transport from ploughed soils is significantly lower than runoff from grasslands.[45]

5.4.3 Drain Flow

The transport of a range of antibacterial substance (i.e., tetracyclines, macrolides, sulfonamides, and trimethoprim) has been investigated using lysimeter and field-based studies in tile-drained clay soils.[38,46,48] Following application of pig slurry spiked with antibiotics to an untilled field, test compounds were detected in drain-flow at concentrations up to a maximum of 613 $\mu g\ l^{-1}$ for oxytetracyline and 36 $\mu g\ l^{-1}$ for sulfachloropyridazine.[38] The spiking concentrations for the test compounds were all similar, so differences in maximum concentrations were likely due to differences in sorption behavior. In a subsequent investigation at the same site[38] in which the soil was tilled, much lower concentrations were observed in the drainflow (i.e., 6.1 $\mu g\ l^{-1}$ for sulfachloropyridazine and 0.8 $\mu g\ l^{-1}$ for oxytetracyline). While the pig slurry used in these studies was obtained from a pig farm where tylosin was used as a prophylactic treatment, this substance was not detected in any drainflow samples—possibly because it is not persistent in slurry.[15]

5.4.4 Uptake into Biota

Veterinary medicines may also be taken up from soil into biota.[49–51] Studies with a range of veterinary medicines[51] showed that florfenicol, trimethoprim, levamisole, diazinon, and enrofloxacin are taken up by plants following exposure to soil at environmentally realistic concentrations of the compounds (Figure 5.3). However, phenylbutazone, oxytetracyline, tylosin, sulfadiazine, and amoxicillin were not detected in plant material. The lack of uptake observed may have been due to factors such as high limits of detection or significant degradation during the study. For example, results for amoxicillin, sulfadiazine, and tylosin could well be explained by their dissipation in soils, with greater than 90% dissipation being observed by the time lettuces were harvested. The results for sulfadiazine contrasted to previous studies into

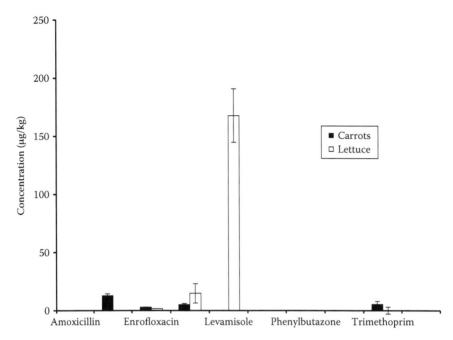

FIGURE 5.3 Mean (± 1 SE) concentrations of veterinary medicines in plants following 3 months exposure (lettuce) and 5 months exposure (carrot) to spiked soil. (Boxall et al.[51] With permission.)

the uptake of sulfonamide antibiotics (sulfamethoxine), where uptake was observed by roots and stems of certain plant species[52,53] following exposure to high concentrations of the study compounds (13 to >2000 mg kg^{-1}). Kumar et al.[50] showed uptake of chlortetracycline from manure amended soils into corn (*Zea mays* L.), green onion (*Allium cepa* L.), and cabbage (*Brassica oleracea* L.). Tylosin was not taken up by the three crops. Chapter 8 of this book presents the results of a study investigating plant uptake of pharmaceuticals, while Chapter 9 provides evidence of detoxification of antibiotics by plant-derived enzymes.

It is generally recognized that chemicals are taken up into plants via the soil pore water. Data for pesticides and neutral organic substances show that root uptake of organic chemicals from soil water is typically related to the octanol-water partition coefficient of the compound.[54,55] Uptake of chemicals by roots is greatest for more lipophilic compounds, whereas polar compounds are accumulated to a lesser extent.[54] Studies of translocation of pesticides into shoots indicate that uptake into shoots (and, hence, aboveground plant material) is related to Log K_{ow} by a Gaussian curve distribution.[54,55] Maximum translocation is observed at a Log K_{ow} around 1.8. More polar compounds are taken up less well by shoots, and uptake of highly lipophilic compounds (Log K_{ow}>4.5) is low. The available data indicate that these relationships may not hold true for veterinary medicines.[51] This is perhaps not surprising as data for other environmental processes (e.g., sorption to soil) indicate that the behavior of veterinary medicines in the environment is poorly related to hydrophobicity but is determined by a range of factors, including H-bonding potential, cation exchange, cation bridging at clay surfaces, and complexation. Through controlled experimental studies it may be possible in the future to begin to understand those factors and processes affecting the uptake of veterinary medicines into plants and to develop modelling approaches for predicting uptake.

5.5   MODELING EXPOSURE IN SOILS

From the information provided above it is clear that the behavior of veterinary medicines in the environment is complex and depends on a range of chemical properties and environmental processes. In order to support the environmental risk assessment process for veterinary medicines, a number of approaches have therefore been developed for predicting concentrations of veterinary medicines in soil and the potential for a medicine to be transported to groundwaters and surface waters.[13,56] These models incorporate many of the fate processes described above and will typically require data on sorption and persistence as input values. Some of the approaches are described below.

In order to harmonize the environmental assessments of veterinary products, the European Federation for Animal Health (FEDESA now IFAH Europe) developed a uniform scheme for calculating predicted environmental concentrations of veterinary medicines in soil following spreading of manure from treated animals.[56] The scheme provides a sequence of standard equations and a database containing information on three major agricultural species: cattle, pigs, and poultry. The database also contains information on the agricultural practices and relevant regulations for various regions within the EU. Inputs to the model are the dose and treatment regime. If information is available on metabolism or degradation, this can be incorporated into the calculation. The model calculated the concentration of the veterinary medicine in animal manure and then uses the nitrogen content of the manure and the maximum spreading rate of manure nitrogen onto land to calculate the maximum quantity of veterinary medicine applied per hectare. The output from the model is a predicted soil concentration. Since the introduction of this model there have been a number of minor modifications and amendments introduced, but the basic premise is that the predicted environmental concentration (PEC) depends on the nitrogen content of the manure and the maximum application rate to land.

The ETox models developed by Montforts[13] predicts concentrations of veterinary medicines using scenarios that are specific to agricultural practices in the Netherlands. The model is more complex than the uniform approach and can be used for medicines that are given internally (e.g., oral and injection treatments) or medicines applied externally (e.g., udder disinfection treatments). A range of input pathways are considered (i.e., direct excretion of dung and urine onto a field, spreading of manure and slurry, and direct spillage onto a field). The following groups of organisms are considered: cows (milk cows, suckling cows, beef cows), pigs (fattening pigs, sows), and poultry (hens, broilers, and turkeys). The outputs from the model include concentrations of the veterinary drug in soil, groundwater, surface waters, and biota.

VetCalc estimates PEC values for soil, groundwater, and surface water for 12 predefined scenarios that were chosen on the basis of the size and importance of their livestock production and its diversity, the range of agricultural practice covered by the scenarios, and the desire to cover three different European climate zones (Mediterranean, Central, and Continental Scandinavian). Each of the scenarios has been ranked in terms of its importance as a scenario for each livestock species. Background information on key drivers for exposure such as treatment regimes (both body weight and nonbody weight related), animal characteristics and husbandry practices, manure characteristics and management regimes, environmental characteristics (soil, hydrology, weather), agricultural practices, and chemical parameters are provided within the model databases.

5.6   OCCURRENCE IN THE SOIL ENVIRONMENT

In recent years a number of studies have begun to explore the occurrence of veterinary medicines in the soil environment resulting from normal agricultural practices.[57–59] These studies have focused on antibacterial medicines and some parasiticides (Table 5.2). Selected studies have explored the distribution of medicines in the soil profile as well as the persistence over time. While some groups of substances have not been detected in soils (e.g., the macrolides and fluoroquinolones), some classes have been detected at concentrations of tens to hundreds of $\mu g\ kg^{-1}$ (Table 5.2). In some cases the compounds will persist in the soil for prolonged time periods. These differences can usually be explained by the laboratory persistence data and usage and treatment scenarios for the different medicines. The availability of real-world monitoring data allows us to evaluate the performance of the exposure models described previously. Generally, these exposure models will overpredict concentrations of a veterinary medicine in the environment (Figure 5.4) by a number of orders of magnitude. The reason for this is that the scenarios used in the models are highly conservative and that adequate data are not always available to describe the different dissipation processes that determine how much of a medicine will reach the soil.

5.7   CONCLUSION

This paper has reviewed the data available in the public domain on the pathways of veterinary medicines to the soil environment and their subsequent fate and transport. There is clearly a large body of data available on veterinary medicines in the soil environment, and it is timely to begin to further synthesize this information in order to provide a general understanding of the fate and transport of medicines in the soil environment and to develop approaches for predicting how a substance will behave in the soil environment. There are, however, still gaps in the data and in our understanding. Those gaps are outlined in the following paragraphs.

Researchers are still focusing on only a small proportion of the medicines in use (including the avermectins, tetracyclines, sulfonamides, macrolides, and fluoroquinolones). There are many more classes of medicines in use, so it would be worthwhile to prioritize these and begin to develop an understanding of the fate (and effects) of some of the more important classes in the environment.

A large body of data is now available on the effects of the manure matrix on fate and transport, and there is evidence that these matrices can affect transport at the

TABLE 5.2

Environmental Monitoring Data for Veterinary Medicines in Agricultural Soils

Compound	Therapeutic Use	Concentration Detected ($\mu g\ kg^{-1}$)	LOD ($\mu g\ kg^{-1}$)	Country	Reference
Chlortetracycline	Antibiotic	9.5	0.7	Germany	42
		26.4	1	Germany	42
		41.8	1	Germany	42
		39	2	Germany	37
		4	—	US	58
Ciprofloxacin	Antibiotic	nd	1	UK	48
Enrofloxacin	Antibiotic	nd	1	UK	48
Ivermectin	Endectocide	2	10.2	US, UK	58
		46			48
Lincomycin	Antibiotic	98.5	—	US	59
			1.26	UK	48
Monensin	Coccidiostat	1.08	—	Canada	61
Oxytetracycline	Antibiotic	8.6	0.7	Germany	42
		254	—	US	59
		7	5	Italy	57
		305	—	UK	48
Sulfadiazine	Antibiotic	0.8	—	UK	48
Sulfadimidine	Antibiotic	15	—	Germany	60
Sulfamethazine	Antibiotic	2	0.5	Germany	37
Tetracycline	Antibiotic	12.3	0.7	Germany	42
		32.2	1	Germany	42
		39.6	11	Germany	42
		295	—	Germany	37
		9		US	59
Tilmicosin	Antibiotic	<10	—	US	59
Trimethoprim	Antibiotic	0.5	—	UK	48
Tylosin	Antibiotic	trace	0.2	Germany	42
		<2	—	US	59
		<10	10	Italy	57

field scale. However, we do not yet fully understand the mechanisms causing these differences in behavior and their implications in terms of environmental risk.

Information on the formation and fate of veterinary medicine metabolites is sparse. It is likely that most metabolites will be less toxic than the parent compounds; however, they may be more mobile and more persistent than their associate parent. Further work on model compounds and their metabolites would help identify the circumstances under which metabolites deserve more detailed attention during risk assessment.

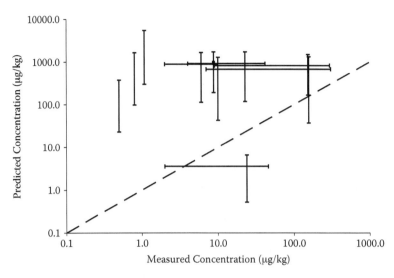

FIGURE 5.4 Comparison of predicted environmental concentrations obtained using exposure models listed in the EU Technical Guidance Document with ranges of measured concentrations in agricultural soils.

The uptake of veterinary medicines into plants is now being considered. It appears that uptake cannot be modelled using existing plant uptake models for pesticides and industrial chemicals. Further work to understand the mechanisms of uptake of veterinary medicine into plants is therefore warranted.

From the information available it appears that inputs from herd or flock treatments are probably the most significant in terms of environmental exposure. However, the relative significance of novel routes of entry to the environment from livestock treatments, such as wash-off following topical treatment and farmyard runoff, and aerial emissions, have not generally been considered.

Monitoring data are available for a range of veterinary medicines in soils. The studies have generally focused on parasiticides and the antibiotics. Once full datasets are obtained on the usage, properties, and effects of other chemical groups, further targeted monitoring should be performed to determine concentrations in the environment. These data could then be used along with the existing data to further evaluate current risk assessment exposure models.

REFERENCES

1. Boxall, A.B.A. The environmental side effects of medication. *EMBO Reports.* 5(12), 1110, 2004.
2. Boxall, A.B.A. et al. Are veterinary medicines causing environmental risks? *Environmental Science and Technology* 37(15), 2003, 286A.
3. Jørgensen, S.E. and Halling-Sørensen, B. Drugs in the environment. *Chemosphere*, 40, 691, 2000.
4. Sarmah, A.K., Meyer, M.T., and Boxall, A.B.A. A global perspective on the use, sales, exposure pathways, fate and effects of veterinary antibiotics (VAs) in the environment. *Chemosphere* 65(5), 725, 2006.

5. Boxall, A.B.A. et al. Veterinary medicines in the environment. *Reviews in Environmental Contamination and Toxicology* 180, 1, 2004.
6. Sommer, C. et al. Ivermectin excreted in cattle dung after sub-cutaneous injection or pour-on treatment: concentrations and impact on dung fauna. *Bulletin of Entomological Research*, 82, 257, 1992.
7. Magnussen, J.D. et al. Tissue residues and metabolism of avilamycin in swine and rats. *J. Agric. Food Chem.* 39, 306, 1999.
8. Stout, S.J. et al. Maduramycin α: characterisation of 14C-derived residues in turkey excreta. *J. Agric. Food Chem.* 39, 386, 1991.
9. Chiu, S.H.L. et al. Absorption, tissue distribution, and excretion of tritium-labelled ivermectin in cattle, sheep and rat. *J. Agric. Food Chem.*, 38, 2072, 1990.
10. Donoho, A.L. Metabolism and residue studies with actaplanin. *Drug Metab. Rev.* 18, 163, 1987.
11. Floate, K. et al. Faecal residues of veterinary parasiticides: non-target effects in the pasture environment. *Annual Reviews in Entomology* 50, 153, 2005.
12. Halling-Sørenson, B. et al. Worst-case estimations of predicted environmental soil concentrations (PEC) of selected veterinary antibiotics and residues used in Danish agriculture. In Kümmerer, K. (Ed.), Pharmaceuticals in the Environment — Sources, Fate, Effects and Risks. Springer-Verlag, Germany, 2001.
13. Montforts, M.H.M.M. Environmental risk assessment for veterinary medicinal products. Part 1: other than GMO-containing and immunological products. RIVM Report 601300001, RIVM, Bilthoven, The Netherlands, 1999.
14. Blackwell, P.A. et al. An evaluation of a lower tier exposure assessment model for veterinary medicines. *J. Agric. Food Chem.* 53(6), 2192, 2005.
15. Loke, M.L. et al. Stability of tylosin A in manure containing test systems determined by high performance liquid chromatography. *Chemosphere* 40, 759, 2000.
16. Teeter, J.S. and Meyerhoff, R.D. Aerobic degradation of tylosin in cattle, chicken and swine excreta. *Environmental Research*, 93, 45, 2003.
17. Kolz, A.C. et al. Degradation and metabolite production of tylosin in anaerobic and aerobic swine-manure lagoons. *Water Environment Research* 77(1), 49, 2005.
18. Halling-Sørenson, B. et al. Occurrence, fate and effect of pharmaceutical substances in the environment—A review. *Chemosphere*, 36, 357, 1998.
19. Strong, L. and Wall, R. Effects of ivermectin and moxidectin on the insects of cattle dung. *Bull. Entomol. Res.*, 84, 403, 1994.
20. McCracken, D.I. The potential for avermectins to affect wildlife. *Vet. Parasitol.* 48, 273, 1993.
21. Ter Laak, T.L., Gebbink, W.A., and Tolls, J. Estimation of sorption coefficients of veterinary medicines from soil properties. *Environ. Toxicol. Chem.* 25(4), 933, 2006.
22. Ter Laak, T.L., Gebbink, W.A., and Tolls, J. The effect of pH and ionic strength on the sorption of sulfachloropyridazine, tylosin and oxytetracycline to soil. *Environ. Toxicol. Chem.* 25(4), 904, 2006.
23. Strock, T.J., Sassman, S.A., and Lee, L.S. Sorption and related properties of the swine antibiotic carbadox and associated n-oxide reduced metabolites. *Environ. Sci. Technol.* 39, 3134, 2005.
24. Sassman, S.A. and Lee, L.S. Sorption of three tetracyclines by several soils: assessing the role of pH and cation exchange. *Environ. Sci. Technol.* 39, 7452, 2005.
25. Jones, A.D., et al. Factors influencing the sorption of oxytetracycline to soils. *Environ. Toxicol. Chem.* 24(4), 761, 2005.
26. Boxall, A.B.A. et al. The sorption and transport of a sulphonamide antibiotic in soil systems. *Toxicology Letters* 131, 19, 2002.
27. Thiele-Bruhn, S. and Aust, M.O. Effects of pig slurry on the sorption of sulfonamide antibiotics in soil. *Archives of Environmental Contamination and Toxicology* 47(1), 31, 2004.

28. Ingerslev, F. and Halling Sorensen, B. Biodegradability of metronidazole, olaquindox and tylosin and formation of tylosin degradation products in aerobic soil/manure slurries. *Chemosphere* 48, 311, 2001.

29. Gilbertson, T.J. et al. Environmental fate of ceftiofur sodium, a cephalosporin antibiotic: role of animal excreta in its decomposition. *J. Agric. Food Chem.* 38, 890, 1990.

30. Halley, B.A., VandenHeuvel, W.J.A., and Wislocki, P.G. Environmental side effects of the usage of avermectins in livestock. *Vet. Parasitol.* 48, 109, 1993.

31. Halley, B.A., Jacob, T.A., and Lu, A.Y.H. The environmental impact of the use of ivermectin: environmental effects and fate. *Chemosphere* 18, 1543, 1989.

32. Wolters, A. and Steffens, M. Photodegradation of antibiotics on soil surfaces: laboratory studies on sulfadiazine in an ozone-controlled environment. *Environ. Sci. Technol.* 39, 6071, 2005.

33. Blackwell, P.A., Kay, P., and Boxall, A.B.A. The dissipation and transport of veterinary antibiotics in a sandy loam soil. *Chemosphere* 67, 292, 2007.

34. Aga, D.S., Goldfish, R., and Kulshrestha, P. Application of ELISA in determining the fate of tetracyclines in land-applied livestock wastes. *Analyst* 128, 658, 2003.

35. Kreuzig, R. and Holtge, S. Investigations on the fate of sulfadiazine in manured soil: laboratory experiments and test plot studies. *Environmental Toxicology and Chemistry*, 24, 771, 2005.

36. Burkhard, M. et al. Surface runoff and transport of sulfonamide antibiotics on manured grassland. *Journal of Environmental Quality* 34, 1363, 2005.

37. Hamscher, G. et al. Different behaviour of tetracyclines and sulfonamides in sandy soils after repeated fertilization with liquid manure. *Environ. Toxicol. Chem.* 24(4), 861, 2005.

38. Kay, P., Blackwell, P., and Boxall, A. Fate and transport of veterinary antibiotics in drained clay soils. *Environ. Toxicol. Chem.* 23, 1136, 2004.

39. Kay, P., Blackwell, P.A., and Boxall, A.B.A. A lysimeter experiment to investigate the leaching of veterinary antibiotics through a clay soil and comparison with field data. *Environmental Pollution* 134, 333, 2005.

40. Kay, P., Blackwell, P.A., and Boxall, A.B.A. Column studies to investigate the fate of veterinary antibiotics in clay soils following slurry application to agricultural land. *Chemosphere* 60(4), 497, 2005.

41. Kay, P., Blackwell, P.A., and Boxall, A.B.A. Transport of veterinary antibiotics in overland flow following the application of slurry to land. *Chemosphere* 59(7), 951, 2005.

42. Hamscher, G. et al. Substances with pharmacological effects including hormonally active substances in the environment: identification of tetracyclines in soil fertilised with animal slurry. *Dtsch. tierärztl. Wschr.* 107, 293, 2000.

43. Carlson, J.C. and Mabury, S.A. Dissipation kinetics and mobility of chlortetracycline, tylosin, and monensin in an agricultural soil in Northumberland County, Ontario, Canada. *Environ. Toxicol. Chem.* 25, 1, 2006.

44. Hirsch, R. et al. Occurrence of antibiotics in the aquatic environment. *Sci. Tot. Environ.* 225, 109, 1999.

45. Kreuzig, R. et al. Test plot studies on runoff of sulfonamides from manured soils after sprinkler irrigation. *Environ. Toxicol. Chem.* 24, 777, 2005.

46. Gupta, S. et al. Antibiotic losses in runoff and drainage from manure applied fields. USGS-WRRI 104G National Grant, 2003.

47. Stoob, K., Singer, H.P., Mueller S.R., Schwarzenbach, R.P., and Stamm, C.H. Dissipation and transport of veterinary sulphonamide antibiotics after manure application to grassland in a small catchment. *Environ. Sci. Technol.*, in press.

48. Boxall, A. et al. Targeted Monitoring Study for Veterinary Medicines. Environment Agency R&D Technical Report, Bristol, UK, Environment Agency, 2006.

49. Migliore, L., Cozzolino, S., and Fiori, M. Phytotoxicity and uptake of enrofloxacin in crop plants. *Chemosphere* 52, 1233, 2003.
50. Kumar, K. et al. Antibiotic uptake by plants from soil fertilized with animal manure. *Journal of Environmental Quality* 34(6), 2082, 2005.
51. Boxall, A.B.A. et al. Uptake of veterinary medicines from soils into plants. *J. Agric. Food Chem.* 54(6), 2288, 2006.
52. Migliore, L. et al. Effect on plants of sulfadimethoxine used in intensive farming (Panicum miliacaum, Pisum sativum and Zea Mays). *Agriculture Ecosystems and Environment* 52, 103, 1995.
53. Migliore, L. et al. Effect of sulphadimethoxine contamination on barley (Hordeum distichum L, Poaceae, Lilipidosa). *Agricultural Ecosystems and Environment* 60, 121, 1996.
54. Briggs, G., Bromilow, R., and Evans, A. Relationships between lipophilicity and root uptake and translocation of non-ionised chemicals by barley. *Pesticide Science* 13, 495, 1982.
55. Burken, J. and Schnoor, J. Predictive relationships for uptake of organic contaminants by hybrid polar tress. *Environ. Sci. Technol.* 32, 3379, 1998.
56. Spaepen, K.R.I. et al. A uniform procedure to estimate the predicted environmental concentration of the residues of veterinary medicines in soil. *Environ. Toxicol. Chem.* 16, 1977, 1997.
57. De Liguoro, M. et al. Use of oxytetracycline and tylosin in intensive calf farming: evaluation of transfer to manure and soil. *Chemosphere* 52, 203, 2003.
58. Nessel, R.J. et al. Environmental fate of ivermectin in a cattle feedlot. *Chemosphere*, 18, 1531, 1989.
59. Zilles, J. et al. Presence of macrolide-lincosamide-streptogramin B and tetracycline antimicrobials in swine waste treatment processes and amended soil. *Water Environ. Res.*, 77, 57, 2005.
60. Christian, T. et al. Determination of antibiotic residues in manure, soil and surface waters. *Acta Hydrochim. Hydrobiol.* 31(1), 36, 2003.
61. Donoho, A.L. Biochemical studies on monensin. *J. Anim. Sci.* 58, 1528, 1984.

6 Sorption and Degradation of Selected Pharmaceuticals in Soil and Manure

Nadia Carmosini and Linda S. Lee

Contents

6.1 INTRODUCTION

Over the past two decades, the number of livestock and poultry farms in the United States has decreased by approximately half, while the number of animal units being raised has increased by about 10%.[1] This shift toward fewer but increasingly larger industrialized farms, termed "confined" or "concentrated animal feeding operations" (CAFOs), has resulted in concomitant increases in pharmaceutical use and manure generated per unit land area. During their lifespan, roughly 60 to 80% of commercial livestock are treated with antibiotics as therapeutic, prophylactic, or growth promoting agents,[2] and much of the ingested dose is excreted unchanged or as active metabolites.[3] Hormones are also used for growth promotion and reproductive control. In addition, hormones and their metabolites are produced and excreted naturally, with more than 50 tons of reproductive hormones being released annually by farm animals in the United States.[4] As a result, the 130 billion pounds of manure produced every year represent an expansive source of pharmaceutical contamination.

CAFOs typically store animal wastes in an outdoor lagoon, underground pit, or litter storage facility. For example, approximately 23% of swine operations in

139

the United States use lagoons; 57% use below-ground pits; and the remaining 20% employ other storage techniques, such as manure piles.[5] Lagoons and pits contain liquid, sludge, and solids that are periodically pumped out and injected or surface applied to agricultural land as an economical disposal strategy that capitalizes on the waste's fertilizer value. Most CAFOs pump down or dewater lagoons three to four times annually (58%), while others dewater less that twice a year (16%) or never at all (26%).[5] During these variable residence times, concentrations of excreted residues in the liquid phase may dissipate as a result of anaerobic degradation and sorption to particulates. After land application, aerobic degradation and sorption to soils further reduce the persistence and mobility of pharmaceuticals in the environment. In a similar manner, municipal wastewater treatment plants generate pharmaceutical-laden biosolids and wastewater effluents, which are disposed by land application and used to meet irrigation demands in arid regions. Recent surveys of organic contaminants in wastewater effluents and biosolids have reported a suite of antibiotics and other pharmaceuticals, albeit at concentrations substantially lower than those in livestock and poultry manures (less than µg/kg vs. mg/kg levels).[6,7]

Concern over potential negative impacts of biologically active pharmaceuticals on nontarget organisms and ecosystem health has prompted many laboratory-scale and a few field-scale assessments of their environmental fate. Several reviews of this information have been published in the past 5 years, especially for veterinary pharmaceuticals. Most recently, Lee et al.[8] summarized what is known on the occurrence, environmental fate, ecological impacts, and analytical techniques associated with veterinary antibiotics and hormones. Khanal et al.[9] reviewed the removal rates and mechanisms for natural estrogens in wastewater treatment systems. Sarmah et al.[10] compiled data on veterinary antibiotic use and sales trends for several countries, including the United States, the European Union, New Zealand, Kenya, Canada, Japan, China, and Russia. Kumar et al.[11] summarized veterinary antibiotic occurrence, excretion rates, and subsequent environmental fate, including sorption, degradation, and transport. Preceding these reviews, Tolls[12] and Thiele-Bruhn[13] provided a review of the sorption and presence of veterinary antibiotics in soils, respectively. In this chapter we will emphasize the studies that have been particularly instrumental in gleaning information on the processes that determine the sorption behavior of antibiotics, as well as present an overview of emerging research on the degradation of antibiotics and hormones in the environment.

6.2 ASSESSING CONTAMINANT FATE AND TRANSPORT IN SOIL ENVIRONMENTS

The potential for antibiotics and hormones to contaminate and adversely impact the environment is directly influenced by their mobility and persistence in animal wastes and soils. As a result, many studies have examined the dissipation and partitioning of these contaminants in environmental matrices, such as soil, manure, organic matter, clays, metal oxides, and oxyhydroxides. For the most part, either batch equilibrium or column displacement techniques have been used. The interpretation and application of data from these experiments requires careful consideration of the experimental design employed. In batch equilibrium experiments, multiple concentrations of a sorbate are equilibrated with a constant amount of sorbent in an electrolyte solution (e.g., 5 mM $CaCl_2$). Ideally, experiments should be conducted using a sorbent mass to solution volume ratio and contaminant concentration range that ensure $50 \pm 25\%$ sorption of the applied compound when near-equilibrium conditions are attained. This degree of partitioning facilitates accurate quantification of sorbate concentrations in both the solid (C_s, mg/kg or mmol/kg) and aqueous (C_w, mg/L or mmol/L) phases. These concentrations are plotted against one another to construct a sorption isotherm, which can be described with one of three commonly used equilibrium based models: (1) Linear equation, $K_d = C_s/C_w$, where K_d (L/kg) is the linear distribution coefficient; (2) Freundlich equation, $C_s = K_f C_w^N$, where K_f ($mg^{1-N} L^N kg^{-1}$ or $mmol^{1-N} L^N kg^{-1}$) is the Freundlich sorption coefficient and N (unitless) is a measure of isotherm nonlinearity; and (3) Langmuir equation, $C_s = C_{s,max} K_L C_w /(1+K_L C_w)$, where K_L (L/mg or L/mmol) is the Langmuir affinity coefficient and $C_{s,max}$ is the maximum monolayer adsorption capacity (mg/kg or mmol/kg). Comparisons among sorbents can be made independent of linearity using nonlinear model coefficients by estimating a concentration-specific sorption coefficient (K_d^*) at an equilibrium concentration (C_w) that falls within the experimental range: $K_d^* = K_f C_w^{N-1}$.

Direct quantification of the contaminant in both phases is critical for assessing potentially reversible sorption and ensuring that the loss of the parent compound from solution is due only to sorption. If biotic and abiotic transformations, volatilization, sorption to labware, and other loss processes are significant and attributed to sorption, inflated estimates of sorption will be obtained. Methods commonly used to inhibit biodegradation include sterilization of the sorbent by ^{60}Co irradiation, autoclaving, and filtration, or incorporating a dilute concentration of an antimicrobial, such as sodium azide (NaN_3) or mercuric chloride ($HgCl_2$), in the batch reactor.[14,15] When these steps are not taken, batch experiments can be used to assess biotic transformations in conjunction with sorption as long as concentrations of the parent compound and metabolites in both sorbed and solution phases are determined. To a certain extent, biotic transformations may occur even after systems have been sterilized due to extracellular enzymes that remain active.

In column displacement studies, the movement of a contaminant front or pulse through a packed or intact soil column is evaluated over time. Information on the chemical's partitioning between the soil and aqueous phases is derived from a plot of the outflow concentration vs. time, called a breakthrough curve, which is compared to the breakthrough curve for a nonsorbing, conservative tracer (e.g., Cl^-, Br^-). Compared to batch experiments, column studies may more closely mimic the spatial and temporal processes of a field scenario, particularly when intact soil cores are used. Contaminants moving through a soil profile may not persist in space and time for a sufficiently long period of time to reach equilibrium; thus, observations under flow conditions can provide additional insights into the processes that affect a chemical's persistence and mobility in soils. However, most column studies are conducted under saturated steady-state flow conditions, which are far not representative of vertical transport through the vadose zone where unsaturated transient flow is the norm. Also, even column studies performed with intact cores often cannot reflect the effect of field-scale heterogeneity on fate and transport phenomena or adequately represent the magnitude of macropore flow at the field scale.[16]

One of the most challenging aspects of batch, column, and field-scale experiments is the accurate quantification of low analyte concentrations in complex sample matrices. Many spectroscopic techniques, such as ultraviolet-visible (UV-VIS), fluorescence, and mass spectrometry, are often plagued by positive or negative matrix effects, necessitating the use of internal standards or matrix-matched standards. Finally, the use of sorption data in environmental fate models will only accurately describe chemical partitioning in other systems with similar physicochemical properties (e.g., pH, electrolyte concentration and composition, solute concentration range, sorbent composition). For example, the magnitude of sorption exhibited by a pure clay sorbent may differ from that of the clay within a soil due to the presence of other sorption domains and the interactions between those domains (e.g., organic matter coatings on clay surfaces).

6.3 ANTIBIOTICS

6.3.1 SORPTION BY SOIL

To date, research on the fate of pharmaceuticals is dominated by work on veterinary antibiotics in response to concerns that their extensive use in livestock production is contributing to bioactive residues in the environment. Although their physicochemical properties are highly variable, most antibiotics are moderately water soluble and exhibit low octanol-water partition coefficients (K_{ow}) due to the presence of polar functional groups (e.g., -C=O, -OH, -COOH, -NO$_2$, -NH$_2$, -CN). Some of these functionalities are ionizable, resulting in compounds that exist as either neutral or charged species (e.g., cations, anions, zwitterions) in proportions dependent on pH conditions. While neutral molecules partition to solid phases via relatively weak van der Waals and electron donor-acceptor interactions, charged species can interact with charged sorbents (e.g., organic matter, clays, metal oxides, and oxyhydroxides) through stronger electrostatic mechanisms, such as cation-exchange, cation-bridging, and complexation. Table 6.1 summarizes the structures, class assignments, molecular weights, acid dissociation constants (pK$_a$), and usage information for the antibiotics discussed in this chapter.

For antibiotics possessing a positive charge, cation exchange has emerged as an important sorption mechanism. In work conducted three decades before the present surge of interest in antibiotics, Porubcan et al.[17] examined the sorption of clindamycin and tetracycline (TC) by montmorillonite clay over a broad pH range (1.5 to 11). X-ray diffraction analysis showed that the clay's interlayer spacing increased under low pH conditions where the antibiotics' cationic species predominated. In contrast, the spacing remained unchanged at high pH where clindamycin is neutral and TC is predominantly anionic (+-- or 0--). Infrared radiation (IR) analyses also showed that clindamycin decreased the intensity of the water absorption band, which was attributed to the replacement of hydrated exchangeable cations in the clay's interlayer by clindamycin. Under neutral to moderately alkaline conditions, IR spectral shifts indicated that sorption of the TC zwitterions (+--0 and +--) occurred by cation exchange as well as by the formation of a complex between the interlayer Ca^{2+} cations and the antibiotic's carbonyl group. For the uncharged clindamycin molecule,

TABLE 6.1
Properties of Selected Antibiotics

Antibiotic	Class	Molecular Weight (g mol^{-1})	pKa	Usage
Clindamycin	Lincosamide	424.98	7.6[a]	Human and veterinary therapeutic
Tetracycline	Tetracycline	444.44	3.3[b] 7.68[b] 9.3[b]	Human and veterinary therapeutic; Livestock growth promoter and prophylactic
Oxytetracycline	Tetracycline	460.44	3.27[b] 7.32[b] 9.11[b]	Human and veterinary therapeutic; Livestock growth promoter and prophylactic; Fruit production; Aquaculture

(Continued)

TABLE 6.1.
(Continued)

Antibiotic	Class	Molecular Weight (g mol⁻¹)	pKa	Usage
Chlortetracycline	Tetracycline	478.89	3.30[b] 7.44[b] 9.27[b]	Human and veterinary therapeutic; Livestock growth promoter and prophylactic
Ciprofloxacin	Fluoroquinolone	331.35	5.90[c] 8.89[c]	Human therapeutic
Enrofloxacin	Fluoroquinolone	359.40	5.94[d] 8.70[d]	Veterinary therapeutic

Name	Structure	Class	MW	pKa	Use
Flumequin		Fluoroquinolone	261.25	6.5[e]	Veterinary therapeutic
Oxolinic Acid		Quinolone	261.23	6.9[e]	Veterinary therapeutic; Aquaculture
Sarafloxacin		Fluoroquinolone	385.37	4.1[e] 6.8[e]	Veterinary therapeutic; Aquaculture

(Continued)

TABLE 6.1.
(Continued)

Antibiotic	Class	Molecular Weight (g mol⁻¹)	pka	Usage
Monensin	Ionophore	670.88	4.2ᶠ	Livestock growth promoter and prophylactic
Lasalocid	Ionophore	590.80	2.6ᶠ	Livestock growth promoter and prophylactic

Name	Structure	Class	Molecular weight	pKa	Use
Salinomycin		Ionophore	751.01	4.45[g]	Livestock growth promoter and prophylactic
Tylosin		Macrolid	916.11	7.7[h]	Veterinary therapeutic; Livestock growth promoter and prophylactic
Erythromycin		Macrolid	733.94	8.88[h]	Human and veterinary therapeutic; Livestock growth promoter and prophylactic

(Continued)

TABLE 6.1.
(Continued)

Antibiotic	Class	Molecular Weight (g mol⁻¹)	pKa	Usage
Roxithromycin	Macrolid	837.05	9.28^g	Human therapeutic
Carbadox	Quinoxaline	262.22	$<0i9.61^i$	Livestock growth promoter and prophylactic
Olaquindox	Quinoxaline	263.25	$None^g$	Livestock growth promoter and prophylactic

Compound	Class	MW	pK_a	Use
Sulfamethazine	Sulfonamide	278.32	2.28[i] 7.42[j]	Human therapeutic; Livestock growth promoter and prophylactic
Sulfamethoxazole	Sulfonamide	253.27	1.85[j] 5.29[j]	Human and veterinary therapeutic
Sulfapyridine	Sulfonamide	249.29	2.30[k] 8.43[k]	Human therapeutic
Tiamulin	Pleuromutilin	493.74	9.51[g]	Veterinary therapeutic and prophylactic

(Continued)

TABLE 6.1.
(Continued)

Antibiotic	Class	Molecular Weight (g mol⁻¹)	pKa	Usage
Triclosan 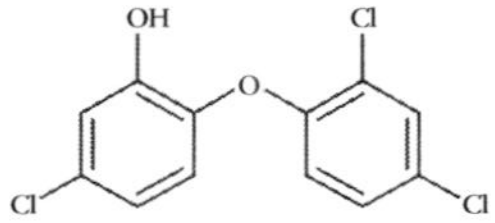	Na	289.54	7.99[l]	Commercial disinfectant
Chlorophene 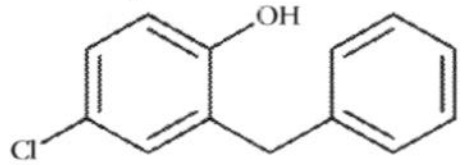	Na	218.68	9.96[l]	Commercial disinfectant

[a] Ref 17
[b] Ref 22
[c] Ref 36
[d] Ref 38
[e] Ref 42
[f] Ref 43
[g] Estimated with Marvinsketch™ (http://www.chemaxon.com/marvin/index.html) (Accessed August 11, 2007)
[h] Ref 85
[i] Ref 44
[j] Ref 86
[k] Ref 87
[l] Ref 56

sorption was attributed to relatively weak physical processes (e.g., H-bonding, van der Waals forces). No sorption was observed for either antibiotic at high pH conditions (pH 11).

Recently, the importance of sorption by cation exchange has been documented extensively in studies on tetracycline interactions with model clays, soils, and humic materials under varying pH and ionic strength conditions.[18-23] The range of the three pK_a values ($\approx$3.3 to 9.3, see Table 6.1) for tetracyclines results in large shifts in the proportion of the cation (+00), zwitterion (+--0), and predominantly anionic species (+--) over the environmentally relevant pH range. Figueroa et al.[18] effectively described pH effects on oxytetracycline (OTC) sorption using a model that included cation exchange plus surface complexation of the +--0 zwitterion species. Previous work on less complicated organic bases (e.g., quinoline and aromatic amines) has shown that when cation exchange dominates contaminant binding to a sorbent, K_d or $K_d{}^*$ can be normalized with the soil's cation exchange capacity (CEC) as follows: $K_{CEC}^* = K_d/CEC$, where K_{CEC}^* (L/equiv) is the CEC normalized distribution coefficient.[24,25] Likewise, Figueroa et al.[18] derived an empirical model using species-specific sorption coefficients normalized to pH-dependent CEC ($K_{CEC}^{+00}, K_{CEC}^{+-0}, K_{CEC}^{+--}$) that are weighted by the pH-dependent fraction of each species ($f_{+00}, f_{+--0}, f_{+--}$) as follows: $K_{CEC}^* = K_{CEC}^{+00} f_{+00} + K_{CEC}^{+-0} f_{+-0} + K_{CEC}^{+--} f_{+--}$, which fit the data well for OTC sorption by Na-saturated montmorillonite in 10 mM sodium bicarbonate buffer.[18] Estimated values for K_{CEC}^{+00} (70 800 L/equiv) were 20 times higher than K_{CEC}^{+-0} (3 500 L/equiv), indicating that the cation contributes the most to OTC sorption even under pH conditions where the zwitterion species predominates in solution. To explain the sorption of the +--0 zwitterion, Figueroa et al.[18] proposed a complexation mechanism involving positively charged protons on the montmorillonite surfaces. The proposed hypothesis is supported by Kulshrestha et al.'s [19] Fourier transformed (FT)-IR analysis of Na-montmorillonite equilibrated with OTC at pH 5.0, which showed peak shifts consistent with coordination of the antibiotic's carbonyl group with interlayer cations or hydrogen bonding with hydroxyl groups of water coordinated to interlayer cations.

Sassman and Lee[22] employed the same model to describe the sorption of OTC, TC, and chlortetracycline (CTC) by eight soils varying in pH, type and amount of clay, CEC, anion exchange capacity (AEC), and OC content. The model fit the data well across all soils (pH range of 3.8 to 7.49) except for a gibbsite-rich soil with high AEC. Values for K_{CEC}^{+00} were one to two orders of magnitude greater than K_{CEC}^{+-0} values (e.g., 6.43 × 10⁶ and 1.46 × 10⁵ L/equiv, respectively, for OTC in 0.01 N $CaCl_2$). Ter Laak et al.[23] reported similar trends in a study on OTC sorption by two agricultural soils.

In contrast, work by Jones et al.,[26] which examined OTC sorption by 30 soils at a buffered pH of 5.5, found CEC to be a weak predictor of sorption ($r^2 = 0.24$). However, the CEC values reported and used in the correlation were measured at the soils' natural pH, which ranged from 3.6 to 7.5, and not the operational CEC values at the buffered isotherm pH of 5.5. CEC values can differ substantially over the pH range of interest for soils containing significant amounts of sorbents with variable or pH-dependent charge properties (e.g., kaolinitic clays, organic matter, metal oxides, and hydroxides).[27,28] For example, CEC increased from 2 to 6 $cmol_c$/kg with increasing pH of 3 to 6 for a highly weathered oxic soil and 10 to 35 $cmol_c$/kg in the 4 to 7 pH range for a high organic matter soil (9% organic compound [OC]).[29] Thus, the operational CEC would be higher than the reported CEC if the natural soil pH was lower than the buffered pH of 5.5. Alternately, for soils with pH values higher than 5.5, the operational CEC would be lower than the reported CEC. These discrepancies may have masked the true contribution of CEC to the sorption of OTC in the Jones et al.[26] study.

Additional experiments on electrolyte composition effects on the sorption of tetracyclines also support a cation exchange mechanism. When cation exchange is a controlling process, decreases in the concentration of competing inorganic cations (e.g., 0.01 to 0.001 N $CaCl_2$) or a reduction in the selectivity of the inorganic cation (e.g., substituting K^+ for Ca^{2+}) are expected to increase solute sorption. Indeed, Sassman and Lee[22] observed that reductions in CTC sorption with electrolyte composition, after accounting for induced pH-shifts, generally followed the trend: 0.001 N $CaCl_2$ > 0.01 N KCl > 0.01 N $CaCl_2$. Ter Laak et al.[23] found that increases in ionic strength decreased sorption of OTC in two agricultural soils with a more pronounced effect induced by $CaCl_2$ compared to NaCl. Figueroa et al.[18] also observed that an increase in ionic strength from 10 mM to 510 mM by the addition of Na^+ decreased $K_{CEC}^{+\,00}$ by a factor of almost 13 and $K_{CEC}^{+\,-\,0}$ by a factor of about 2 for OTC sorption to Na-montmorillonite. An increase in ionic strength from 0.01 M to 0.1 M by the addition of Na^+ reduced the sorption of TC by Elliott soil humic acid by 50%.[21]

Alternatively, for model clays, Ca^{2+} has been shown to enhance solute sorption by serving as a cation bridge between negatively charged tetracycline molecules and clay surfaces. Under alkaline conditions, OTC sorption coefficients for Ca-saturated montmorillonite were higher than either Na-saturated or untreated montmorillonite.[18] Tetracycline has also been shown to have a high affinity for the positively charged surfaces of metal-hydrous oxides, with K_f and N values of 150 mol^{1-N} kg^{-1} L^N and 0.95 for Al-hydrous oxides and 59.1 mol^{1-N} kg^{-1} L^N and 0.85 for Fe-hydrous oxides, respectively, measured at an ionic strength of 0.01 M and pH 5.4.[30] (Units for K_f were erroneously reported to be unitless by Gu and Karthikeyan.[30]) Fourier transform infrared spectroscopy indicated that the predominantly zwitterion TC (+-0) interacted with the metal oxides via the tricarbonylamide and carbonyl functional groups. The formation of the surface complex subsequently resulted in the dissolution of the metal-hydrous oxides and aqueous concentrations of 2:1 metal to TC complexes. Other studies have shown that tetracyclines form complexes with divalent cations, such as Ca^{2+}, Cu^{2+}, Mn^{2+}, Mg^{2+}, and Zn^{2+} at multiple chelating sites.[31,32] Since some animals are fed rations with high concentrations (g/kg range) of inorganic cations for growth promotion (e.g., Cu^{2+} and Zn^{2+}),[33-35] elucidating the role of complexing metals is particularly important in predicting the fate of antibiotics at CAFO-impacted sites.

Complexation with cations has also emerged as an important sorption mechanism for fluoroquinolone antibiotics, which consist of a bicyclic aromatic ring skeleton with a carboxylic acid at position 3, keto group at position 4, and a basic N-moiety at position 7.[36-38] Gu and Karthikeyan[39] examined the binding of cipro-floxacin (CIP) ($pK_{a1} = 6.16$, $pK_{a2} = 8.62$) to Al- and Fe-hydrous oxides. CIP is an important third-generation fluoroquinolone reserved for human use, and it is also the primary metabolite of enrofloxacin, a veterinary antibiotic that differs structurally from CIP by a single ethyl group. Sorption of CIP by Fe-hydrous oxide increased from pH 4 ($K_d \approx 400$ L/kg), where the compound exists predominantly as a cation, to a maximum at pH 7 ($K_d \approx 2000$ L/kg), where it is predominantly a zwitterion. Further increases to pH 10 reduced sorption as a result of increasingly repulsive interactions between the negatively charged antibiotic and sorbate. The same trend was observed for CIP sorption to Al-hydrous oxide; however, the maximum K_d was less than 400 L/kg. Analysis by FT-IR indicated this was attributable to differences in the types of antibiotic-metal complexes formed. CIP and Fe formed a strong bidentate complex between the metal and the deprotonated β-keto acid. In contrast, CIP formed only a monodentate complex between the deprotonated carboxylate group and an Al atom. Earlier work by Nowara et al.[40] that measured low sorption by soil of decarboxylated enrofloxacin relative to enrofloxacin and other quinolone derivatives also provides support for a cation-bridging sorption mechanism.

Cation exchange and cation bridging also contribute to the binding of antibiot-ics to aqueous dissolved organic carbon (DOC), which may enhance their mobility through a soil profile or in surface runoff. Using a dialysis membrane technique, MacKay and Canterbury[41] quantified the binding of OTC by solutions of Aldrich humic acid with and without amendments of Al^{3+}, Fe^{3+}, and Ca^{2+}. DOM binding coefficients (K_{DOC}, L/kg DOC) ranged from 5500 to 250,000 L/kg and increased in the presence of Al^{3+} and Fe^{3+}, which were hypothesized to promote the formation of ternary complexes between OTC and humic acid ligands. In contrast, K_{DOC} values decreased to 2980 L/kg with the addition of Ca^{2+}, suggesting that Ca^{2+} competed with OTC for cation exchange sites on the humic acid. However, sorption coeffi-cients for TC by Elliott soil humic acid (23 mmol Ca^{2+}/kg DOC) were increased from approximately 4000 L/kg OC to 5000 L/kg DOC by Ca^{2+} amendments (8333 mmol Ca^{2+}/kg DOC).[21]

Holten Lützhøft et al.[42] measured K_{DOC} values in the same range for the binding of fluoroquinolones by Aldrich humic acid (3000 to 200,000 L/kg DOC). Sorption of flumequin and oxolinic acid increased markedly between pH 3 and 6 but remained constant with further increases in pH. Anticipated electrostatic repulsion between the negatively charged antibiotic and the humic acid did not reduce sorption as pH increased. These results suggest a cation bridging mechanism since flumequin and oxolinic acid posses a single ionizable -COOH ($pK_a = 6.4$ and 6.9, respectively), which can bind to negatively charged sites on the humic acid only via a cation bridge. In contrast, the binding of sarafloxacin to humic acid was high under acidic condi-tions (59,000 L/kg DOC at pH 3) and less sensitive to changes in pH. Binding to humic acid at low pH conditions was attributed to cation exchange via sarafloxacin's positively charged amine group ($pK_a = 8.6$).

Although strong electrostatic interactions govern the sorption of many ionizable antibiotics, hydrophobic partitioning has also been shown to contribute to the sorp-tion of others. Sassman and Lee[43] recently found that linear isotherms adequately modeled the sorption by eight soils of two ionophores, monensin and lasalocid. Ionophores are relatively large molecules comprised of a backbone consisting of tetrahydropyran or tetrahydrofuran groups with multiple carboxylic acid and ester groups. Log K_{oc} values ranged from 2.1 to 3.8 for monensin and 2.9 to 4.2 for lasalocid and generally decreased with increasing soil pH (pH range 4.2 to 7.5), which is expected since as carboxylic acid groups are deprotonated under alkaline conditions. Interestingly, the carboxyl and ether O atoms in the molecules can chelate environmentally relevant cations (e.g., Na^+, K^+, Ca^{2+}, Mg^{2+}), which may increase the apparent hydrophobicity of the molecules and possibly alter their sorption and mobility by reducing their net charge. Elucidating interactions with divalent metals, especially those frequently incorporated in animal feeds, will be necessary for understanding the fate of these antimicrobials in the environment.

Hydrophobic partitioning also appears to contribute to sorption by soils of tylosin, a basic macrolid antibiotic ($pK_a = 7.7$), and two of its metabolites, tylosin-D and tylosin A-aldol.[15] Sorption isotherms measured for six soils were nonlinear with Freundlich model N values of 0.42 to 0.80. Comparisons among soils made by examining K_d^* values estimated at an aqueous equilibrium concentration of 2×10^{-4} mmol/L indicated that sorption was lowest for a sandy soil (2.23 L/kg) and highest for a clay soil (5520 L/kg). As with cationic antibiotics, sorption was strongly correlated to CEC ($r^2 = 0.70$), soil surface area ($r^2 = 0.91$), and clay content ($r^2 = 0.86$). A lesser but moderately positive correlation was also observed with OC content ($r^2 = 0.41$), indicating some contribution from hydrophobic forces. Given that CEC originates from both OC and clays, independent contributions of OC and clay domains cannot be definitively separated. However, in a natural soil, evidence for hydrophobic forces was provided from the high mass recovery (95 ± 14%) attained for most soils using only methanol to recover the sorbed fraction. Only an acidic soil (pH of 4.4) required an extraction solution containing NH_4^+, which promoted cation exchange of the antibiotic and improved recovery to 98% from 26 to 50% attained with methanol alone.

Hydrophobic forces have also been shown to play a role in the sorption of carbadox, a quinoxaline antibiotic added to the feed of starter pigs in the United States to prevent dysentery and improve feed efficiency. Strock et al.[44] investigated the sorption of carbadox and its N-oxide reduced metabolites (N4 oxide, N1 oxides, and bis-desoxycarbadox) by several soils, a sediment, and homoionic smectite and kaolinite clays. Both carbadox and bis-desoxycarbadox have been identified as potential mutagens,[45] prompting the need for information on the environmental fate of these novel aromatic N-oxide compounds. Sorption appeared well correlated to soil OC content (e.g., log $K_{oc} = 3.98 \pm 0.18$ for CBX). However, sorption was enhanced by the presence of K^+ relative to Ca^{2+}, competitive sorption by the metabolites was observed, and sorption by clay minerals was large (e.g., log $K_d \approx 5$ for montmorillonite) and inversely correlated to surface change density. In the absence of a clay surface, hydrophobic type forces dominated as evidenced by increasing K_{ow} values and reverse-phase chromatographic retention times with the loss of oxygen from the aromatic nitrogens. Therefore, although sorption was generally well described by the OC domain, specific interactions with clays also contributed significantly to the sorption of carbadox and its metabolites.

Compared to the antibiotic classes discussed thus far, sulfonamide antibiotics sorb to soils, clays, and organic materials to a lesser extent. The low sorption of sulfonamides is expected because of their high polarity and net charge (neutral and negatively charged) at environmentally relevant pH conditions. The pK_a of the anilinic amine of sulfonamides is typically ≤ 2.3; thus, its cationic species will only exist at pH values less than 4.5, which is approaching the lower boundary of pH conditions typical of the natural environment, whereas pK_a values for the deprotonation of the amide group (pK_a values from 5.5 to 8, see Table 6.1) are within the typical soil pH range. Gao and Pedersen[46] examined the sorption of sulfamethazine, sulfamethoxazole, and sulfapyradine to montmorillonite and kaolinite clays as a function of pH, ionic strength, and exchangeable cation composition. In the pH range 5 to 7, where the neutral species dominates, sorption was relatively insensitive to pH and ionic composition and essentially linear with approximated K_d values in the 10 to 16 L/kg range. At pH <5, sorption of sulfonamides increased with decreasing pH and decreased with increasing ionic strength, which are characteristic trends of sorption by the protonated anilinic amine. At pH >7, sorption decreased to near zero with increasing pH as the fraction of the anionic species increased. The magnitude of sorption and the pH effect were dependent on clay charge density. Similar trends in sorption ($K^{+0} >$ $K^{00} > K^{0-}$) and with pH have been reported for the sorption of sulfathiazole to compost, manure, and humic acid.[47] Sorption of a suite of sulfonamides by whole soils and soil fractions exhibited a range of sorption nonlinearity ($N = 0.75$ to 1.21) and correlations to OC appeared dependent on the composition of organic matter indicating the occurrence of some site-specific interactions.[48,49] In model systems, Bialk et al.[50] exemplified enzyme-mediated oxidative cross-coupling of sulfonamides with some simple model humic constituents as well as acid birnessite (MnO_2)-mediated transformation of certain sulfonamides, which may be important pathways for reducing the mobility and persistence of these antibiotics in the environment.

6.3.2 DEGRADATION IN MANURE AND SOIL

Lee et al.[8] and Kumar et al.[11] have recently reviewed the literature on the degradation of antibiotics under various environmental conditions. Since these summaries were published, a handful of additional studies have emerged, and the discussion below will focus on these new results or information that was not contained in previous reviews. Several papers have examined the degradation of common growth promoters in anaerobic manure slurries since their stability under these conditions will be a major determinant for their release into the environment. Over a 180-d anaerobic incubation in liquid swine manure, first-order half-lives were estimated for erythromycin, roxithromycin, and salinomycin of 41 d, 130 d, and 6 d, respectively, whereas no degradation of tiamulin was observed.[51] The dissipation of tylosin was studied in anaerobic slurries of swine manure obtained from open and covered lagoons for which 90% disappearance times of 40 h and 310 h, respectively, were calculated using a two-compartment, first-order model.[52] Addition of NaN_3 (50 g/L), intended to eliminate biodegradation, increased these times to 90 h and 500 h, respectively. In contrast, aerating the systems in either the presence or absence of NaN_3 reduced the 90% disappearance times for the open and covered lagoons to 12 h and ≤ 32 h, respectively, in a manner that followed a first-order degradation model. The authors attributed that the dissipation of tylosin primarily to irreversible sorption to solids and aeration induced increases in pH, which enhanced base-catalyzed reactions. In another study on tylosin, calculated first-order half-lives in aerobic cattle, chicken, and swine manure slurries ranged from 6.2 to 7.6 d.[53]

New studies have reported on the dissipation of various antibiotics in soils. Sassman and Lee[43] investigated the degradation of monensin and lasalocid in ^{60}Co irradiated and nonsterile laboratory soil microcosms with and without manure amendment. In nonsterile soils, less than 25% of either ionophore remained after 8 d with no significant effect from manure amendment. In the sterile soils, 40% of the applied monensin degraded abiotically and became nonextractable within 4 d, whereas appreciable lasalocid loss occurred only in the unsterilized microcosms. This indicates that abiotic processes contribute significantly to monensin dissipation, although the complete absence of microbial activity in the soil microcosms was not confirmed. In a field study in Ontario, Canada, the disappearance of tylosin, CTC, and monensin in a sandy-loam agricultural soil followed first-order kinetics with half-lives of 6.1 d, 21 d, and 3.8 d, respectively.[54] The addition of dairy cow manure (approximately 27 tons/hectare) to the fields decreased the half-life of tylsoin to 4.1 d, whereas no effect was observed for CTC and monensin. These field-based data for monensin are similar to half-lives measured by Sassman and Lee[43] for monensin (1.2 to 1.9 d) and lasalocid (1.4 to 4.1 d) in soil microcosms. They are also comparable to half-lives of just over 3 weeks measured for tylsoin in soil microcosms.[15] In contrast, significantly longer half-lives were measured for CTC (20 to 42 d) and tylosin (40 to 86 d) in a Danish field study with sandy loam and sandy agricultural soils.[55]

Recently, several studies have investigated the transformation by manganese oxide (δ-MnO$_2$) of several antibacterial agents and related model compounds, including fluoroquinolones, quinoxalines, triclosan, chlorophene, and OTC.[56–59] Naturally occurring manganese oxide minerals (e.g., birnessite) in soils and sediments have a high reduction potential (1.23 V) and can play an important role in the transformation of organic compounds. Departure from pseudo-first-order kinetics was reported in each study, likely due to changes to the MnO$_2$ surface properties over the course of the experiments.[59] To facilitate comparisons of reaction kinetics and calculations of reaction order, Zhang and Huang[56–58] used initial rates during the first few hours where pseudo-first-order kinetics were approximated. Except for flumequine for which no transformation was observed, all tested fluoroquinolones degraded in the presence of MnO$_2$ at pH 6 with initial reaction rate constants (k_{init}) ranging from 0.54 h^{-1} to 1.11 h^{-1}.[57] Triclosan and chlorophene degraded in reagent water (pH 5) with k_{init} values of 1.74 h^{-1} and 2.67 h^{-1}, respectively.[56] Quinoxaline transformations occurred more slowly. Carbadox was oxidized with a k_{init} value of 0.3 h^{-1}, whereas desoxycarbadox reacted with two orders of magnitude slower and no measurable loss of olaquindox occurred.[58] Rubert and Pedersen[59] also measured a relatively slow k_{init} value for OTC (4×10^{-4} μM$^{-0.5}$ h^{-1}).

All studies reported that reaction rates were inversely related to pH conditions, which is consistent with the decreasing oxidative power of MnO$_2$ with increasing pH. In addition, rates are also limited by the ability of the antibacterial to adsorb to the MnO$_2$ surface. Oxidation of the organic compound requires the formation of a precursor complex with the surface of the MnO_2, and the ability to form this complex will vary among the different pH-dependent species of the ionizable compounds.[57,58] Differences in transformation rates among related compounds of a particular anti-bacterial class were attributed to specific functional groups and described in detail by the authors. In related work, Bialk et al.[50] found that selected sulfonamides also underwent significant transformation (20 to 65%) in the presence of MnO_2. In addition, the presence of certain model substituted phenols, intended to simulate humic materials, enhanced transformation rates by promoting covalent cross-coupling of the sulfonamides to the model phenols. Goethite has also been shown to oxidatively transform fluoroquinolones, albeit at rates approximately three orders of magnitude lower than MnO_2, reflecting the lower redox potential (0.63 V) of the mineral.[60]

6.4   HORMONES

6.4.1   SORPTION BY SOIL AND SEDIMENT

The occurrence of natural and synthetic estrogens and androgens in surface waters has sparked interest by the research community and the general public because of their potential to elicit adverse ecological effects at concentrations in the low ng/L range.[61,62] It is estimated that livestock and poultry annually excrete 49 tons of estrogens, 4.4 tons of androgens, and 279 tons of gestagens.[4] Natural steroidal estrogens from human and animal wastes are also released in the effluents of conventional wastewater treatment plants at moderately high ng/L concentrations.[63,64]

Whether estrogenic or androgenic, naturally excreted or manufactured, reproductive hormones and their primary metabolites share similar physicochemical properties. They are moderately hydrophobic, poorly soluble in water, and do not possess a charge in manure wastes or soil environments (e.g., estradiol $pK_a = 10.71$).[65] Reported log K_{ow} and aqueous solubility values range from 3 to 4 and 13 to 41 mg/L, respectively.[8] Several studies have examined the sorption of these compounds to sediments or soils, and as with other hydrophobic organic contaminants, interactions with the OC domain appear to drive this process. Lee et al.[66] measured the simultaneous sorption and transformation of three reproductive hormones and their metabolites in several Midwestern soils. Concentrations of the parent compounds and metabolites were measured directly in the aqueous and soil phases to generate sorption isotherms that, in general, fit the linear model well. Near equilibrium conditions in the batch reactors were achieved within a few hours. Average log K_{oc} values for 17β-estradiol (E2), 17α-ethynylestradiol (EE2), and testosterone were 3.34, 2.99, and 3.34, with standard deviations of ≤0.18 log units. Log K_{oc} values in the same range (2.77 to 3.69) were also obtained by Das et al.[67] and by Casey et al.[68,69] Recent work by Khan et al.[70] on the widely used livestock growth promoter 17β-trenbolone acetate and its primary metabolites, 17α-trenbolone and trendione, shows that these compounds sorb in a manner similar to natural steroid hormones. Average log K_{oc} values for 17β-trenbolone (3.10 ± 0.28) and trendione (3.42 ± 0.26) were consistent with reported aqueous solubilities and log K_{ow} values, indicating that soil OC is a primary sorption domain.

There is also evidence that processes other than hydrophobic partitioning to OC contribute to sorption of steroid hormones. Several researchers have reported slight-to-moderate nonlinear sorption isotherms (N <1).[66,71] Competition for binding sites has also been observed, suggesting a contribution by site-specific interactions.[71,72] In work on river sediments, Lai et al.[71] found a strong correlation (r^2 = 0.86 to 0.94) between sorption of estrogens and total OC content. In addition, sorption increased with increasing compound hydrophobicity, as indicated by K_{ow} values: mestranol > EE2 > E2 > estrone (E1) > estriol (E3). However, substantial sorption by an Fe-oxide, intended to simulate a sediment lacking OC, was also observed. This sorption was attributed to ion exchange interactions between the oxide's surface hydroxyl groups and the estrogens polar phenolic groups. Recently, Shareef et al.[73] also reported considerable sorption of E1 and EE2 to kaolinite, goethite, and montmorillonite. In addition, sorption by montmorillonite increased by a factor of two with a pH increase from 8 to 10, and desorption was low. Although it was hypothesized that sorption was driven by electrostatic interactions between the negatively charged hormones and water molecules in the interlayer spaces of the mineral, it is unlikely that this process would lead to low desorption. The same experiment with mestranol, which lacks an ionizable phenolic functional group, followed a similar sorption pattern as EE2 and E1, further negating the electrostatic interaction hypothesis. In each of these cases, sorption to the minerals was estimated by difference rather than by direct quantification of the sorbed concentration. Thus, loss of mass from solution may have resulted from surface-catalyzed transformations or instability in alkaline conditions rather than sorption.

The presence of DOC can result in lower-than-expected sorption of hormones

due to the association of the hormones as follows: $K_d^{obs} = \dfrac{K_d}{1 + K_{DOC}[DOC]}$ where

K_d^{obs} is the apparent sorption coefficient in the presence of DOC and [DOC] is the concentration of DOC (kg/L). High log K_{DOC} values for the binding of E2, EE2, and E3 (4.13 to 4.86) have been reported for commercial humic and fulvic acids and from sediment-derived DOC.[74,75] Interestingly, log K_{DOC} values were greater than log K_{oc} values and were not correlated to log K_{ow}, which differs from the trends observed for other more strongly hydrophobic compounds.[76] Yamamoto et al.[74] proposed that hydrogen bonding and interactions between π-electrons of estrogenic compounds and the DOC augmented sorption beyond what would be predicted from simple hydrophobic partitioning interactions.

Recently, Kim et al.[77] measured log K_{oc} values for testosterone (6.18 to 6.80) and androstenedione (6.04 to 6.92) that are significantly higher than previously reported values and are inconsistent with the moderate log K_{ow} values and solubilities of these compounds. Sorption isotherms were slightly nonlinear with Freundlich model N values of 0.70 to 0.90 and equilibrium was achieved over several weeks, which differs from the apparent rapid approach to equilibrium reported by others.[66,68,69,71] Although degradation was inhibited with NaN_3, active extracellular enzymes may have resulted in some compound loss. Sorption was calculated by difference and neither the conservation of mass nor the absence of metabolites were confirmed. Transformations of steroid hormones under apparently sterile conditions have been reported elsewhere[78] and may have occurred during the long equilibration times leading to inflated log K_{oc} estimates.

6.4.2 DEGRADATION IN MANURE AND SOIL

Relatively few studies have examined dissipation rates of steroid hormones in manure during storage and after land application. Hakk et al.[79] measured decreases in water-soluble estrogenic activity (83 to 13 ng estradiol equivalents/g) and androgenic activity (115 to 11 ng testosterone equivalents/g) in poultry manure compost piles over a 139-d period. Dissipation followed first-order kinetics with loss rate constants of 0.01 d^{-1} and 0.015 d^{-1}, respectively. Water-extractable hormones were still detectable after 4.6 months of composting.

Colucci et al.[78] and Colucci and Topp[80] reported somewhat longer dissipation rate constants for E2 (0.07 to 3.12 d^{-1}) and EE2 (0.1 to 0.37 d^{-1}) in moist agricultural soil microcosms. These rates correspond to half-lives between 0.5 d and 7.7 d. Transformation of E2 to E1 occurred in both autoclaved and nonsterile soil, whereas EE2 persisted in sterilized soil. In nonsterile soils, both E2 and E1 formed nonextractable residues in a few days (56 to 91%), and mineralization to CO_2 was minimal (15%) over the 61-d experiment. Half-lives ranging from a few hours to a few days were also reported for E2, EE2, and testosterone by Lee et al.[66] Dissipation rates varied with hormone and soil or sediment type. However, all three hormones exhibited the longest half-lives for a sandy sediment with the lowest OC content, which was likely the soil with lowest fertility. Half-lives for 17β-trenbolone in moist soils are in the same range (0.5 to ≥2 d); however, half-lives increased when applied soil concentrations were increased from 1 to 10 mg/kg.[70]

Das et al.[67] used a saturated flow-interruption column technique to estimate degradation coefficients for testosterone, E2, and their metabolites in an agricultural surface soil, freshwater sediment, and two sands. Estimated rate constants ranged from 0.003 h^{-1} to 0.015 h^{-1} for testosterone and 0.0003 h^{-1} to 0.075 h^{-1} for E2, with degradation for the metabolites occurring more quickly than the parent compounds. Rates were observed to decrease with increasing column life, which was attributed to nutrient depletion over time since additions of NH_4^+ restored degradation. Estimated sorption mass-transfer constants calculated during flow were at least an order of magnitude higher than the degradation rate constants. This indicates that although sorption by soils of the hormones and metabolites was substantial (4 to 142 L/kg), degradation was not impeded by sorption, and near-equilibrium conditions could be assumed within a short time period, which is consistent with other batch experiments.[66,68,71] Schiffer et al.[81] also used saturated columns to monitor the transformation of 17β-trenbolone to 17α-trenbolone. Conversion was higher in a surface Ap horizon soil (1.6% OC) compared to a subsurface Bt horizon soil (0.6% OC), which may be attributed to potentially greater microbial activity inherent with higher OC soils and is in agreement with observations by Das et al.[67] Likewise, Khan et al. (unpublished data) observed faster degradation in a higher OC soil for trenbolone compared to a sandy soil. They also noted no differences between the two isomers and that degradation was faster at lower applied hormone concentrations of 1 mg/kg soil (half-lives of <0.5 d) compared to 10 mg/kg (half-lives up to 2.5 d). The primary metabolite trendione exhibited longer half-lives of 3 d to 2 weeks.

Two studies have measured dissipation and mineralization rates in soils amended with different types of animal litter.[82,83] Lucas and Jones[82] found that addition of sheep and cattle wastes aged from 7 d to 2 years reduced half-lives for E2 from a range of 5 to 25 d to 1 to 9 d, with mineralization rates being largely independent of manure age and animal type. However, mineralization occurred more rapidly in manure-amended soils compared to those that were amended with urine only. Degradation was fit with an exponential decay model with two decay constants, one to describe a rapid CO_2 production phase stemming from immediate catabolic processes and a second slower rate attributed to anabolic uptake and subsequent turnover of the microbial biomass. Since mineralization to CO_2 was measured, transformations to less estrogenically active metabolites are expected to be much faster. Jacbosen et al.[83] also found that complete conversion of E2 to E1 and complete mineralization to CO_2 were enhanced when swine manure slurry was added to a loam soil (10% v/w). Testosterone transformation to androstenedione was also accelerated with swine manure, although mineralization to CO_2 was suppressed for a short period (<6 d). Essentially no transformation was observed in sterile conditions over a 96-h period.

Despite apparently high sorption and rapid degradation rates, residues of steroid hormones applied to soils may escape attenuation if they move through preferential flow paths. In an intact soil core experiment with a pulse application of radiolabelled E2 and testosterone, Sangsupan et al.[84] found that as much as 27% and 42%, respectively, of the applied hormones leached out. In five of the six soil columns, peak hormone and Cl⁻ breakthrough concentrations occurred simultaneously, indicating movement through preferential flow paths. Schiffer et al.[81] also found that small amounts of 17β-trenbolone and melengestrol acetate leached through packed soil columns within one pore volume, and subsequent breakthrough occurred earlier than predicted from sorption isotherm data.

6.5 CONCLUSION

As the world's human population grows and global affluence increases, the demand for food animals can be expected to rise. Current usage data indicate that antibiotics and hormones are likely to play a major role in meeting this demand, particularly as developing countries modernize their livestock and poultry industries to meet food requirements.[10] As a result, the pressure on agricultural lands to accommodate more animal and human wastes can be expected to increase concomitantly.

At this time there are significant gaps in our understanding of the fate and environmental impacts of the large quantities of pharmaceutical residues present in the billions of pounds of animal wastes produced annually. The primary concern with antibiotic use in animal production is the persistence of excreted antibiotic-resistant bacteria or the development of new antibiotic-resistant bacteria. For hormones, low concentration residues have been implicated with endocrine disruption resulting in alterations in vertebrate reproductive endocrinology and physiology. Laboratory-scale studies generally indicate that rapid degradation rates and high sorption to soils should minimize the persistence and mobility of these emerging contaminants. Nevertheless, many of the compounds least expected to exist in the environment based on laboratory-based sorption and degradation experiments are among the most frequently detected. This calls to attention the limited degree to which laboratory experiments can fully and accurately evaluate processes at the field scale. In addition, relatively few studies have evaluated the routes to surface waters, which will vary with local hydrology and climate conditions.

To better assess the real contribution of food animal production and associated waste management practices on antibiotic and hormone inputs into the environment, field-scale systematic studies are needed to address (1) chemical persistence in the field after land application; (2) the relative contribution of runoff events, tile drainage, and leaching on the actual quantities of pharmaceuticals released to water sources, which will vary with region and time after land application; and (3) correlation between time after application and rainfall events on chemical loadings to aquatic systems. Research is also needed to understand how manure storage or composting parameters can be optimized toward reducing manure-borne pharmaceutical concentrations prior to land application, and how application methods and timing of applications affect potential loadings to aquatic systems.

REFERENCES

1. Gollehon, N. et al., Confined animal production and manure nutrients, Agriculture Information Bulletin No. 771. Resource Economics Division, Economic Research Service. USDA, 2001.
2. USEPA, Proposed regulations to address water pollution from concentrated animal feeding operations, Report No. EPA 833-F-00-016, Office of Water, Washington, DC, 2000.
3. Addison, J.B., Antimicrobials in sediments and run-off waters from feedlots, *Resid. Rev.*, 92, 1, 1984.
4. Lange, I.G. et al., Sex hormones originating from different livestock production systems: fate and potential disrupting activity in the environment, *Anal. Chim. Acta.*, 473, 27, 2002.
5. USDA, Part III: Reference of swine health and environmental management in the United States 2000, Report No. #N361.0902, USDA:APHIS:VS, CEAH, National Animal Health Monitoring System, Fort Collins, CO, 2002.
6. Kinney, C.A., Furlong, E.T., Werner, S.L., and Cahill, J.D., Presence and distribution of wastewater-derived pharmaceuticals in soil irrigated with reclaimed water, *Environ. Toxicol. Chem.*, 25, 317, 2006.
7. Kinney, C.A. et al., Survey of organic wastewater contaminants in biosolids destined for land application, *Environ. Sci. Technol.*, 40, 7202, 2006.
8. Lee, L.S. et al., Agricultural contributions of antimicrobials and hormones on soil and water quality, *Adv. Agron.*, 93, 1, 2007.
9. Khanal, S.K. et al., Fate, transport, and biodegradation of natural estrogens in the environment and engineered systems, *Environ. Sci. Technol.*, 40 (21), 6537, 2006.
10. Sarmah, A.K., Meyer, M.T., and Boxall, A.B.A., A global perspective on the use, sales, exposure pathways, occurrence, fate and effects of veterinary antibiotics (VAs) in the environment, *Chemosphere*, 65, 725, 2006.
11. Kumar, K., Gupta, S.C., Chander, Y., and Singh, A.K., Antibiotic use in agriculture and its impact on the terrestrial environment, *Adv. Agron.*, 87, 1, 2005.
12. Tolls, J., Sorption of veterinary pharmaceuticals in soils: a review, *Environ. Sci. Technol.*, 35, 3397, 2001.
13. Thiele-Bruhn, S., Pharmaceutical antibiotic compounds in soils—A review, *J. Plant Nutr. Soil Sci.*, 166, 145, 2003.
14. Boxall, A.B.A. et al., The sorption and transport of a sulfphonamide antibiotic in soils systems, *Toxicol. Lett.*, 131, 19, 2002.

15. Sassman, S.A., Sarmah, A.K., and Lee, L.S., Sorption of tylosin A, tylosin D and tylosin A-aldol in soils, *Environ. Toxicol. Chem.*, Accepted, 26, 2007.

16. Haws, N.W., *Integrated flow and transport processes in subsurface-drained agricultural fields*, Ph.D. Dissertation, Purdue University, 2003.

17. Porubcan, L.S., Serna, C.J., White, J.L., and Hem, S.L., Mechanism of adsorption of clindamycin and tetracycline by montmorillonite, *J. Pharm. Sci.*, 67, 1081, 1978.

18. Figueroa, R.A., Leonard, A., and Mackay, A.A., Modeling tetracycline antibiotic sorption to clays, *Environ. Sci. Technol.*, 38, 476, 2004.

19. Kulshrestha, P., Giese Jr., R.F., and Aga, D.S., Investigating the molecular interactions of oxytetracycline in clay and organic matter: insights on factors affecting its mobility in soil, *Environ. Sci. Technol.*, 38, 4097, 2004.

20. Pils, J.R.V. and Laird, D.A., Sorption of tetracycline and chlortetracycline on K- and Ca-saturated soil clays, humic substances, and clay-humic complexes, *Environ. Sci. Technol.*, 41, 1928, 2007.

21. Gu, C., Karthikeyan, K.G., Sibley, S.D., and Pedersen, J.A., Complexation of the antibiotic tetracycline with humic acid, *Chemosphere*, 66, 1494, 2007.

22. Sassman, S.A. and Lee, L.S., Sorption of three tetracyclines by several soil types, *Environ. Sci. Technol.*, 39, 7452, 2005.

23. Ter Laak, T.L., Gebbink, W.A., and Tolls, J., The effect of pH and ionic strength on the sorption of sulfachloropyridazine, tylosin, and oxytetracycline to soil, *Environ. Toxicol. Chem.*, 25, 904, 2006.

24. Lee, L.S. et al., Initial sorption of aromatic amines to surface soils, *Environ. Toxicol. Chem.*, 16, 1575, 1997.

25. Zachara, J.M., Ainsworth, C.C., Felice, L.J., and Resch, C.T., Quinoline sorption to surbsurface materials: role of pH and retention of the organic cation, *Environ. Sci. Technol.*, 20, 620, 1986.

26. Jones, A.D., Bruland, G.L., Agrawal, S.G., and Vasudevan, D., Factors influencing the sorption of oxytetracycline to soils, *Environ. Toxicol. Chem.*, 24, 761, 2005.

27. Oorts, K., Vanlauwe, B., Pleysier, J., and Merckx, R., A new method for the simultaneous measurement of pH-dependent cation exchange capacity and pH buffering capacity, *Soil Sci. Soc. Am. J.*, 68, 1578, 2004.

28. Hyun, S. and Lee, L.S., Hydrophilic and hydrophobic sorption of organic acids by variable-charge soils: effect of chemical acidity and acidic functional group, *Environ. Sci. Technol.*, 38, 5413, 2004.

29. Hyun, S., Lee, L.S., and Rao, P.S.C., Significance of anion exchange in pentachlorophenol sorption, *J. Environ. Qual.*, 32, 966, 2003.

30. Gu, C. and Karthikeyan, K.G., Interaction of tetracycline with aluminum and iron hydrous oxides, *Environ. Sci. Technol.*, 39, 2660, 2005.

31. Tongaree, S., Flanagan, D.R., and Poust, R.I., The interaction between oxytetracycline and divalent metal ions in aqueous and mixed solvent systems, *Pharm. Dev. Technol.*, 4, 581, 1999.

32. Williamson, D.E. and Everett Jr., G.W., A proton nuclear magnetic resonance study of the site of metal binding in tetracycline, *J. Am. Chem. Soc.*, 97, 2397, 1975.

33. Case, C.L. and Carlson, M.S., Effect of feeding organic and inorganic sources of additional zinc on growth performance and zinc balance in nursery pigs, *J. Anim. Sci.*, 80, 1917, 2002.

34. Arias, V.J. and Koutsos, E.A., Effects of copper source and level on intestinal physiology and growth of broiler chickens, *Poultry Sci.*, 85, 999, 2006.

35. Apgar, G.A., Kornegay, E.T., Lindemann, M.D., and Notter, D.R., Evaluation of copper sulfate and a copper lysine complex as growth promoters for weanling swine, *J. Anim. Sci.*, 73, 2640, 1995.

36. Drakopoulos, A.I. and Ioannou, P.C., Spectrofluorometric study of the acid-base equilibria and complexation behavior of the fluoroquinolone antibiotics ofloxacin, norfloxacin, ciprofloxacin and perfloxacin in aqueous solution, *Anal. Chim. Acta.*, 354, 197, 1997.

37. Brown, S.A., Fluoroquinolones in animal health, *J. Vet. Pharmacol. Ther.*, 19, 1, 1996.

38. Lizondo, M., Pons, M., Gallardo, M., and Estelrich, J., Physicochemical properties of enrofloxacin, *J. Pharm. Biomed. Anal.*, 15, 1845, 1997.

39. Gu, C. and Karthikeyan, K.G., Sorption of the antimicrobial ciprofloxacin to aluminum and iron hydrous oxides, *Environ. Sci. Technol.*, 39, 9166, 2005.

40. Nowara, A., Burhenne, J., and Spiteller, M., Binding of fluoroquinlone carboxylic acid derivatives to clay minerals, *J. Ag. Food Chem.*, 45, 1459, 1997.

41. MacKay, A.A. and Canterbury, B., Oxytetracycline sorption to organic matter by metal-bridging, *J. Environ. Qual.*, 34, 1964, 2005.

42. Holten Lützhøft, H.-C., Vaes, W.HJ., Freidig, A.P., Halling-Sørensen, B., and Hermens, J.L.M., Influence of pH and other modifying factors on the distribution behavior of 4-quinolones to solid phases and humic acids studied by "negligible-depletion" SPME-HPLC, *Environ. Sci. Technol.*, 34, 4989, 2000.

43. Sassman, S.A. and Lee, L.S., Sorption by soils of veterinary ionophore antibiotics: monensin and lasalocid, *Environ. Toxicol. Chem.*, Accepted, Vol 26, September 2007.

44. Strock, T.J., Sassman, S.A., and Lee, L.S., Sorption and related properties of the swine antibiotic carbadox and associated N-oxide reduced metabolites, *Environ. Sci. Technol.*, 39, 3134, 2005.

45. Freedom of Information Summary NADA 041-061, http://www.fda.gov/cvm/FOI/520. htm, 1998, (accessed January 26, 2007).

46. Gao, J. and Pedersen, J.A., Adsorption of sulfonamide antimicrobial agents to clay and minerals, *Environ. Sci. Technol.*, 39, 9509, 2005.

47. Kahle, M. and Stamm, C., Sorption of the veterinary antimicrobial sulfathiazole to organic materials of different origin, *Environ. Sci. Technol.*, 41, 132, 2007.

48. Thiele, S., Adsorption of the antibiotic pharmaceutical compound sulfapyridine by a long-term differently fertilized loess Chernozem, *J. Plant Nutr. Soil Sci.*, 163, 589, 2000.

49. Thiele-Bruhn, S., Seibicke, T., Schulten, H.-R., and Leinweber, P., Sorption of sulfonamide pharmaceutical antibiotics on whole soils and particle-size fractions, *J. Environ. Qual.*, 33, 1331, 2004.

50. Bialk, H.M., Simpson, A.J., and Pedersen, J.A., Cross-coupling of sulfonamide antimicrobial agents with model humic constituents, *Environ. Sci. Technol.*, 39, 4463, 2005.

51. Schlüsener, M.P., von Arb, M.A., and Bester, K., Elimination of macrolids, tiamulin, and salinomycin during manure storage, *Arch. Environ. Contam. Toxicol.*, 51, 21, 2006.

52. Kolz, A.C. et al., Degradation and metabolite production of tylosin in anaerobic and aerobic swine-manure lagoons, *Water Environ. Res.*, 77, 49, 2005.

53. Teeter, J.S. and Meyerhoff, R.D., Aerobic degradation of tylosin in cattle, chicken, and swine excreta, *Environ. Res.*, 93, 45, 2003.

54. Carlson, J.C. and Mabury, S.A., Dissipation kinetics and mobility of chlortetracycline, tylosin, and monensin in an agricultural soil in Northhumberland County, Ontario, Canada, *Environ. Toxicol. Chem.*, 25, 1, 2006.

55. Halling-Sørensen, B. et al., Dissipation and effects of chlortetracycline and tylosin in two agricultural soils: a field-scale study in southern Denmark, *Environ. Toxicol. Chem.*, 24, 802, 2005.

56. Zhang, H. and Huang, C.-H., Oxidative transformation of triclosan and chlorophene by manganese oxides, *Environ. Sci. Technol.*, 37, 2421, 2003.

57. Zhang, H. and Huang, C.-H., Oxidative transformation of fluoroquinolone antibacterial agents and structurally related amines by manganese oxide, *Environ. Sci. Technol.*, 39, 4474, 2005.

58. Zhang, H. and Huang, C.-H., Reactivity and transformation of antibacterial N-oxides in the presence of manganese oxide, *Environ. Sci. Technol.*, 39, 593, 2005.

59. Rubert IV, K.F. and Pedersen, J.A., Kinetics of oxytetracycline reaction with a hydrous manganese oxide, *Environ. Sci. Technol.*, 40, 7216, 2006.

60. Zhang, H. and Huang, C.-H., Adsorption and oxidation of fluoroquinolone antibacterial agents and structurally related amines with goethite, *Chemosphere*, 66, 1502, 2007.

61. Oberdorster, E. and Cheek, A., Gender benders at the beach: Endocrine disruption in marine and estuarine organisms, *Environ. Toxicol. Chem.*, 20, 23, 2001.

62. Jobling, S., Nolan, M., Tyler, C.R., Brighty, G., and Sumpter, J.P., Widespread sexual disruption in wild fish, *Environ. Sci. Technol.*, 32, 2498, 1998.

63. Belfroid, A.C. et al., Analysis and occurrence of estrogenic hormones and their glucuronides is surface water and waste water in the Netherlands, *Sci. Total. Environ.*, 225, 101, 1999.

64. Ying, G.-G., Kookana, R.S., and Ru, Y.-J., Occurrence and fate of hormone steroids in the environment, *Environ. Int.*, 28, 545, 2002.

65. Lewis, K.M. and Archer, R.D. pKa values of estrone, 17β-estradiol, and 2-methoxyestrone, *Steroids*, 34, 485, 1979.

66. Lee, L.S., Strock, T.J., Sarmah, A.K., and Rao, P.S.C., Sorption and dissipation of testosterone, estrogens, and their primary transformation products in soils and sediments, *Environ. Sci. Technol.*, 37, 4098, 2003.

67. Das, B.S., Lee, L.S., Rao, P.S.C., and Hultgren, R.P., Sorption and degradation of steroid hormones in soils during transport: column studies and model evaluation, *Environ. Sci. Technol.*, 38, 1460, 2004.

68. Casey, F.X.M. et al., Sorption, mobility, and transformation of estrogenic hormones in natural soil, *J. Environ. Qual.*, 34, 1372, 2005.

69. Casey, F.X.M., Larsen, G.L., Hakk, H., and Simunek, J., Fate and transport of 17β-estradiol in soil-water systems, *Environ. Sci. Technol.*, 37, 2400, 2003.

70. Khan, B., Sassman, S.A., and Lee, L.S., Sorption and degradation, SETAC North America 26th Annual Meeting, November 13–17, 2005, Baltimore, MD.

71. Lai, K.M., Johnson, K.L., Scrimshaw, M.D., and Lester, J.N., Binding of waterborne steroid estrogens to solid phases in river and esturine systems, *Environ. Sci. Technol.*, 34, 3890, 2000.

72. Yu, Z. and Huang, W., Competitive sorption between 17α-ethinyl estradiol and naphthalene/phenanthrene by sediments, *Environ. Sci. Technol.*, 39, 4878, 2005.

73. Shareef, A., Angove, M.J., Wells, J.D., and Johnson, B.B., Sorption of bisphenol A, 17α-ethynylestradiol and estrone to mineral surfaces, *J. Colloid Interface Sci.*, 297, 62, 2006.

74. Yamamoto, H., Liljestrand, H.M., Shimizu, Y., and Morita, M., Effects of physical-chemical characteristics on the sorption of selected endocrine disruptors by dissolved organic matter surrogates, *Environ. Sci. Technol.*, 37, 2646, 2003.

75. Bowman, J.C., Zhou, J.L., and Readman, J.W., Sediment-water interactions of natural oestrogens under estuarine conditions, *Mar. Chem.*, 77, 263, 2002.

76. McCarthy, J.F. and Jimenez, B.D., Interactions between polycyclic aromatic hydrocarbons and dissolved humic material: Binding and dissociation, *Environ. Sci. Technol.*, 19, 1072, 1985.

77. Kim, I., Yu, Z., Xiao, B., and Huang, W., Sorption of male hormones by soils and sediments, *Environ. Toxicol. Chem.*, 26, 264, 2007.

78. Colucci, M.S., Bork, H., and Topp, E., Persistence of estrogenic hormones in agricultural soils: 1. 17β-estradiol and estrone, *J. Environ. Qual.*, 30, 2070, 2001.

79. Hakk, H., Millner, P., and Larsen, G., Decrease in water-soluble 17β-estradiol and testosterone in composted poultry manure with time, *J. Environ. Qual.*, 34, 943, 2005.

80. Colucci, M.S. and Topp, E., Persistence of estrogenic hormones in agricultural soils: II. 17α-ethynylestradiol, *J. Environ. Qual.*, 30, 2077, 2001.

81. Schiffer, B. et al., Mobility of the growth promoters trenbolone and melengestrol acetate in agricultural soil: column studies, *Sci. Total Environ.*, 326, 225, 2004.
82. Lucas, S.D. and Jones, D.L., Biodegradation of estrone and 17β-estradiol in grassland soils amended with animal wastes, *Soil Biol. Bioch.*, 38, 2803, 2006.
83. Jacobsen, A.-M., Lorenzen, A., Chapman, R., and Topp, E., Persistence of testosterone and 17β-estradiol in soils receiving swine manure or municipal biosolids, *J. Environ. Qual.*, 34, 861, 2005.
84. Sangsupan, H.A. et al., Sorption and transport of 17β-estradiol and testosterone in undisturbed soil columns, *J. Environ. Qual.*, 35, 2261, 2006.
85. McFarland, J.W. et al., Quantitative structure-activity relationships among macroli antibacterial agents: *In vitro* and *in vivo* potency against *Pasteurella multocida*, *J. Med. Chem.*, 40, 1340, 1997.
86. Lin, C.E., Chang, C.C., and Lin, W.C., Migration behavior and separation of sulfonamides in capillary zone electrophoresis, III. Citrate buffer as a background electrolyte, *J. Chromatogr. A*, 768, 105, 1997.
87. Lin, C.E., Chang, C.C., and Lin, W. C., Migration behavior and separation of sulfonamides in capillary zone electrophoresis, 2. Positively charged species at low pH, *J. Chromatogr. A*, 759, 203, 1997.

7 Mobility of Tylosin and Enteric Bacteria in Soil Columns

Keri L. Henderson,
Thomas B. Moorman, and Joel R. Coats

Contents

7.1 INTRODUCTION

The production of swine, cattle, and poultry raised for human consumption represents a significant portion of the U.S. agricultural economy. To maximize production, producers regularly use antibiotics as supplements in animal feed and water to increase weight gain and prevent diseases among their livestock. In swine, for example, it is estimated that antibiotics are used for disease prevention and growth promotion in more than 90% of starter feeds, 75% of grower feeds, 50% of finishing feeds, and 20% of sow feeds, and equally relevant numbers are seen in beef cattle production (Hayes et al., 1999; USDA APHIS Swine 2000 and COFE). It has been well documented that measurable quantities of these antibiotics are excreted, often in original form, in feces and urine of livestock (FAO/WHO, 1991). Livestock waste, containing antibiotics, is often used as fertilizer for farm fields or pastures and may result in nonpoint source pollution of ground or surface waters (Loke et al., 2000). Although antibiotic residues have been studied extensively in tissues and excrement, we are only beginning to understand the environmental fate of antibiotics and their metabolites once the excreta reaches soil and water environments.

Recently, antibiotics, including the veterinary antibiotic tylosin, which is described in this study, were found in 48% of 139 stream waters tested in 30 states, according to the U.S. Geological Survey (Kolpin et al., 2002). Antibiotics entering the environment could potentially alter bacterial populations and their activity in sediment and water, thus affecting biodegradation, nutrient cycling, and water quality. In addition, there is concern that antibiotics in the environment may induce antibiotic resistance, resulting in adverse human health effects. Certainly, there is significant evidence for development of antibiotic resistance within animals and in the excretion of antibiotic-resistant bacteria in manure (Beaucage et al., 1979; Aarestrup et al., 1997; Kelley et al., 1998). Much less is known about the ability of low concentrations of antibiotics to induce resistance in the environmental microbial population or to provide selective pressure for maintenance of antibiotic resistance genes among microorganisms, although the transfer of antibiotic resistance from agricultural settings to humans has been reported (Oppegaard et al., 2001).

7.1.1 TYLOSIN

Tylosin is a macrolide antibiotic with activity against gram-positive and certain gram-negative bacteria, including *Staphylococcus, Listeria, Legionella,* and *Enterococcus.* It has little activity against gram-negative enteric bacteria such as *E. coli.* Tylosin is used exclusively in veterinary applications and is closely related to erythromycin, which has an important role in public health. Tylosin consists of four major factors: tylosin A, B, C, and D (Figure 7.1); each of the factors is biologically active, with tylosin A being most active and most prevalent in medicinal and feed formulations (Teeter and Meyerhoff, 2003). Tylosin acts in bacteria by binding to the 50S ribosome subunit, which leads to inhibition of protein synthesis. Sensitive bacteria are inhibited by as little as 500 µg/L. Tylosin is used as a growth promoter applied in swine feed and as a therapeutic product in swine and cattle. Tylosin is a common antibiotic used internationally in swine, cattle, and poultry production as both a therapeutic and a prophylactic (Massé et al., 2000; Rabølle and Spliid, 2000). In swine production, tylosin is among the three antibiotics that accounted for the majority (78.8%) of disease prevention. Tylosin was the most used antibiotic at 31.3% of

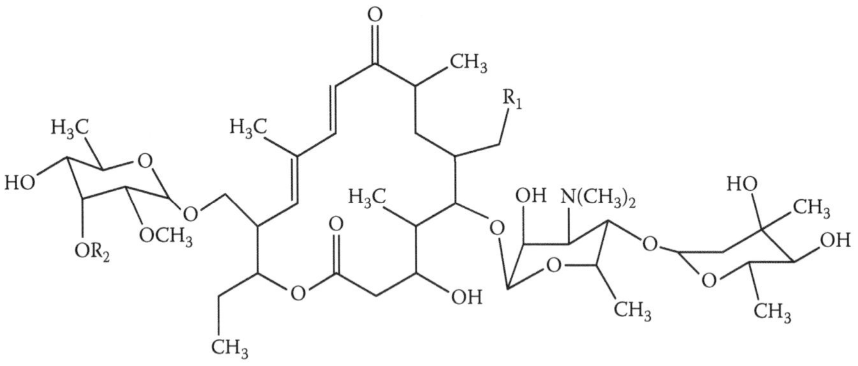

FIGURE 7.1 Chemical structure of tylosin including factors: A (R_1=CHO, R_2=CH$_3$); B (TYL A minus mycinose); C (R_1=CHO, R_2=H); and D (R_1=CH$_2$OH, R_2=CH$_3$).

swine production facilities surveyed (Bush and Biehl, 2001). It has been shown that tylosin is transformed in the animal from tylosin A to tylosin D, which is a change from an aldehyde to an alcohol on the macrolide ring. However, tylosin D may be converted back to its original form in excreta (FAO/WHO, 1991). Concentrations of tylosin in swine feed range from 10 to 100 g tylosin/ton feed for growth promotion purposes (Elanco Animal Health Tylan® Premix product label).

Tylosin was listed in the top ten most frequently detected antibiotics in surface water from 1999 to 2000 (Kolpin et al., 2002). Boxall et al. (2003) identified tylosin as a key pharmaceutical of interest in the environment. Several studies have shown that this antibiotic may have an affinity for clay particles and organic matter in soil, as well as the organic components of manure, which could affect its ability to degrade (Rabølle and Spliid, 2000; Kolz et al., 2005). Sorption to soil and manure components may affect its bioavailability. Huang et al. (2001) described tylosin as one of the most likely water contaminants from agricultural runoff. Due to its sorption characteristics, it is believed that tylosin would be transported with sediment during a runoff event (Davis et al., 2006). Very few studies have evaluated mobility and degradation in the environment in the presence of a manure substrate (Rabølle and Spliid, 2000; Kay et al., 2004; 2005); however, these studies assessed only total tylosin residues and did not quantify tylosin metabolites. Sorption of chemicals onto solid phases, such as soil, sediment, or manure, is extremely important because it could affect the fate and impact of these substances in that environment. An understanding of the degradation and fate of veterinary antibiotics in soil is important because of widespread use of the compounds in livestock production in the United States, and the concurrent application of manure to land. Agricultural lands typically contain subsurface tile drain networks, which may drain directly into streams and other surface water bodies. Based on the lack of data regarding the leaching ability of tylosin factors in soil, one objective of the present study was to address these data gaps by quantifying tylosin residues (specifically tylosin A and tylosin D) in leachate from tylosin applied to soil columns in a manure slurry.

7.1.2 ENTERIC BACTERIA

Two genera of enteric bacteria were selected for use in the present study. *Escherichia coli* are gram-negative, rod-shaped members of γ-Proteobacteria. *Enterococcus sp.* are gram-positive cocci. Both *E. coli* and enterococcus inhabit the gastrointestinal tract of many mammals, including livestock, and are excreted from the animals and found in manure (Schleifer and Kilpper-Balz, 1987; Schaechter, 2000). Both organisms are potentially pathogenic and can develop resistance to antibiotics and have the potential to transfer resistance genes to other bacteria (Wegener et al., 1999; Ochman et al., 2000). These bacteria are also used as fecal indicator species for water quality assessments (Molina, 2005). Because of these characteristics, it is important to understand the survival and mobility of these microorganisms in the environment. Several researchers have reported *E. coli* surviving up to 8 weeks, and enterococcus survival ranging from 35 to >200 d in soils, depending on soil texture, amount and type of manure applied, temperature, and competition with indigenous soil microorganisms (Cools et al., 2001; Lau and Ingham, 2001; Andrews et al., 2004; Entry et al., 2005; Johannessen et al., 2005). A study examining the mobility of enteric bacteria in soil indicated 2 to 6% of the inoculated enterococcus leached through soil columns; however, the bacteria were applied directly to the top of the soil rather than in a manure slurry (Celico et al., 2004). Soupir et al. (2006) reported enterococcus as being highly mobile in runoff from a simulated heavy rainfall event; different types of manure were tested and counts ranged from 6000 to 187,000 cfu/100 mL.

As very little information is available on the fate of bacteria excreted in manure once the manure is applied to soil, particularly in the presence of drug residues, another objective of the present study was to determine the survival, movement, and antibiotic resistance of enteric bacteria in undisturbed soil columns.

7.2 MATERIALS AND METHODS

7.2.1 PRELIMINARY TRIAL

Twenty intact soil cores of Tama series soil were collected from an agricultural field near Grinnell, Iowa. The field had not received manure application for over 20 years, thereby reducing the likelihood of background contamination of antibiotics in the present study. The soil was a loam, containing 46% sand, 36% silt, and 18% clay. Soil cores (10-cm diameter × 30-cm depth) were collected using a Giddings soil core apparatus (Giddings Machine Co., Windsor, Colorado). Soil columns were immediately taken to the lab and were saturated from the bottom with 5mM $CaSO_4$ for 48 h. Soil columns were placed in shelving units and rested on funnels plugged with glass wool and filled with washed sea sand (Fisher Scientific, Pittsburgh, Pennsylvania), so that the bottom of the soil column rested firmly on the sand. Soil columns were allowed to drain, and those with drainage times of 24 to 48 h were chosen for the experiment and were randomly divided among treatment and control groups.

Fresh hog manure was collected from hogs on an antibiotic-free diet (Iowa State University Swine Nutrition Farm, Ames, Iowa). Twelve-gram aliquots of manure were prepared and 10 mL ultrapure water was added to each aliquot to make a field-representative manure slurry. The manure slurry was spiked with 60 μg tylosin tartrate in a methanol carrier to reach a tylosin concentration of 5 ppm in manure. Next, 2×10^8 gfp-labeled ampicillin-resistant *E. coli* 0157:H7 B6914 were added to the slurry. These organisms were selected for the preliminary trial because of their ease of detection, relevance, and availability. The slurry was then poured onto the surface of the columns. The five untreated (control) columns received 20 mL ultrapure water, equivalent to the moisture addition of the treated columns. After application the top 2 cm of the columns were raked with a sterilized spatula to simulate the incorporation of manure into soil that would occur during manure application in the field. These methods are similar to those described by Saini et al. (2003).

Sand was wetted with ultrapure water prior to leaching events. Forty-eight hours after application, a 5-cm "rainfall" in the form of 410 mL 5 mM $CaSO_4$ was applied to the column drop-wise over 2.5 to 3.5 h. Leachate was collected from the bottom of the columns for 48 h, then immediately analyzed for tylosin and *E. coli* 0157:H7 B6914.

The concentration of tylosin in leachate was determined using enzyme-linked immunosorbent assay kits (ImmunoDiagnostic Reagents, San Diego, California) in which the concentration was correlated to absorbance at 405 nm using a THER-

MOmax microplate reader with SOFTmax Pro V3.0 software (Molecular Devices, Sunnyvale, California).

The presence of *gfp*-labeled *E. coli* O157:H7 in leachate was measured using a most-probable number (MPN) technique (IDEXX, Westbrook, Maine). Total coliform bacteria were also enumerated using MPN.

The remaining leachate was concentrated using solid-phase extraction cartridges (Waters Oasis® HLB, Milford, Massachusetts). Cartridges were conditioned with 5 mL acetonitrile followed by 5 mL of 10% acetonitrile. Following sample retention, cartridges were rinsed three times with 10 mL distilled water and 5 mL 10% acetonitrile. Tylosin was eluted from the cartridges with 2 mL of 98:2 acetonitrile:glacial acetic acid. Extracts were analyzed for total tylosin, tylosin A, and tylosin D using high performance liquid chromatography with tandem mass spectrometry (LC/MS/MS) with a gradient of ammonium acetate pH 4.0 : acetonitrile over 35 min at 40°C on a Zorbax SD-C18 4.6 × 250 mm column (Agilent Technologies, Santa Clara, California). Mass spectrometry was used for analysis of tylosin factors based on mass, as reference standards for each factor were not available. Mass spectrometry methods were similar to those described by Kolz et al. (2005). The limit of detection was 0.5 µg/L.

7.2.2 MAIN STUDY

Fifty-six soil columns were collected at the same site previously described. These columns were divided among seven treatment groups: control, manure only, manure plus *Enterococcus*, manure plus tylosin, tylosin plus *Enterococcus*, and manure plus tylosin only, and *Enterococcus*. *Enterococcus* was chosen for the main study because of its greater susceptibility to tylosin compared to *E. coli*, as one of the objectives of the present study was to evaluate development of resistance by manure-associated bacteria in the soil columns. Similar methods were employed as those previously described, except the saturation and preleaching components were not performed, and all soil columns were utilized, resulting in eight replicates per treatment group. Additionally, we increased the concentration of tylosin and the amount of manure applied to better represent real-world applications. Thirty grams of fresh hog manure was spiked with tylosin in an acetone carrier to reach a concentration of 50 µg/g tylosin in the manure.

Both enterococcus (which was inoculated) and *E. coli* (from manure) were measured in leachate water after each rain event, and results are expressed as cells per 100 mL of leachate water. A few samples were not sufficiently diluted and thus saturated the MPN panel, resulting in an MPN value that underestimates the true concentration. These values were used in the data analysis. Control columns (no manure) and columns treated with manure and *Enterococcus* leached no *E. coli*.

7.3 RESULTS AND DISCUSSION

7.3.1 PRELIMINARY TRIAL

Following a single rain event, analysis of leachate using LC/MS/MS revealed total tylosin residues up to 2.8 ng/mL, with a mean concentration of 0.8 ng/mL (se =

0.3). We found similar results when using immunoassay; 0.6 ng/mL was detected in leachate. When examining specific tylosin factors using LC/MS/MS, we found that tylosin A accounted for approximately 22% of the total tylosin residues; this corresponds to a concentration of 0.2 ng/mL. Tylosin D, another major factor, was detected at 0.5 ng/mL, or 65% of the total residues. This result is quite interesting considering the composition of the tylosin applied to the top of the column; tylosin D only accounted for approximately 10% in the formulation applied. Finding such different proportions in the leachate implies a differential metabolism or a differential mobility between tylosin A and D. It is possible that tylosin D is more stable or more mobile than tylosin A. Further studies are needed to elucidate this phenomenon.

Additionally, tylosin D has only 35% of the antibacterial activity as tylosin A, so the differences could have implications for low-level effects on soil microbial communities (Teeter and Meyerhoff, 2003). Each of the other tylosin factors (B and C), and even metabolites, can possess antimicrobial activity, which contributes to a complex situation in soil and water with respect to biological activity. Likewise, some analytical methods, e.g., ELISA residue quantification, also are differentially less sensitive to some factors or metabolites and more sensitive to others. The major advantage to the high performance liquid chromatography (HPLC) or LC/MS method is the specificity, but the ELISA method is faster, cheaper, and can detect very low concentrations in aqueous samples (Hu et al., 2006).

Following MPN testing of leachate, total coliform bacteria were highly variable in manure-treated columns, with a range of 1 to 644 CFU/mL in three of five treatments. One control column had 1 cell/mL in the leachate, indicating external sources of bacteria, which could include wildlife from the area in which the columns were collected. *E. coli* O157:H7 B6914 were detected in the leachate of two of the five treated columns; they also ranged from 1 to 644 cells/mL (the limit of detection for the assay).

Although <0.1% of the gfp-labeled cells and tylosin residues applied to the top of the column were detected in the leachate in this study, these results do indicate the ability of tylosin and some bacteria to move in an agronomic soil.

7.3.2 MAIN STUDY

After four rain events, less than one-third of the treated columns leached detectable amounts of tylosin, with the average concentration at <1 ng/mL in leachate. These results were similar to those found in the preliminary trial.

There were no apparent differences in the *E. coli* leaching from any of the manure treatments; therefore, the *E. coli* were averaged over these treatments for each leaching period (Figure 7.2). *E. coli* (from manure) were detected in all leachates, but the maximum mean concentrations were in the second and third leachates. The decline in *E. coli* concentration seen in the fourth leachate is likely due to decreasing survival in the soil and washout from the column.

Enterococcus also leached from the soil columns but in numbers far exceeding those observed for *E. coli* (Figure 7.3). In addition, the number of organisms was dependent upon treatment. Soil treated with manure leached no *Enterococcus* in the first rain event but averaged 5199 cells/100 mL in the second leaching, then declined

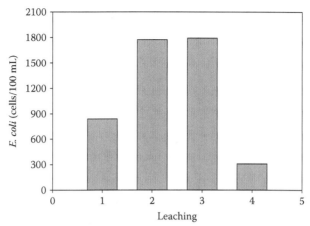

FIGURE 7.2   Average *E. coli* in leachate water from intact soil columns treated with swine manure.

to less than 400 cells/100 mL in the third and fourth rain events. The manure plus tylosin (MT) treatment was similar in magnitude and pattern to the manure-only treatment in the leaching of *Enterococcus*, suggesting that the tylosin was not active toward the *Enterococcus* in this soil/manure environment. *Enterococcus* added to manure (MB) also resulted in bacteria being leached, as did the MTB treatment. Tylosin plus *Enterococcus* without manure (TB) treatment resulted in much fewer *Enterococcus* being leached, reaching a maximum average of 30 cells/100 mL at the third leaching. It is possible that the tylosin was more available to inhibit the microbes in this treatment compared with the similar manure-containing treatment (MTB). *Enterococcus* was detected in leachate from 4 of 23 untreated soil columns (controls) and in 2 of 12 leachates from columns treated only with tylosin, indicating minimal input of enterococcus from the soil.

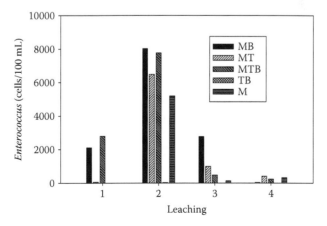

FIGURE 7.3   Average *Enterococcus* in leachate water from soil columns treated with swine manure alone (M), manure plus *Enterococcus* (MB), manure plus tylosin (MT), manure plus tylosin and *Enterococcus* (MTB), or tylosin and *Enterococcus* without manure (TB).

Putative tylosin-resistant *Enterococcus* were not recovered in the controls or the tylosin treatment (no manure or *Enterococcus* added), which was expected. Only trace levels (2 cells/100 mL) of tylosin-resistant *Enterococcus* were recovered in the second leaching from the manure-treated columns. Resistant *Enterococcus* were recovered in both the first and second leachates from the MB treatment at levels of 1955 and 779 cells/100 mL. These detections were in the absence of exposure to tylosin or other antibiotics and may be explained by a natural level of resistance in the population. Leaching of tylosin-resistant *Enterococcus* in the MT and MTB treatments was also observed (3228 cells/100 mL in MT second leaching and 137 cells/100 mL in MTB second leaching), but no tylosin-resistant bacteria were found in leachate from the TB columns. Thus, it can be concluded that tylosin-resistant bacteria were only leached from manure-treated columns, but that there was no obvious effect of tylosin. It is possible those resistant organisms were present in the manure or that some organisms in the manure developed resistance to tylosin during the study. The results of the tylosin plus *Enterococcus* treatment (TB) could be due to the poor survival of *Enterococcus* in the absence of manure, the effect of tylosin, or both these factors.

The pattern of leaching was the same for both bacteria (*E. coli* and *Enterococcus*) over time, with peak concentrations coming from the second and third leachings. Greater concentrations of *Enterococcus* were seen in column leachates compared to *E. coli*, particularly at the second leaching. There was no obvious effect of tylosin on the prevalence of tylosin-resistant *Enterococcus*, but manure treatment resulted in elevated levels of resistant *Enterococcus*. This could be due to the presence of indigenous tylosin-resistant *Enterococcus* already present in the swine manure. The movement of both the indicator bacteria and antibiotics is likely due to macropores in this well-structured soil. The transport of these agents illustrates their mobility.

Examination of current literature reveals a small number of comparable studies. Rabølle and Spliid (2000) performed a leaching study with tylosin in packed soil cores of two soil types: a sandy loam and a sandy type. The K_d values described for those soils were 128 and 10.8, respectively, and desorption was reported at 13 and 26%. After one "rain" event, approximately 70% of the tylosin was recovered in the top 20 cm of the 48-cm soil columns, and no tylosin was detected in the leachate from either soil type; however, it should be noted that the limit of detection reported in the study was 7 µg/L, compared to 0.5 µg/L reported in this study. Using the sorption coefficient data from this study, Tolls (2001) stated tylosin would be low to slightly mobile in most soils, if comparing to pesticide sorption and mobility data; these results are similar to our findings in a loam soil.

Freundlich partition coefficients for tylosin in silty clay loam, sand, and manure ranged from 1000 to 2000 (Clay et al., 2005); desorption was found to be <0.2% in the same soils. The concentrations used in these batch sorption studies were 23 to 200 mg/kg, similar to the concentrations of tylosin in the manure applied to our soil columns. These sorption and desorption values may also be useful in a comparison to the conditions in our soil.

A field-scale study performed on tylosin mobility in a clay loam soil determined that up to 6% of applied tylosin was present in runoff water from an agricultural soil (Oswald et al., 2004). The same study also examined the effect of manure on tylosin mobility and found increased runoff potential of tylosin (up to 23%) in manured treatments; this was likely due to the greatly decreased infiltration of the applied rainfall. Infiltration was reduced by approximately 85% in manured treatments. This information may be important in identifying factors that affected leaching of tylosin in our study.

Finally, Saini et al. (2003) found increased survival of an *E. coli* strain when manure in which they were residing was incorporated into the soil. Additionally, they reported that most bacteria leached from soil columns after the first rain event and that increased time between application of manure and the rainfall event resulted in decreased leaching of the bacteria, which could be a result of decreased survival. These results are different from our finding of higher numbers of bacteria leaching from the second and third rain events. Saini et al. (2003) reported 3.4 to 4.5 log CFU/100 mL in the leachate from the rain event 16-d postapplication; 1 to 10% of the applied inoculum was detected in the leachate, regardless of time between application and first rain event (Saini et al., 2003). Additionally, Recorbet et al. (1995) found that survival of bacteria in soil may be attributed to colonization of clay fractions in soil, which could provide protection from stressors, including environmental contaminants, and that preferential flow in soil columns may be extremely important, which is in agreement with our results.

7.4   CONCLUSION

The goal of the present study was to evaluate the mobility and degradation of tylosin and the mobility of enteric bacteria in undisturbed agronomic soil columns. Results from the present study indicate a low amount of mobility of tylosin in a loam soil, with an average of 0.8 ng/mL total tylosin detected in the leachate from multiple simulated rainfall events. Tylosin D was the predominant factor present in the leachate. Microbiological analysis of the leachate revealed that enteric bacteria were frequently present in the leachate at numbers exceeding the suggested water quality criteria of 126 cells/100 mL for *E. coli* and 33 cells/100 mL for enterococcus (U.S. EPA, 2003). It is likely that preferential flow played a role in the transport of tylosin and bacteria through the soil profile in the intact soil columns.

Numerous monitoring studies have now detected very low residues of antibiotics in surface water. The current study has endeavored to take some first steps toward understanding how the compounds move to surface water, how long they persist in soil, and what transformation processes and products are evident in environmental matrices. Many questions remain unanswered, the most intriguing of which is "What is the significance of the low concentrations of antibiotic residues in the environment?" Much work is still needed to answer questions regarding the significance of pharmaceuticals in the environment.

ACKNOWLEDGMENTS

The authors would like to thank Beth Douglass, USDA-ARS, for her technical assistance throughout the project and Dingfei Hu for his assistance with HPLC analysis. Funding for this project was provided in part by the Center for Health Effects of Environmental Contamination and a USDA-CSREES NRI grant. This work is part of the Iowa Agricultural and Home Economics Experiment Station Projects No. 5075 and 5091.

REFERENCES

Aarestrup, F.M., Nielsen, E.M., Madsen, M., and Engberg, J. 1997. Antimicrobial susceptibility patterns of thermophilic *Campylobacter* spp. from humans, pigs, cattle and broilers in Denmark. *Antimicrob. Agents Chemother.* 41:2244–2250.

Andrews, R.E., Johnson, W.S., Guard, A.R., and Marvin, J.D. 2004. Survival of enterococci and Tn*916*-like conjugative transposons in soil. *Can. J. Microbiol.* 50:957–966.

Beaucage, C.M., Fox, J.G., and Whitney, K.M. 1979. Effect of long-term tetracycline exposure (drinking water additive) on antibiotic-resistance of aerobic gram-negative intestinal flora of rats. *Am. J. Vet. Res.* 40:1454–1457.

Boxall, A.B.A., Fogg, L.A., Kay, P., Blackwell, P.A., Pemberton, E.J., and Croxford, A. 2003. Prioritisation of veterinary medicines in the UK environment. *Toxicol. Lett.* 142:207–218.

Bush, E.J., and Biehl, L.G. 2001. Use of antibiotics and feed additives by U.S. pork producers. 2001 *Proceedings of U.S. Animal Health Association.* Nov. 7, 2001. Hershey, PA.

Celico, F., Varcamonti, M., Guida, M., and Naclerio, G. 2004. Influence of precipitation and soil on transport of fecal enterococci in fracture limestone aquifers. *Appl. Environ. Microbiol.* 70:2843–2847.

Clay, S.A., Liu, Z., Thaler, R., and Kennouche, H. 2005. Tylosin sorption to silty clay loam soils, swine manure, and sand. *J. Environ. Sci. Health B.* 40:841–850.

Cools, D., Merckx, R., Vlassak, K., and Verhaegen, J. 2001. Survival of *E. coli* and *Enterococcus* spp. derived from pig slurry in soils of different texture. *Appl. Soil Ecol.* 17:53–62.

Davis, J.G., Truman, C.C., Kim, S.C., Ascough II, J.C., and Carlson, K. 2006. Antibiotic transport via runoff and soil loss. *J. Environ. Qual.* 35:2250–2260.

Entry, J.A., Leytem, A.B., and Verwey, S. 2005. Influence of solid dairy manure and compost with and without alum on survival of indicator bacteria in soil and on potato. *Environ. Pollut.* 138:212–218.

FAO/WHO. 1991. Tylosin. *Residues of some veterinary drugs in animals and foods.* 38th Meeting of the Joint FAO/WHO Expert Committee on Food Additives. pp. 109–127.

Hayes, D.J., Jensen, H.H., Backstrom, L., and Fabiosa, J. 1999. Economic Impact of a Ban on the Use of Over-the-Counter Antibiotics. *Center for Agricultural and Rural Development, Iowa State University, Ames, IA. Staff Report 99-SR 90.* http://www.card.iastate.edu/publications/texts/99sr90.pdf.

Hu, D., Henderson, K.L., and Coats, J.R. 2006. Environmental fate of tylosin and analysis of immunological cross-reactivity among tylosin-related compounds. Presented at 231st National Meeting of the American Chemical Society, Atlanta, GA, (accessed on March 26, 2006).

Huang, C., Renew, J., and Smeby, K. 2001. Assessment of potential antibiotic contaminants in water and preliminary occurrence analysis. 22nd Annual Meeting of the Society of Environmental Toxicology and Chemistry. Baltimore, MD.

Johannessen, G.S., Bengtsson, G.B., Heier, B.T., Bredholt, S., Wasteson, Y., and Rorvik, L.M. 2005. Potential uptake of *Escherichia coli* O157:H7 from organic manure into crisphead lettuce. *Appl. Environ. Microbiol.* 71:2221–2225.

Kay, P., Blackwell, P.A., and Boxall, A.B.A. 2004. Fate of veterinary antibiotics in a macroporous tile drained clay soil. *Environ. Toxicol. Chem.* 23:1136–1144.

Kay, P., Blackwell, P.A., and Boxall, A.B.A. 2005. A lysimeter experiment to investigate the leaching of veterinary antibiotics through a clay soil and comparison with field data. *Environ. Pollut.* 134:333–341.

Kelley, T.R., Pancorbo, O.C,. Merka, W.C., and Barnhart, H.M. 1998. Antibiotic resistance of bacterial litter isolates. *Poultry Sci.* 77:243–247.

Kolpin, D.W., Furlong, E.T., Meyer, M.T., Thurman, E.M., Zaugg, S.D., Barber, L.B., and Buxton, H.T. 2002. Pharmaceuticals, hormones and other organic wastewater contaminants in US Streams, 1999–2000: A national reconnaissance. *Environ. Sci. Technol.* 36:1202–1211.

Kolz, A.C., Ong, S.K., and Moorman, T.B. 2005. Sorption of tylosin onto swine manure. *Chemosphere* 60:284–289.

Lau, M.M., and Ingham, S.C. 2001. Survival of faecal indicator bacteria in bovine manure incorporated into soil. *Lett. Appl. Microbiol.* 33:131–136.

Loke, M.L., Ingerslev, F., Halling-Sorensen, B., and Tjornelund, J. 2000. Stability of tylosin A in manure containing test systems determined by high performance liquid chromatography. *Chemosphere* 40:759–765.

Massé, D.I., Lu, D., Massé, L., and Droste, R.L. 2000. Effect of antibiotics on psychrophilic anaerobic digestion of swine manure slurry in sequencing batch reactors. *Bioresource Technol.* 75:205–211.

Molina, M. 2005. Temporal and spatial variability of fecal indicator bacteria: implications for the application of MST methodologies to differentiate sources of fecal contamination. U.S. Environmental Protection Agency National Exposure Research Laboratory Research Abstract. http://epa.gov/nerl/research/2005/g2-2.html. Accessed April 30, 2007.

Ochman, H., Lawrence, J.G., and Groisman, E.A. 2000. Lateral gene transfer and the nature of bacterial innovation. *Nature* 405:299–304.

Oppegaard, H., Steinum, T.M., and Wateson, Y. 2001. Horizontal transfer of a multi-drug resistance plasmid between coliform bacteria of human and bovine origin in a farm environment. *Appl. Environ. Microbiol.* 67:3732–3734.

Oswald, J.K., Trooien, T.P., Clay, S.A.., Liu, Z., and Thaler, R. 2004. Assessing potential transport of tylosin in the landscape. *Proc. South Dakota Acad. Sci.* 83:39–46.

Rabølle, M., and Spliid, N.H. 2000. Sorption and mobility of metronidazole, olaquindox, oxytetracycline and tylosin in soil. *Chemosphere* 40:715–722.

Recorbet, G., Richaume, A., and Jocteur-Monrozier, L. 1995. Distribution of a genetically-engineered Escherichia coli population introduced into soil. *Lett. Appl. Microbiol.* 21:38–40.

Saini, R., Halverson, L.J., and Lorimor, J.C. 2003. Rainfall timing and frequency influence on leaching of *Escherichia coli* RS2G through soil following manure application. *J. Environ. Qual.* 32:1865–1872.

Schaechter, M. 2000. *Escherichia coli*, general biology. In Encyclopedia of microbiology, 2nd ed., vol. 2, Lederberg, J. Ed. 260–269. San Diego: Academic Press.

Schleifer, K.H., and Kilpper-Balz, R. 1987. Molecular and chemotaxonomic approaches to the classification of streptococci, enterococci, and lactococci: a review. *Syst. Appl. Microbiol.* 10:1–19.

Soupir, M.L., Mostaghimi, S., Yagow, E.R., Hagedorn, C., and Vaughan, D.H. 2006. Transport of fecal bacteria from poultry litter and cattle manures applied to pastureland. *Water, Air, Soil Pollut.* 169:125–136.

Teeter, J.S., and Meyerhoff, R.D. 2003. Aerobic degradation of tylosin in cattle, chicken, and swine excreta. *Environ. Res.* 93:45–51.

Tolls, J. 2001. Sorption of veterinary pharmaceuticals in soils: a review. *Environ. Sci. Technol.* 35:3397–3406.

USDA APHIS Swine 2000 survey. http://www.aphis.usda.gov/vs/ceah/cahm/Swine/Swine2000/antibiotics2.PDF. Accessed June 20, 2002.

USDA APHIS Cattle on Feed Evaluation. http://www.aphis.usda.gov/vs/ceah/cei/antiresist. antibiouse.pdf. http://www.aphis.usda.gov/vs/ceah/cahm/Beef_Feedlot/Cofdes1.pdf. Accessed February 2, 2004.

U.S. Environmental Protection Agency. 2003. Bacterial water quality standards for recreational waters (freshwater and marine waters) status report. EPA-823-R-03-008. http://www.epa.gov/waterscience/beaches/local/statrept.pdf. Accessed April 30, 2007.

Wegener, H.C., Aarestrup, F.M., Jensen, L.B., Hammerum, A.M., and Bager, F. 1999. Use of antimicrobial growth promoters and *Enterococcus faecium* resistance to therapeutic antimicrobial drugs in Europe. *Emerg. Infect. Dis.* 5:329–335.

8 Plant Uptake of Pharmaceuticals from Soil
Determined by ELISA

Rudolf J. Schneider

Contents

8.1 INTRODUCTION

Plants take their nutrients from the air and air moisture (fog, dew, and rainfall), and are able to take up dissolved compounds from soil water to live and grow. Many plants live in symbiosis with soil fungi (mycorrhiza), which themselves are able to uptake and deliver specific solutes to their host. Unfortunately, in the process of taking up essential plant nutrients and energy sources, plants also take up heavy metals, pesticides, and other bioavailable pollutants that are undesirable for human consumption. For example, it has been recognized that there are mechanisms that lead to the uptake of insecticides such as DDT in nontargeted plants. Having learned the lessons on the potential of passive uptake of xenobiotics into the crops, there is concern that plants would also take up pharmacologically active compounds from contaminated soil, thus accumulating unwanted pharmaceuticals in edible parts of crops.

Knowledge on nutrient uptake, such as that of nitrate, phosphate, sulfate, ammonium, and potassium, is well established and is the basis of understanding and managing plant production. Similarly, knowledge on the uptake of soil-applied pesticides is also huge because it is essential for designing and assessing the efficacy of crop protection agents. However, studies on plant uptake of pharmaceuticals are limited to date. Therefore, this paper describes results of experiments that study plant uptake of two sulfonamide antimicrobials: sulfamethazine and sulfamethoxazole—the first being a veterinary drug and the latter used in human therapy. Model experiments examining degradation, leaching, and uptake were performed. Concentrations in water, soil, and plants were analyzed using two selective enzyme immunoassays (ELISA).

8.2 BACKGROUND

8.2.1 CODEPOSITION OF VETERINARY ANTIBIOTICS WITH MANURING PRACTICES

Millions of chemical compounds are present in the atmosphere, mostly of natural origin, but thousands of them are of xenobiotic nature. By wet and dry deposition, often adsorbed on aerosol particles and traveling over large distances, they are deposited onto the soil surface. For most plants the top soil layer is the most important interface with their environment, providing support and protection of the roots and, at the same time, water, nutrients, and some other low-molecular weight compounds.

There are other intentional depositions of chemical substances that are traditionally carried out, especially the application of fertilizer or pesticides on agricultural fields. The applications of plant residues such as straw and mulch and of nutrient-rich residues from livestock breeding as dung and liquid manure are in many regions of the world widely used to fertilize the soil in order to fill up the soil's nutrient reservoirs. In the 1980s researchers became aware that this practice could lead to the input of pharmaceutical compounds into soils used for plant production.

After oral or parenteral administration of pharmaceuticals, such as veterinary antibiotics or growth promoters, these compounds may undergo natural metabolism, in which they are degraded to a major or minor degree, depending on their chemical properties. Pharmaceuticals can be ranked among the more persistent substances,

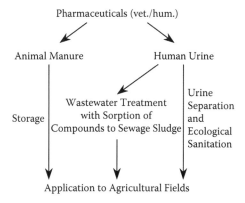

FIGURE 8.1 Pathways of pharmaceuticals to agricultural soils.

since they have been developed to remain active in the body for some time without deactivation or degradation.[1] Many antimicrobial substances are excreted unchanged, with a share from 50 to 70% for penicillin G and 70 to 90% for tetracyclines.[2] Therapeutic doses of tetracyclines (ca. 40 mg kg^{-1} d^{-1}) can result in manure concentrations of 200 mg kg^{-1} during treatment.[3]

The stability of some groups of pharmaceuticals in manure, especially antimicrobials, has been the topic of many studies.[4–7] It was found that during storage of the manure, some residues remain stable for a long period of time. Meyer et al.[8] found tetracyclines in concentrations between 5 and 870 μg L^{-1} in all of 13 studied swine manure storage tanks. Other studies found concentrations of tetracycline up to 66 g m^{-3} in swine manure with a degradation of less than 5% after 7 weeks.[5] Sulfonamides are stable in manure until application, too,[9] especially with nonoptimal conditions (e.g., cold weather, anaerobic conditions),[10,11] but higher temperatures and adaptation of the microorganisms enhance degradation.[12] Metabolites (e.g., the glucuronide of chloramphenicol or N-4-acetylsulfamethazine) can be cleaved in manure and thus transformed back to the active ingredients, such that their concentrations may increase with time.[13] As manure is usually collected and stored for several weeks, only compounds that resist degradation for this period will enter the soil compartment when manure is applied to soil (Figure 8.1).

8.2.2 PHARMACEUTICALS FOR HUMAN USE IN SOIL

Wastewater of human origin may also contain residues of pharmaceuticals, especially when they come from hospitals, which obviously have a larger consumption in pharmaceuticals such as antibiotics. Some substances are excreted in relatively high amounts, while on the other hand there are compounds like sulfamethoxazole that are metabolized up to 85%; hence, only 15% of the original substance that can be found in urine or feces[2] will enter wastewater. The fate of pharmaceuticals during wastewater treatment is discussed in Part III of this book.

Wastewater in urbanized regions is usually transported via subterranean sewer systems to wastewater treatment plants. It should be noted that less polar pharmaceuticals may adsorb to sewage sludge during wastewater treatment. In countries where agricultural use of sewage sludge is still an option, this may lead to an input of these compounds into the soil environment (see Figure 8.1). Some countries, some of which have scattered settlement in rural areas or in alpine regions, started opting for complementing "decentralized wastewater treatment" pursuing "ecological sanitation" (see also: http://www.ecosan.org/; http://www.ecosanres.org/). Ecological sanitation is based on source separation of feces from urine (e.g., in "no-mix" toilets and waterless urinals) so that after some basic treatment to eliminate fecal pathogens (often simple storage is sufficient) the solid components can be composted together with organic garbage and the urine used separately. These low-cost and low-maintenance systems are especially interesting for developing countries where more than 80% of sewage is discharged without any treatment, wasting the nutrients nitrogen (nitrate, ammonia), phosphate, and potassium, which can be used in sustainable agriculture like organic farming, closing nutrient cycles while protecting surface water and groundwater from eutrophication.

It is clear that human urine may contain residues of pharmaceuticals and that a sanitation concept including urine recycling will lead to an input of these compounds into the top soil layer of agricultural fields or will limit the usability of the urine. The experiments described in this contribution were carried out with urine collected in a urine tank from a small closed-loop sanitation project (Water Mill Museum "Lambertsmühle," Burscheid, Germany) where acceptance, efficiency, storage, and fertilization studies have been undertaken.[14]

8.2.3 FATE OF PHARMACEUTICALS IN SOIL

Reviews on the fate of veterinary antibiotics,[15] fluoroquinolones,[16] and sulfonamides[17] appeared recently. It is also discussed in Chapter 5 of this book. Compounds that have entered the pedosphere disappear with time. Dissipation can be subdivided in the concurrent processes sorption/binding, degradation, leaching, and plant uptake. The share of these processes depends predominantly on the properties of the compounds and the soils and the relative time scale of each process.

Physicochemical parameters that are important in predicting the fate of contaminants in soil have been collated for antibiotics in the review articles by Tolls[18] and Thiele-Bruhn.[19] Chapter 6 in this book also provides an overview of sorption and degradation of selected pharmaceuticals in soil and manure. In all these reviews it is apparent that the sorption of pharmaceuticals in soil can vary widely. For instance, partition coefficients between soil and water (K_d) for sulfonamides in diverse soils range from 0.9 to 10, suggesting low sorption, while tetracyclines show values between 417 and 1026, indicating that tetracyclines are strongly adsorbed. Streptomycin adsorbs preferentially to the clay fraction in soil.[11] This also holds true for the fluoroquinolones.[20–22] Adsorption of sulfonamide antimicrobials is highly dependent on soil pH and ionic strength.[23] Sorption of tetracyclines to clays showed differences for K- and Ca-saturation due to cation bridging, with humic substances being an additional factor.[24] Kinetics also have to be taken into account.[25] With some compounds (e.g., virginiamycin) soil metabolites form that have to be accounted for.[26]

Nondegraded residues in the top soil layers can be dislocated by surface run-off or interflow and preferential flow.[27] Usually the governing process with polar compounds is leaching. In sandy soils tetracyclines can leach to lower levels, but to a higher extent this happens with sulfonamides, eventually even into groundwater.[28,29] Sulfamethoxazole is relatively mobile in soil and has been detected in groundwater (up to 0.47 μg L[−1]).[8] For another sulfonamide, sulfachlorpyridazine, sorption coefficients in soil have been determined that imply preferential leaching with the drainage water.[30] The sulfonamide sulfapyridine has been found to adsorb stronger in moist soils than in dry ones.[31] In a lysimeter study it has been shown that a breakthrough of sulfadiazine through soil columns was only slightly retarded against a water tracer.[32]

8.2.4 FATE OF PHARMACEUTICALS IN PLANTS

8.2.4.1 Uptake

There are a number of studies indicating that pharmaceutical residues that are reversibly adsorbed to soil may be taken up by plants.[33] In early experiments using hydroponic culture, as well as in pot experiments with chlortetracycline and oxytetracycline, the effect of stimulated nitrogen uptake shown by wheat and corn plants grown on a sandy loam demonstrated the uptake and activity of the residues within the plants.[34,35] In a greenhouse experiment, corn took up lasalocid and monensin.[36] In laboratory experiments it has been demonstrated that uptake of sulfadimethoxin in sorghum, pea, and corn influence their development and is dependent on bioaccumulation of these species; bioaccumulation is more pronounced for C-4 plants (sorghum and corn) than for C-3 plants (pea, ryegrass).[9] From hydroponic culture (300 mg L[−1] Sulfadimethoxin) plants incorporated the sulfonamide until final concentrations of 180 to 2000 mg kg[−1]. Roots of corn and sorghum accumulated much more active ingredient than the shoots. Similar results were obtained from rye, carrot, corn, sorghum, and pea in field trials.[37] Enrofloxacin was also accumulated in μg g[−1] amounts.[38]

Plant uptake is governed by many factors, and it has been studied with many compounds other than pharmaceuticals because of its potential for phytoremediation.[39] Even genotype can be important to assess the uptake of contaminants.[40] Roots such as carrots and potato tubers can take up very high amounts of pollutants.[41] Mycorrhization is also an important factor as has been shown for the uptake of atrazine.[42] Even desorption-resistant organic compounds can be taken up into the roots, to some extent, from sediments, which means that compound properties sometimes do not modulate uptake and that root sorption may be the dominant mechanism vs. translocation in the plant for nonpolar compounds.[43] The class of xenobiotics whose uptake has been intensively studied is pesticides, especially herbicides that may affect nontarget plants.[44,45]

Uptake of pharmaceuticals may also influence plant development.[46] It is in part not clear if the negative effects on plants originate from a direct damage on the plant by the ingredients (phytotoxicity) or if the antimicrobial action on soil microorganisms is responsible for the damage by affecting the plant-microorganism symbiosis.[47–49] Antibiotics in the soil may influence plant development indirectly by disrupting soil communities: the decay in the number of soil bacteria leads to a lack of feed for soil fauna (protozoa, nematodes, microarthropods) and finally influences soil functions: plant residues are decomposed slower, denitrification is slower, and therefore nutrients are recycled more slowly.[50] Risk assessments for the uptake of pharmaceuticals in edible portions of crops suggest that with the allergenic potential and long-term effects of antibiotics, the risk is not negligible.[51,52]

8.2.4.2 Detoxification

Plants are not without defense against xenobiotic compounds, which they (passively) take up. There is a three-step detoxification mechanism that acts on foreign compounds, once termed the "green liver."[53,54] Nucleophilic substances are conjugated to glutathione, a process intensively studied with herbicides.[55,56] Glutathione conjugates are the transport forms of many xenobiotics; compartmentation is potentially the final purpose of this process.[57] Meanwhile, this pathway has also been found for the detoxification of chlortetracycline in corn[58]; Chapter 9 of this book is dedicated to that mechanism.

8.2.5 ANALYSIS OF PHARMACEUTICAL RESIDUES IN SOIL AND PLANTS

8.2.5.1 Extraction

Sulfonamides are highly water soluble (sulfamethazine: 1500 mg L^{-1}, sulfamethoxazole: 610 mg L^{-1}), allowing efficient extraction using methanol or methanol/water mixtures as extraction solvents. On the other hand, it is well known that methanol coextracts a huge variety of organic soil constituents, and this is also true for plants. liquid chromatography with tandem mass spectrometry (LC-MS/MS) may therefore be hampered, especially by giving rise to suppression during ionization, which leads to underestimations of the concentrations in the extracts.

Methanol has a relatively low bowling point of ca. 64°C, but with the method of pressurized fluid extraction,[59] which can be performed with commercially available accelerated solvent extraction (ASE) equipment, organic solvents can be used at temperatures up to 200°C by holding, for example, a pressure of 20 MPa in high pressure cells.[60] With polar herbicides extraction times have been reduced dramatically (18 h Soxhlet versus <22 min ASE) and a reduction of solvent use (300 mL versus <80 mL) was also achieved.[59] Samples that contain considerable amounts of water, like plants, can be handled in a better way by addition of diatomaceous earth to the samples (e.g., Extrelut™, Merck).[60] Aged residues of sulfonamide antibiotics could be exhaustively extracted by a mixture of aqueous buffer and acetonitrile at elevated temperature.[61]

8.2.5.2 Determination of Antimicrobial Residues

Residues of pharmaceuticals in plant and soil samples are usually analyzed by chromatographic methods coupled to mass spectrometric (LC/MS) methods.[62] For food analysis, especially with sulfonamides in milk, immunoassays have been used for almost two decades (e.g., Märtlbauer et al.[63]). In fact, class specific assays have been developed for this application.[64,65] For the analysis of pharmaceuticals in plants or soils by immunoassay, only a small number of literature exists. One example is the determination of tetracycline with ELISA kits by Aga and coworkers[66] and our study on the determination of sulfamethazine in soil extracts[13] using the antibodies prepared by Fránek et al.[67]

8.3 MATERIALS AND METHODS

All chemicals were used at least in analytical grade. Pharmaceutical standards sulfamethazine (sulfadimidin) and sulfamethoxazole for the construction of calibration curves were prepared from reference substances obtained from Riedel-de Haën and were of 99.7% and 99.9% purity, respectively. The expression "ultrapure water" in the text refers to water deionized by reverse osmosis and subsequent passage through a Milli-Q® water purification system. Methanol was high performance liquid chromatography (HPLC) grade.

8.3.1 SULFONAMIDES AND PLANTS STUDIED

Sulfamethazine (SMZ, in Europe frequently named "sulphadimidine") and sulfamethoxazole (SMX) have been chosen from the sulfonamide antimicrobials because much data on their properties and behavior exist. The chemical structures of these compounds are show in Figure 8.2. pK_a values are 2.4 and 7.4 and 1.8 and 6.0 for sulfamethazine and sulfamethoxazole, respectively.[61] Sulfamethazine is a frequently used veterinary antibiotic, and in a previous study it has been detected in soil in concentrations up to 17 µg kg^{-1} soil.[13] As it is frequently detected in swine manure it cannot be excluded that the soils studied had been in contact with the chemical before. Sulfamethoxazole is a standard antibiotic against common cold. It has been found frequently in freshwater monitoring.[68] As a model plant for uptake Italian ryegrass (*Lolium multiflorum italicum*, type: Turilo), a forage grass plant, was chosen.

8.3.2 MODEL SOILS

Soils of different composition were selected for the studies. Two of the soils originated from the long-term field experiments at the University of Bonn[69] and at Bayer

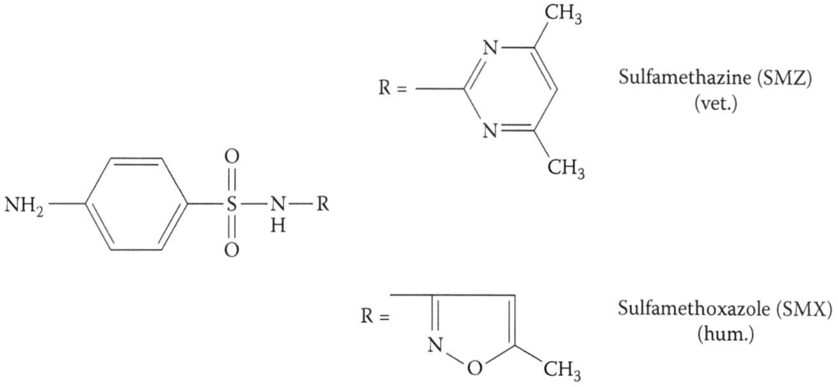

FIGURE 8.2 Chemical structures of the compounds studied.

TABLE 8.1
Model Soil Characteristics

Soil	A	B	C	D
Soil Type	Loamy silt	Sandy silt	Loam	Sand
Texture	lU	sU	L	S
Sampling Depth	0–30 cm	0–15 cm	0–15 cm	0–30 cm
Status	Dry	Dry	Fresh*	Dry
pH	6.3	5.8	6.5	Not determined
Organic Carbon Content	1.2%	2.7%	1.7%	<0.1 %

* For the laboratory experiments, soil C was also air dried and all the soils were sieved through a 2-mm sieve.

Crop Science Corporation facilities at Burscheid. All soils had been taken from the top soil layer at least 1 year before the experiments, air dried and sieved through a sieve with 5-mm pore width, except soil C which was used freshly. Table 8.1 provides soil data.

8.3.3 URINE

Urine was collected from the model project on ecological sanitation described in Section 8.2.2 and was between 1 and 3 months of age; pH was 9. Urine spiking was performed at IWW Water Center, Mülheim/Ruhr, Germany. Eight other pharmaceuticals were present in the spiking solution but were not analyzed in this project. Their effect on the course of the plant experiment and on the analytical accuracy has been determined to be negligible in separate experiments (data not shown).

8.3.4 SOIL EXTRACTION

Soil samples (equivalent to 10.0 g dry mass) were subjected to ASE using an ASE® 200 Extractor (Dionex). The 11-mL extraction cells were used, with the frits of the cell heads protected against clogging by a glass fiber filter and ca. 1 g of diatomaceous earth (Isolute® HM-N, Part. No. 9800-1000, Separtis) on top of the soil sample. The extracts were collected in 40-mL collection vials with Teflon-lined silicon septa. The parameters of extraction were: preheating 10 min, heating of the cell 5 min, stationary extraction 8 min, pressure 140 bar, temperature 80°C, solvent methanol : ultrapure water 9 : 1 v/v, rinsing solvent volume 50% of cell volume, two extraction cycles, purge solvent from cell with nitrogen (5.0) for 60 sec.

After extraction, about 15 to 20 mL of extract were obtained, depending on the water contents of the soil sample and its pore volume. The extracts were evaporated to almost dryness in a water bath (30°C) by nitrogen (5.0) from stainless steel needles. The larger part of methanol evaporated first and an azeotrope of methanol and water formed reducing largely evaporation speed. The residue is almost completely aqueous. It was resuspended in 5.00 mL of ultrapure water and treated in an ultrasonic bath for 30 sec to completely dissolve the active ingredient. A portion of this aqueous extract was filtered through a membrane syringe filter (GHP Acrodisc 13 mm, 0.45 μm, Gelman). An aliquot of 100 μL was diluted with 9.90 mL of pure water by a factor of 100 and this sample was stored in brown 10-mL glass vials with plastic lid and a Teflon-coated rubber liner in the refrigerator at 4°C until measurement.

In order to determine the recoveries of soil extraction, a preliminary miniaturized incubation experiment with 10 g of soil and two low spiking levels (8 and 16 μg kg^{-1} equivalent to ca. 1 kg ha^{-1} application rate) was used. The soil was put in a 10-mL glass vial, 1.6 mL of urine (a typical "application rate") were added, and the open vials incubated for 4 weeks at 20°C and 80% relative air humidity in the dark. The soil was kept at 14% of its maximum water holding capacity. Each variant was run in duplicate.

8.3.5 PLANT EXTRACTION

The plant biomass was deep frozen immediately after harvest. The water contents of the frozen grass were not determined or accounted for. However, in preliminary experiments it was shown that it did not exceeded 5% of the original mass. For extraction, the deep-frozen plants were put in a porcelain mortar, covered completely by liquid nitrogen, and manually ground with a porcelain pestle to obtain a homogeneous sample. This sample could be easily subdivided into smaller portions after evaporation of the nitrogen.

About 5 g of the homogenate were weighed into a short 100-mL glass measuring cylinder. Then 50.0 mL of pure water were added, and the mixture was homogenized by a rotary mixer at high speed (Ultra-Turrax®) for 60 sec. About 5 mL of the supernatant were filtered through a fluted paper filter and subsequently through a membrane filter (pore width 0.45 μm). Finally, 100 μl of the filtrate were diluted with 9.90 mL of pure water (factor 100). These samples were directly measured by ELISA, and the results were calculated back with the respective factors to a mass concentration of mg pharmaceutical per kg of fresh mass.

8.3.6 IMMUNOASSAY

The immunoassays used follow the principle of direct ELISA. The general procedure has been described elsewhere.[70] The polyclonal anti-sulfamethazine ("sulphadimidine") antibodies were from Fránek et al.[67] Antibodies against sulfamethoxazole were produced beforehand at the University of Bonn. Table 8.2 summarizes the performance data of the ELISAs. Selectivity in the context of the model experiments performed means that the cross-reactivity of the antibodies to the other respective analyte had been checked and found to be negligible.

Immunoassays often show matrix effects, especially when soil extracts are analyzed. Interference in ELISA is often attributed to coextracted humic substances, which may lead to overestimation of target analyte concentrations. Because the affinity of humic acids and other interferents can be considered inferior to the analyte's affinity to the antibody, it is in many cases possible to eliminate the influence of matrix by dilution of the samples, provided that the assay has low detection limits.

The high application rates in the model experiments for degradation and leaching allowed the extracts to be diluted by a factor of 100 to 10,000. In these measurements

TABLE 8.2
Immunoassays Performance Data

Antibody	Limit of Detection			Average Variance (n = 3)	Selectivity
	Leachate	Soil	Plant		Cross-Reactivity
	µg L^{-1}	mg kg^{-1}	mg kg^{-1}	%	%
anti-sulfamethazine	5	0.005	0.1	21	SMX <0.1%
anti-sulfamethoxazole	10	0.01	0.1	17	SMZ <0.1%

no matrix effects have been observed previously (data not shown). There were some concerns as to whether the urine components in the leachate would interfere and affect the accuracy of the ELISA measurements. Therefore an experiment was carried out on recoveries of spiked urine concentrations, including dilutions of the samples.

8.3.7 LABORATORY STUDY: SORPTION AND MICROBIAL DEGRADATION IN SOIL

In an incubation experiment of the active ingredients with different soils the processes of sorption and degradation compete with each other. This study was performed in a controlled environment in the climate chamber. After preliminary miniaturized recovery experiments it was concluded that soil moisture could be decisive, and therefore the four soils were studied at three different moisture levels. The experiments were performed in "microcosms," 250-mL Erlenmeyer flasks with a narrow neck that were closed by a lid formed of aluminum foil. The flasks were filled with 100.0 g of the respective soils. The maximum water-holding capacity was determined with the bulk soils, and 40, 60, and 90% of that value were maintained readjusting the weight with some drops of ultrapure water. Incorporation of the compounds into the soil was achieved by spiking purified sea sand with a methanolic solution of the substances and letting the solvent evaporate to yield a final mass concentration of ca. 1000 mg kg^{-1} (actual concentrations obtained were 996 mg kg^{-1} SMZ and 1003 mg kg^{-1} SMX). After homogeneous incorporation of 1.00 g of the sea sand into the 100.0 g of soil the resulting soil concentration was 10 mg kg^{-1}. This provided a thorough distribution of the compounds in the soil and a slow desorption. The "fertilization" by urine was done by adding 8 mL of urine to each flask (based on a manure application rate equivalent to 140 kg nitrogen per ha).

The flasks were kept in a climate chamber at 20°C and a relative air humidity of 80% in the dark. Each application was done in triplicate plus one negative control. Sampling was done after 3, 6, and 9 weeks: the soil in the flask was homogenized and mixed by a stainless steel spatula, ca. 10 g of equivalent dry mass of soil was withdrawn. Extraction was performed as described in Section 8.3.4. Problems occurred at 90% maximum water-holding capacity: the soil sample was almost always covered by water, hence, it can be assumed that there were anaerobic conditions in this soil at later stages of the experiment.

8.3.8   GREENHOUSE STUDY: LEACHING VERSUS UPTAKE

In this study the concurrent processes of leaching out of the root zone and plant uptake from the root zone were examined. Degradation, which also takes place, could not be accounted for directly in this study. In this experiment urine, spiked with the two sulfonamides, was applied as a fertilizer on Italian ryegrass. The vegetation experiment was carried out in a greenhouse. Due to limited capacity only three soils (A, C, and D) could be used. The pots were located on a carriage on rails and could be exposed to direct sunlight and wind, while it was also possible to protect them from heavy rainfall. In contrast to the laboratory experiments, the containers could not be kept on a constant moisture level. Water demand is higher in pot experiments than in the open land because the containers surrounding the soil are exposed to higher ambient air temperatures and soil irradiation gives rise to an increased evapotranspiration from the soil surface.

The plants were grown in "Mitscherlich" containers. These containers are made of sheet iron coated with white enamel, which prevents sorptive processes. The containers also possess a lid on the bottom that retains the soil while allowing leachate to drain into a collection dish below. Mitscherlich containers have a surface of 0.03 m^2 and hold 10 L of soil; each soil was filled into the pots to a final weight of 6.00 kg dry mass. Four replicates were prepared on each variant (three replicates and one negative control). About 1 g of Italian ryegrass was sown onto the soil and slightly incorporated into the first centimeter of soil.

A successful experiment could only be performed in a second run. It had been planned that the application of the compounds to the soil would be done via spiked sea sand at the beginning of the experiment. Unfortunately, the high concentrations prevented the grass from growing. A reduction in dose and in the number of compounds (initially 10 compounds should be applied and analyzed) did not give the desired result.

Therefore, the application of the pharmaceuticals to the soil was performed after tillering, about 3 weeks after sowing (grass blades ca. 20 to 25 cm long). An application of sea sand was not suitable at this stage. Thus, the compounds were applied together with the urine. The spiking solution was prepared by IWW Water Center Mülheim/Ruhr, Germany. The sulfonamides were dissolved in acetone at concentrations of 2.26 g L^{-1} of sulfamethazine and 2.41 g L^{-1} of sulfamethoxazole. To avoid direct contact of the solvent with the plant and with soil microorganisms, 25 mL of the spiking solution in acetone was mixed with 480 mL of urine (resulting in a 3.5% solution of acetone in urine). Application took place at a soil moisture of 75% of maximum water-holding capacity. A 56.5-mg portion of SMZ and 60.2 mg of SMX were applied to 6 kg of soil. This gives an initial soil concentration of 9.42 mg kg^{-1} of SMZ and 10.04 mg kg^{-1} of SMX, respectively.

In order to produce significant amounts of leachate, the soils were over-saturated to about 110% of the maximum water-holding capacity twice a week. Certainly these drying/saturation cycles would lead to different microbial populations than in closed container experiments. The drying/saturation cycles are therefore part of a "worst-case" scenario for leaching. After each of these seven intense rainfall events leachate was formed (40 to 960 mL, dependent on the soil), sampled from the collecting dish, weighed, and aliquoted.

At the end of the experiment the whole soil volume was sieved through a 2-mm sieve and reduced to a 200-g laboratory sample. The plants were harvested at two times of the experiment; this means that after a first cut, a second cut was obtained from the same plants that grew without a signal of further growth inhibition. Soil extraction by ASE was performed as described above, the leachate was diluted by a factor of 100 to 10,000, soil extracts by a factor of 1 to 100 (SMZ) and 1 to 10,000 (SMX), respectively.

8.4 RESULTS

8.4.1 ELISA MEASUREMENTS AND EXTRACTION RECOVERY

Table 8.3 shows the results of the recovery experiments of spiked sulfonamides in urine stored for 4 weeks measured with the sulfamethazine ELISA. It can be seen that in direct urine samples an overestimation of the concentrations was observed, giving recovery rates of more than 200%. It can also be seen that the selectivity of the serum for sulfamethazine in the presence of equal concentrations of sulfamethoxazole is sufficient ("recovery" of 2% or less). While a dilution of the samples by a factor of 10 does not solve the problem, a dilution starting from 1:100 and higher can circumvent matrix effects of urine and leads to acceptable recovery rates (mean 75%).

The soil extraction experiment gave an average recovery of 84% for residues aged in a soil for 4 weeks. It was also concluded that with these miniaturized experiments (10 g of soil) it would not be possible to maintain specific soil moisture levels and get a good reproducibility of replicates. Therefore, the next experiments were performed with 100 g of soil.

8.4.2 DEGRADATION AND SORPTION

The results of the laboratory incubation experiments are shown in Figure 8.3. Overall, both sulfonamides can be detected in all soils at all soil moisture levels even after 12 weeks of incubation. In soils A and B no significant (95% level) decline in

TABLE 8.3
Recoveries of Sulfonamides Spiked in Urine (Measured* by SMZ ELISA with and without Dilution)

	Native, Nonspiked Urine (Stored 4 weeks)	Spiked with 100 µg L⁻¹ Sulfamethoxazole	Spiked with 100 µg L⁻¹ Sulfamethazine	Spiked with 100 µg L⁻¹ SMZ/SMX Mix
undiluted	1	2	200	210
diluted 1 : 10	2	1	170	210
diluted 1 : 100	0	1	65	85

* mean of two sets of triplicates from two different runs

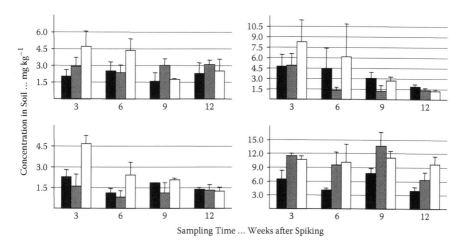

FIGURE 8.3 Mean residual concentration of sulfamethazine in the four soils (up left: A; up right: B; down left: C; down right: D); Shading: soil moisture levels 30% (black), 60% (gray), 90% (white); Sampling intervals represent 3, 6, 9, and 12 weeks after spiking. The initial concentration was 10.0 mg kg^{-1}; Error bars represent one standard deviation.

the concentrations of sulfamethazine was observed. Solely in soil C, which is the freshly taken soil that had not been dried before, a decline was found to a level of 20% of the initial concentration after 3 months. In the sandy soil D both degradation and sorption are lowest. Hence, any concentration higher than 50% of the initial concentration after 3 months can be attributed to high desorption of the residues from this sandy soil. Even with soil C, it is reasonable to assume that not all 80% of the compound lost has been degraded; i.e., only about 50% degraded while 30% is strongly bound to the organic matter of the soil. This bound fraction possesses a lower bioavailability and will be more resistant to degradation.

With sulfamethoxazole the situation is similar (Figure 8.4). There is no clear correlation between the degradation rate and the moisture level. However, the highest moisture level appears to lead to reduced degradation. With 90% water saturation anaerobic conditions can be assumed in the soil, at least temporarily. Under these circumstances sulfamethoxazole seems to be especially difficult to degrade. In the fresh field soil C degradation decreases the levels to about 15% as well as in the alluvial soil B. In the sandy soil D concentrations were determined that are within the error tolerance close to the initial concentration. Sulfamethoxazole does not appear to degrade in sandy soil D. Only the drier variants show some decay. With the incubations at high moisture the compound seems to be adsorbed so weakly that ASE extracts the residues almost quantitatively.

8.4.3 Uptake versus Leaching

8.4.3.1 Growth

Severe growth suppression was observed in the treated grass compared to the control as a signal of the uptake of compounds, but the grass survived, and after 4 weeks a

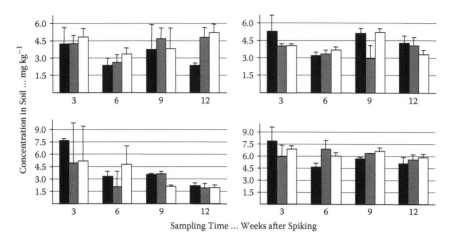

FIGURE 8.4 Mean residual concentration of sulfamethoxazole in the four soils. See legend of Figure 8.3.

first harvest could be performed. The grass continued growing at the same growth rate and 2 weeks later a second cut was obtained.

8.4.3.2 Leaching

By sampling 1 day after every "heavy rain" event, with leachate concentrations of up to 30 mg L^{-1} and determination of amount of leachate together with the respective concentrations, the leached loads of sulfonamides could be calculated. For sulfamethazine the data are presented in Figure 8.5 for the three soils. It can be seen that the main carry-out of the compound takes place after the first rain event with a load of 3 to more than 6 mg. Milligram amounts were also leached on the following days, the sandy soil giving the largest tailing. The loamy silt shows the lowest peak concentration.

8.4.3.3 Plant Uptake

Figure 8.6 shows the residues of free sulfonamides found in Italian ryegrass grown in spiked pot experiments. The controls had mass concentrations below 0.05 mg kg^{-1}. Comparing the three soils in the initial phase, it can be seen that the uptake is dependent on the availability of the antimicrobial in the soil which is best in the sandy soil D. With the course of the experiment, the most available residues (i.e., residues in soil D) had been washed out to a high extent and were therefore not as much available as in soils that showed a lower leaching potential. Therefore, the later uptake was less in the sandy soil due to depletion of the compound in the pore water. Sulfamethazine is more mobile and is therefore taken up first and depleted more quickly, an indication being its almost three times higher aqueous solubility. With the fresh soil C uptake is slow, giving rise to much smaller plant mass concentrations. On the other hand, in the second cut more compounds were taken up but less for sulfamethazine in this case because of its high leaching potential.

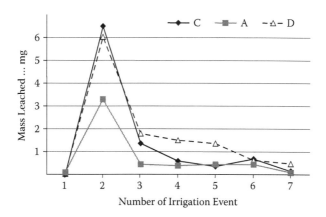

FIGURE 8.5 Breakthrough curve for sulfamethazine with the leachate. Letters represent the soils of Table 8.1.

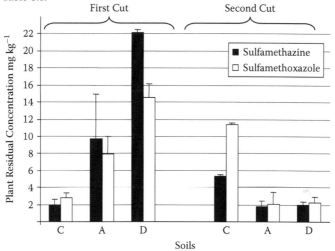

FIGURE 8.6 Residual concentrations (measured in triplicate) in Italian ryegrass grown on three different soils spiked with 10.0 mg kg^{-1} of the sulfonamides sulfamethazine and sulfamethoxazole. Left: grass blades harvested after initial growth; right: concentrations in grass grown after the first cut. Means of three replicates (pots); bars: standard deviations; Soils as in Table 8.1.

8.5 CONCLUSION

It was found that the sulfonamides studied in four soils were not degraded to half of the initial concentration after 3 months. Furthermore, it is possible that the observed decrease in concentrations was not due to metabolism but rather due to poor extraction recovery resulting from the strong binding of the compounds to the soil matrix (minerals and organic matter). Degradation seems to be higher under aerobic conditions when microbial activity is high. This was obvious when fresh soil resulted in higher degradation of the sulfonamides while anaerobic conditions (high water saturation of soils) resulted in lower degradation rates.

Apart from the two sulfonamides for which the concentrations were determined by ELISA and that were applied with a load of 10 mg kg^{-1} (dry mass), which represents an application rate of ca. 20 kg ha^{-1}, there were eight other compounds present in the spiking solution to give a total soil concentration of 60 mg kg^{-1}. This was to present a worst-case scenario.

Leaching is an important pathway for dissipation of sulfonamide residues in soil. Polar compounds such as sulfamethazine are preferentially translocated, and leachate concentrations reach the milligram-per-liter range. Most of the sulfonamides were transported with the first simulated rainfall events, which is similar to the behavior of polar pesticides that often drain to lower levels and even down to groundwater by preferential flow.

For a rough estimation, the total load in the leachate was calculated by summing up the daily loads. This gives an estimate of ca. 6 mg for the loamy silt soil A, ca. 10 mg for the fresh soil C, and ca. 12 mg for the sandy soil D, with its low organic carbon content, respectively. From the 60 mg applied to the container this is 10, 16, and 20%, respectively.

It was found that the polar sulfonamides can be taken up into forage grass. The plant concentrations were in the milligram-per-kilogram range as in some references with high amounts taken up from hydroponics.[9,38] On an average 200 to 500 g of grass had been harvested, and thus between 5 and 20% of the total mass applied has been abstracted from soil certainly under conditions that represent a worst-case scenario with a very high initial soil concentration. On the other hand, plant uptake competes with leaching and degradation so that in consequence uptake declined after some weeks. When leaching and uptake begin to cease, the missing share of 60 to 85% of the applied mass will become degraded or stay adsorbed. The laboratory experiment showed that degradation does not take place to that extent. Therefore, the formation of a more or less available reservoir sorbed in the soil has to be considered. Altogether, this preliminary study shows some aspects of pharmaceutical compounds behavior in soil. It has to be emphasized again that the study assumed a "worst-case" scenario. Sulfonamides are highly polar pharmaceuticals, and very high amounts of the test compounds were used in the study.

Our experiments on plant uptake of selected pharmaceuticals from soil in pot experiments that followed a distinct worst-case scenario underline that, at least with some pharmaceuticals, a combination of factors (i.e., a relatively low degradation rate and a relatively high tendency to leach and not to adsorb) contribute to a non-negligible plant uptake of these chemicals. Under field conditions and with lower concentrations, results may be quite different from the laboratory studies presented here. However, the laboratory studies demonstrated the risks of the agricultural use of urine and manure that contain pharmaceuticals, and results suggest that the risks cannot be ignored.

ELISA has been used for monitoring the concentration of single compounds in model experiments. It could be shown that ELISA is able to deliver the data necessary to assess the behavior of these chemicals fast and at low cost. Therefore, matrix-tolerant immunoassays could be a valuable tool in performing tests that might be necessary for the registration of chemicals or for gathering data on chemicals in the European Registration Evaluation, Authorization, and Restriction of Chemicals (REACH) process.

ACKNOWLEDGMENTS

This work was carried out at the Institute of Crop Science and Resource Conservation (INRES), Department of Plant Nutrition of the Faculty of Agriculture of the University of Bonn. I would like to thank Mrs. Melanie Mucha for valuable laboratory assistance, Mohamed Hashim for help in the lab, and Dr. Thorsten Christian for developing the anti-sulfamethoxazole antibodies. Supply of anti-sulfamethazine antibodies by Dr. Milan Fránek, Veterinary Research Institute, Brno, Czech Republic, is gratefully acknowledged. Part of this work was supported by the Ministry of Environment, Nature Conservation, Agriculture and Consumer Protection of the Federal State of North Rhine-Westphalia, Germany. The IWW Water Center at Mülheim/Ruhr, Germany, is thanked for preparing spiked sea sand samples. Thanks to Dr. Joachim Clemens (INRES) for coordinating the project.

REFERENCES

1. Langhammer, J.P., Büning-Pfaue, H., Winkelmann, J., and Körner, E., Chemotherapeutical residues and resistance in post-partum sows during herd treatment [in German], *Tierärztl. Umsch.*, 43, 375, 1988.
2. Hirsch, R., Ternes, T., Haberer, K., and Kratz, K.-L., Occurrence of antibiotics in the aquatic environment, *Sci. Total Environ.*, 225, 109, 1999.
3. Kühne, M., Wegmann, S., Kobe, A., and Fries, R., Tetracycline residues in bones of slaughtered animals, *Food Control*, 11, 175, 2000.
4. Kühne, M., Ihnen, D., Möller, G., and Agthe, O., Stability of tetracycline in water and liquid manure, *J. Vet. Med. A*, 47, 379, 2000.
5. Winckler, C. and Grafe, A., Use of veterinary drugs in intensive animal production— evidence for persistence of tetracycline in pig slurry, *J. Soils Sediments*, 1, 66, 2001.
6. De Liguoro, M., Cibin, V., Capolongo, F., Halling-Sørensen, B., and Montesissa, C., Use of oxytetracycline and tylosin in intensive calf farming: evaluation of transfer to manure and soil, *Chemosphere*, 52, 203, 2003.
7. Loke, M.L., Ingerslev, F., Halling-Sørensen, B., and Tjornelund, J., Stability of Tylosin A in manure containing test systems determined by high performance liquid chromatography, *Chemosphere*, 44, 865, 2001.
8. Meyer, M.T., Bumgarner, J.E., Varns, J.L., Daughtridge, J.V., Thurman, E.M., and Hostetler, K.A., Use of radioimmunoassay as a screen for antibiotics in confined animal feeding operations and confirmation by liquid chromatography/mass spectrometry, *Sci. Total Environ.*, 248, 181, 2000.
9. Migliore, L., Brambilla, G., Cozzolino, S., and Gaudio, L., Effect on plants of sulphadimethoxine used in intensive farming (*Panicum miliaceum, Pisum sativum* and *Zea mays*), *Agric. Ecosys. Environ.*, 52, 103, 1995.
10. Richardson, M.L. and Bowron, J.M., The fate of pharmaceutical chemicals in the aquatic environment — a review, *J. Pharm. Pharmacol.*, 37, 1, 1985.
11. Gavalchin, J. and Katz, S.E., The persistence of faecal-borne antibiotics in soil, *J. AOAC Int.*, 77: 481–485, 1994.
12. Wang, Q.-Q., Bradford, S.A., Zheng, W., and Yates, S.R., Sulfadimethoxine degradation kinetics in manure as affected by initial concentration, moisture, and temperature, *J. Environ. Qual.*, 35, 2162, 2006.
13. Christian, T., Schneider, R.J., Färber, H., Skutlarek, D., and Goldbach, H.E., Determination of antibiotic residues in manure, soil and surface waters, *Acta Hydrochim. Hydrobiol.*, 31, 36, 2003.

14. Otterpohl, R., Options for alternative types of sewerage and treatment systems directed to improvement of the overall performance, *Water Sci. Technol.,* 45, 149, 2002.

15. Sarmah, A.K., Meyer, M.T., and Boxall, A.B.A., A global perspective on the use, sales, exposure pathways, occurrence, fate and effects of veterinary antibiotics (VAs) in the environment, *Chemosphere,* 65, 725, 2006.

16. Picó, Y. and Andreu, V., Fluoroquinolones in soil—Risks and challenges, *Anal. Bioanal. Chem.,* 387, 1287, 2007.

17. Sukul, P. and Spiteller, M., Sulfonamides in the environment as veterinary drugs, *Rev. Environ. Contam. Toxcol.,* 187, 67, 2006.

18. Tolls, J., Sorption of veterinary pharmaceuticals in soils: a review, *Environ. Sci. Technol.,* 35, 3397, 2001.

19. Thiele-Bruhn, S., Pharmaceutical antibiotic compounds in soils—a review. *J. Plant Nutr. Soil Sci.,* 166: 145, 2003.

20. Burhenne, J., Ludwig, M., Nikoloudis, P., and Spiteller, M., Photolytic degradation of fluoroquinolone carboxylic acids in aqueous solution. Part I: Primary photoproducts and half-lives, *Environ. Sci. Pollut. Res.,* 4, 10, 1997.

21. Burhenne, J., Ludwig, M., and Spiteller, M., Photolytic degradation of fluoroquinolone carboxylic acids in aqueous solution. Part II: Isolation and structural elucidation of polar photometabolites, *Environ. Sci. Pollut. Res.,* 4, 61, 1997.

22. Marengo, J.R., Kok, R.A., O'Brien, K., Velagaleti, R.R., and Stamm, J.M., Aerobic biodegradation of (C14)-sarafloxacin hydrochloride in soil, *Environ. Toxicol. Chem.,* 16, 462, 1997.

23. Gao, J.A. and Pedersen, J.A., Adsorption of sulfonamide antimicrobial agents to clay minerals, *Environ. Sci. Technol.,* 39, 9509, 2005.

24. Pils, J.R.V. and Laird, D.A., Sorption of tetracycline and chlortetracycline on K- and Ca-saturated soil clays, humic substances, and clay-humic complexes, *Environ. Sci. Technol.,* 41, 1928, 2007.

25. Allaire, S.E., Del Castillo, J., and Juneau, V., Sorption kinetics of chlortetracycline and tylosin on sandy loam and heavy clay soils, *J. Environ. Qual.,* 35, 969, 2006.

26. Weerasinghe, C.A. and Towner, D., Aerobic biodegradation of virginiamycin in soil, *Environ. Toxicol. Chem.,* 16, 1873, 1997.

27. Addison, J.B., Antibiotics in sediments and run-off waters from feedlots, *Res. Rev.,* 92, 1, 1984.

28. Hamscher, G., Pawelzick, H.T., Höper, H., and Nau, H., Different behavior of tetracyclines and sulfonamides in sandy soils after repeated fertilization with liquid manure, *Environ. Toxicol. Chem.,* 24, 861, 2005.

29. Hamscher, G., Sczesny, S., Höper, H., and Nau, H., Determination of persistent tetracycline residues in soil fertilized with liquid manure by HPLC-ESI-MS-MS, *Anal. Chem.,* 74, 1509, 2002.

30. Boxall, A.B.A., Blackwell, P., Cavallo, R., Kay, P., and Tolls, P., The sorption and transport of a sulphonamide antibiotic in soil systems, *Toxicol. Lett.,* 131, 19, 2002.

31. Thiele, S., Adsorption of the antibiotic pharmaceutical compound sulfapyridine by a longterm differently fertilized loess Chernozem, *J. Plant Nutr. Soil. Sci.,* 163, 589, 2000.

32. Wehrhan, A., Kasteel, R., Simunek, J., Groeneweg, J., and Vereecken, H., Transport of sulfadiazine in soil columns—Experiments and modelling approaches, *J. Contam. Hydrol.,* 89, 107, 2007.

33. Jjemba, P.K., The potential impact of veterinary and human therapeutic agents in manure and biosolids on plants grown on arable land: a review, *Agric. Ecosyst. Environ.,* 93: 267, 2002.

34. Batchelder, A.R., Chlortetracycline and oxytetracycline effects on plant growth and development in liquid cultures, *J. Environ. Qual.,* 10, 515, 1981.

35. Batchelder, A.R., Chlortetracycline and oxytetracycline effects on plant growth and development in soil systems, *J. Environ. Qual.,* 11, 675, 1982.

36. King, L.D., Safley, L.M. Jr., and Spears, J.W., A greenhouse study on the response of corn (*Zea mays* L.) to manure from beef cattle fed antibiotics, *Agric. Wastes,* 8, 185, 1983.

37. Migliore, L., Brambilla, G., Casoria, P., Civitareale, C., and Gaudio, L., Effect of antimicrobials for agriculture as environmental pollutants, *Fresenius Environ. Bull.,* 5, 735, 1996.

38. Migliore, L., Cozzolino, S., and Fiori, M., Phytotoxicity to and uptake of enrofloxacin in crop plants, *Chemosphere,* 52, 1233, 2003.

39. Makris, K.C., Shakya, K.M., Datta, R., Sarkar, D., and Pachanoor, D., High uptake of 2,4,6-trinitrotoluene by vetiver grass—Potential for phytoremediation? *Environ. Pollut.,* 146, 1, 2007.

40. Mattina, M.I., Isleyen, M., Eitzer, B.D., Iannucci-Berger, W., and White, J.C., Uptake by cucurbitaceae of soil-borne contaminants depends upon plant genotype and pollutant properties, *Environ. Sci. Technol.,* 40, 1814, 2006.

41. Santos de Araujo, B., Dec, J., Bollag, J.M., and Pletsch, M., Uptake and transformation of phenol and chlorophenols by hairy root cultures of *Daucus carota, Ipomoea batatas* and *Solanum aviculare, Chemosphere,* 63, 642, 2006.

42. Huang, H., Zhang, S., Chen, B.-D., Wu, N., Shan, X.-Q., and Christy, P., Uptake of atrazine and cadmium from soil by maize (*Zea mays* L.) in association with the arbuscular mycorrhizal fungus *Glomus etunicatum, J. Agric. Food. Chem.,* 54, 9377, 2006.

43. Gomez-Hermosillo, C., Pardue, J.H., and Reible, D.D., Wetland plant uptake of desorption-resistant organic compounds from sediments, *Environ. Sci. Technol.,* 40, 3229, 2006.

44. Khan, S.U. and Marriage, P.B., Residues of atrazine and its metabolites in an orchard soil and their uptake by oat plants, *J. Agric. Food Chem.,* 25, 1408, 1977.

45. Balke, N.E. and Price, T.P., Relationship of lipophilicity to influx and efflux of triazine herbicides in oat roots, *Pestic. Biochem. Physiol.,* 30, 228, 1988.

46. Colinas, C., Ingham, E., and Molina, R., Population responses of target and nontarget forest soil organisms to selected biocides, *Soil Biol. Biochem.,* 26, 41, 1994.

47. Halling-Sørensen, B., Sengelov, G., and Tjornelund, J., Toxicity of tetracyclines and tetracycline degradation products to environmentally relevant bacteria, including selected tetracycline-resistant bacteria, *Arch. Environ. Contam. Toxicol.,* 42, 263, 2002.

48. Chander, Y., Kumar, K., Goyal, S.M., and Gupta, S.C., Antibacterial activity of soil-bound antibiotics, *J. Environ. Qual.,* 34, 1952, 2005.

49. Opalinski, K.W., Dmowska, E., Makulec, G., Mierzejewska, E., Petrov, P., Pierzynowski, S.G., and Wojewoda, D., The reverse of the medal: feed additives in the environment, *J. Animal Feed Sci.,* 7, 35, 1998.

50. Migliore, L., Civitareale, C., Cozzolino, S., Casoria, P., Brambilla, G., and Gaudio, L., Laboratory models to evaluate phytotoxicity of sulphadimethoxine on terrestrial plants, *Chemosphere,* 37, 2957, 1998.

51. Kumar, K., Gupta, S.C., Baidoo, S.K., Chander, Y., and Rosen, C.J., Antibiotic uptake by plants from soil fertilized with animal manure, *J. Environ. Qual.,* 34, 2082, 2005.

52. Boxall, A.B.A., Johnson, P., Smith, E.J., Sinclair, C.J., Stutt, E., and Levy, L.S., Uptake of veterinary medicines from soils into plants, *J. Agric. Food Chem.,* 54, 2288, 2006.

53. Sandermann, H. Jr., Plant metabolism of xenobiotics, *Trends Biochem. Sci,* 17, 82, 1992.

54. Sandermann, H. Jr., Higher-plant metabolism of xenobiotics — the green liver concept, *Pharmacogenetics,* 4, 225, 1994.

55. Schröder, P. and Collins, C., Conjugating enzymes involved in xenobiotic metabolism of organic xenobiotics in plants, *Intern. J. Phytoremed.,* 4, 247, 2002.

56. Edwards, R. and Owen, W.J., Comparison of glutathione S-transferases of *Zea mays* responsible for herbicide detoxification in plants and suspension-cultured cells, *Planta,* 169, 208, 1986.

57. Coleman, J.O.D., Blake-Kalff, M.M.A., and Davies, T.G.E., Detoxification of xenobiotics by plants: chemical modification and vacuolar compartmentation, *Trends Plant Sci.,* 2, 144, 1997.

58. Farkas, M.H., Berry, J.O., and Aga, D.S., Chlortetracycline detoxification in maize via induction of glutathione S-transferases after antibiotic exposure, *Environ. Sci. Technol.,* 41, 1450, 2007.

59. Zhu, Y., Yanagihara, K., Guo, F.M., and Li, Q.X., Pressurized fluid extraction for quantitative recovery of acetanilide and nitrogen heterocyclic herbicides in soil, *J. Agric. Food Chem.,* 48, 4097, 2000.

60. Obana, H., Kikuchi, K., Okihashi, M., and Hori, S., Determination of organophosphorus pesticides in foods using an accelerated solvent extraction system, *Analyst,* 122, 217, 1997.

61. Stoob, K., Singer, H.P., Stettler, S., Hartmann, N., Mueller, S.R., and Stamm, C.H., Exhaustive extraction of sulfonamide antibiotics from aged agricultural soils using pressurized liquid extraction, *J. Chrom. A,* 1128, 1, 2006.

62. Miao, X.-S. and Metcalfe, C.D., Determination of carbamazepine and its metabolites in aqueous samples using liquid chromatography-electrospray tandem mass spectrometry. *Anal. Chem.,* 75, 3731, 2003.

63. Märtlbauer, E., Usleber, E., Schneider, E., and Dietrich, R., Immunochemical detection of antibiotics and sulfonamides, *Analyst,* 119, 2543, 1994.

64. Grubelnik, A., Padeste, C., and Tiefenauer, L., Highly sensitive enzyme immunoassays for the detection of β-lactam antibiotics, *Food Agric. Immunol.,* 13, 161, 2001.

65. Korpimäki, T., Brockmann, E.-C., Kuronen, O., Saraste, M., Lamminmäki, U., and Tuomola, M., Engineering of a broad specificity antibody for simultaneous detection of 13 sulfonamides at the maximum residue level, *J. Agric. Food Chem.,* 52, 40, 2004.

66. Aga, D.S., O'Connor, S., Ensley, S., Payero, J.O., Snow, D., and Tarkalson, D., Determination of the persistence of tetracycline antibiotics and degradates in manure-amended soil using enzyme-linked immunosorbent assay and liquid chromatography-mass spectrometry, *J. Agric. Food Chem.,* 53, 7165, 2005.

67. Fránek, M., Kolár, V., Deng, A., and Crooks, S., Determination of sulphadimidine (sulfamethazine) residues in milk, plasma, urine and edible tissues by sensitive ELISA, *Food Agric. Immunol.,* 11, 339, 1999.

68. Kolpin, D.W., Furlong, E.T., Meyer, M.T., Thurman, E.M., Zaugg, S.D., Barber, L.B., and Buxton, H.T., Pharmaceuticals, hormones, and other organic wastewater contaminants in U.S. streams, 1999–2000: a national reconnaissance, *Environ. Sci. Technol.,* 36, 1202, 2002.

69. Scherer, H.W., Goldbach, H.E., and Clemens, J., Potassium dynamics in the soil and yield formation in a long-term field experiment, *Plant Soil Environ.,* 49, 531, 2003.

70. Schneider, C., Schöler, H.F., and Schneider, R.J., A novel enzyme-linked immunosorbent assay for ethynylestradiol using a long-chain biotinylated EE2 derivative, *Steroids,* 69, 245, 2004.

9 Antibiotic Transformation in Plants via Glutathione Conjugation

Michael H. Farkas,
James O. Berry, and Diana S. Aga

Contents

9.1 INTRODUCTION

Pharmaceuticals have been introduced into the environment for decades, via land application of manure from antibiotic-treated livestock and via discharges from wastewater treatment plants, where only very limited removal may take place. As an increasing number of investigators report the occurrence of a wide range of pharmaceuticals in the environment,[1–3] there is a need to focus more research on the advancement of treatment technologies to remediate pharmaceutical pollutants in the environment. Efforts to enhance pharmaceutical remediation have been spurred by various concerns, ranging from the emergence of antibiotic-resistant pathogens that may infect the human population[4] to potential risks associated with the long-term exposure of consumers to crops that accumulate the antibiotics.[5] While most of these concerns have not as of yet been verified, some substantiation of these effects has come to fruition. For instance, research on human embryonic cells exposed to 13 pharmaceuticals at concentrations found in the environment has shown a significant decrease in cell proliferation *in vitro*.[6]

Phytoremediation, the application of plants and their associated microbes to enhance biodegradation of contaminants in the environment, has recently been explored for the remediation of sulfadimethoxine[7] and oxytetracycline[8] antibiotics. The phytoremediation of metals and organic pollutants in soil is an emerging, low-cost technology, but the underlying biochemical mechanisms involved in contaminant uptake, detoxification, and translocation in plants are largely unknown. One well-known detoxification pathway involves phytotransformation of contaminants via glutathione (GSH) conjugation, which is catalyzed by glutathione s-transferase (GST) enzymes. This chapter provides an overview on the role of plant GSTs in phytoremediation of organic contaminants and presents recent work on GST-mediated transformations of environmentally relevant antibiotics.

9.2 PLANT UPTAKE AND PHYTOTOXICITY OF PHARMACEUTICALS

Like many heavy metals and organic contaminants, antibiotics can be taken up by plants and can elicit phytotoxicity in susceptible species. In fact, it has been known for quite some time that chlortetracycline, a highly used growth-promoting antibiotic, is phytotoxic to pinto beans (*Phaseolus vulgaris*).[9] More recent reports have demonstrated the phytotoxic effects of sulfadimethoxine antibiotics toward maize and barley[10] and also that of enrofloxacin antibiotics to several agricultural crops.[11] However, some plant species do not appear to be affected by exposure to antibiotics. For example, in the same study that reported phytotoxicity of chlortetracycline to pinto beans, it was shown that *Zea mays* was unaffected by the antibiotic exposure.[9]

Pharmaceutical-induced phytotoxicity does not appear to cause plant mortality but rather leads to inhibition of plant growth and ultimately lower crop yields. For example, enrofloxacin, a broad-spectrum antibiotic that is used in both human and veterinary medicine, inhibits root growth and leaf development in a variety of crop plants.[11] Statins, which are cholesterol-reducing agents that are highly prescribed for human health, can inhibit 3-hydroxyl-3-methylglutaryl coenzyme A reductase in plants, blocking isoprenoid biosynthesis, which is important for a multitude of endogenous functions.[12]

Direct quantification of the amount of contaminants that accumulate in plant tissues is difficult because of the analytical challenges associated with detecting low levels of analytes within the complex plant biomass. Recently, however, Kumar and coworkers[5] were able to measure concentrations of chlortetracycline that accumulated in maize, cabbage, and green onions at the parts per billion (ppb) range using enzyme-linked immunosorbent assay (ELISA) techniques. Beyond pharmaceuticals, other toxic organic compounds have been shown to accumulate in the fruits of apple and peach trees.[13] Trichloroethylene, a cleaning and degreasing agent widely used for industrial and military purposes in the United States, has been found in the environment at levels as high as 500 ppm. A controlled greenhouse study has shown that within 2 years of exposure, apple and peach trees were able to accumulate as much as 34 ppm.

9.3 BIOAVAILABILITY

The bioavailability of pharmaceuticals is an important factor to consider when investigating their interactions with plants. Generally, the toxicity of a contaminant is directly related to its bioavailability.[14] In soil, the bioavailability of an antibiotic depends on two major factors: (1) the degree to which it adsorbs to the soil and (2) the organisms inhabiting the soil. Sorption of antibiotics to soil is dependent on the soil composition and its chemical characteristics, such as pH and ionic strength. Soil composition can vary with regard to its clay, sand, organic matter, and mineral content, all of which play an important role in antibiotic sorption. For instance, sulfonamide antibiotics adsorb more strongly to clay than to sand, and in general, antibiotics adsorb less tightly to soil minerals than to organic matter.[14] However, the soil's mineral composition plays a role in tetracycline binding. It is known that Ca^{2+} sorption to clay increases oxytetracycline adsorption relative to Na^+. Soil pH must also be factored into antibiotic bioavailability. Sulfonamides and tetracyclines are adsorbed more tightly in the presence of acidic soils, whereas the opposite is true for tylosin (macrolide family of antibiotics).[15] Some antibiotics have hydrophobic characteristics and cannot dissolve in water without additional factors, but it appears that hydrophobicity is not directly related to the strength of sorption of these antibiotics to soil.[16,17] In general, electrostatic forces, complexation, and hydrogen bonding, which are regulated by the soil content and chemistry, are all important in determining antibiotic bioavailability.

The rhizosphere, the area surrounding a plant's roots, is very dynamic and full of microorganisms that play a role in the fate and bioavailability of antibiotics and other contaminants in soil. Root exudates in the rhizosphere may contain reactive oxygen species (ROS) such as H_2O_2, which are produced by plants as a general defense response to oxidative stress. Oxytetracycline has been shown to induce the release of H_2O_2 in hairy root cultures of sunflowers (*Helianthus annuus*),[18] resulting in the inactivation of oxytetracyline via oxidation.

9.4 DETOXIFICATION OF XENOBIOTICS VIA GLUTATHIONE CONJUGATION

Plants have an intricate defense system that is capable of combating a variety of intrusions ranging from pathogens to exogenous chemicals. In fact, the plant's defense system is controlled by numerous biochemical pathways and is capable of producing a physiological response that is pathogen-/xenobiotic-specific. The ability of some plants to detoxify harmful compounds upon uptake via these specific pathways has promoted interest in the area of phytoremediation, which is still in its early stages. As mentioned earlier, GST enzymes are responsible for many of these detoxification reactions involving a large number of xenobiotics found in living systems, including plants.

The GST enzymes are primarily found in the cytosol of plants, mammals, bacteria, fungi, and insects. The GSTs are part of a three-phase detoxification system involved in detoxifying xenobiotics in living organisms. Phase I includes enzymes

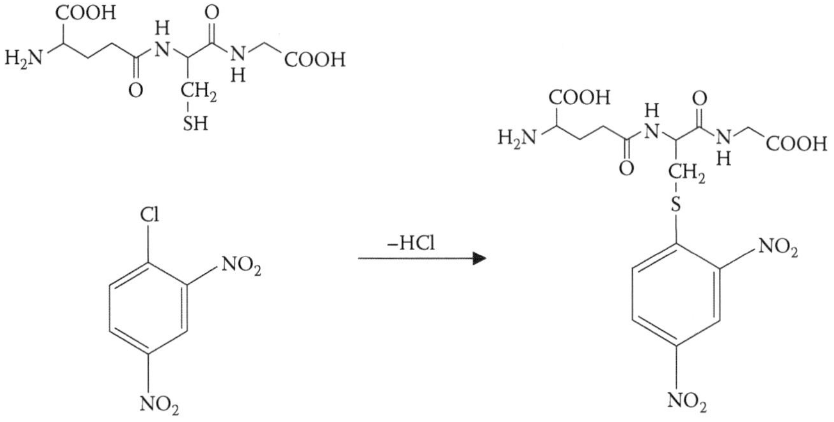

FIGURE 9.1 The reaction between glutathione and 1-chloro-2,4-dinitrobenzene (CDNB) as catalyzed by GST enzyme proceeds via the electrophilic substitution of chlorine atom in CDNB by the sulfur atom of glutathione, producing a dechlorinated conjugate.

such as cytochrome P450 monooxygenases. The purpose of the Phase I enzymes is to introduce reactive functional groups via hydroxylation and epoxidation reactions to xenobiotic compounds.[19] The introduction of functional groups prepares a xenobiotic to be acted upon by Phase II enzymes, of which GSTs are a major component. GSTs detoxify xenobiotics that are typically electrophilic, and they do so by substituting glutathione (GSH) at an electrophilic site, which renders the xenobiotic more polar and more readily translocated.[20] GSH is a tripeptide (γ-glutamyl-cysteinyl-glycine) that is conjugated to many endogenous and xenobiotic compounds (Figure 9.1). GST-mediated conjugations occur very rapidly, and the general mechanism takes place via a nucleophilic attack of the thiol group of GSH on an electrophilic atom in the xenobiotic. The first documented cases of plant GST-mediated detoxification were in the metabolism of herbicides. There are a few cellular mechanisms that render a herbicide selective toward some unwanted species of weeds, but most commonly GSTs are involved in detoxifying herbicides in nontarget species.

Both GSTs and GSH must be already present in abundance (or be induced) for a plant to be able to detoxify a xenobiotic via the glutathione pathway. For example, maize is tolerant to chloracetanilide and chlorotriazine herbicides by using a Type I GST[21,22] to detoxify these chlorinated compounds after plant uptake. Type I GSTs are constitutively expressed in maize; hence, maize plants are inherently able to detoxify atrazine.[21] Type II GSTs, however, are only induced by exogenous chemicals. Safeners are chemicals that are applied to plants that do not have the inherent ability to detoxify a chemical, thus inducing GST expression. For example, maize has a slight tolerance for thiocarbamate herbicides (EPTC), but pretreatment with a safener such as dichlormid or benoxacor greatly increases induction of Type II GSTs that specifically protect the maize from EPTC exposure.[23] While plants are capable of metabolizing xenobiotics using different mechanisms, they may be susceptible if the metabolic pathway they use for detoxification is not efficient. For instance, maize detoxifies atrazine via GST conjugation, which is an efficient mechanism. In contrast, a pea plant metabolizes atrazine slowly and inefficiently via N-dealkylation.[24] The slow rate of atrazine metabolism by pea plants leaves them susceptible to atrazine toxicity.

GST-mediated transformations of xenobiotics is not the only mechanism of detoxification in the Phase II pathway. Glucosyl- and malonyltransferases are Phase II enzymes that add glucose and malonic acid to xenobiotics, respectively.[19] These enzymes serve similar functions as GST with respect to "tagging" a xenobiotic and making it more polar. In Duckweed (*Lemna gibba*), several chlorinated pesticides are transformed using these latter Phase II enzymes.[25] However, unlike GST, these enzymes appear to conjugate at the carboxylic acid and amine R-groups of the pesticide, instead of at the more electrophilic chlorine atoms.

9.5   GLUTATHIONE S-TRANSFERASES: STRUCTURE, FUNCTION, AND EVOLUTION

The GST enzymes are hetero- and homodimeric, with an average molecular weight of around 50 kDa, and serve a variety of functions. GSTs are encoded by a large and diverse gene family. In plants, this family is divided into five classes based on sequence identity. These classes include: phi, tau, theta, zeta, and lambda, in which theta and zeta have homology to the mammalian classes. Genomic analysis of *Arabidopsis thaliana* has located at least 48 GST genes, with the tau and phi classes being most abundant. Each monomer of the GST dimer is composed of two binding sites. The more internal binding site (G site) is responsible for binding glutathione (GSH).[24] The G site is a conserved group of amino acids located in the N terminal domain of the polypeptide. The C terminal domain contains the binding site for the hydrophobic substrate (H site). This region is much more variable in terms of amino acid sequence relative to other GSTs, which is not unexpected when keeping in mind the large number of compounds GSTs can bind.

Electrophilic xenobiotics are particularly deleterious to living organisms, because they can be cytotoxic or genotoxic.[19] There exist two types of these electrophiles: soft and hard. An example of each type of electrophile that is typically found in a xenobiotic includes alkene groups (carbon-carbon double bonds) and halogens, respectively. GSTs can mediate the conjugation of both types of electrophiles. GSH conjugation "tags" a xenobiotic for further processing. Processing of the "tagged" xenobiotic is considered as Phase III of the detoxification pathway, and the end result for the xenobiotic differs depending on the organism. In plants, the GSH-conjugated xenobiotic is either stored in the cell's vacuole or else it is sent to the apoplast (area outside of the cell).[26] Some evidence does exist, which suggests that GSH-labeled xenobiotics are exuded back into the soil after processing of GSH.[27] To transport the xenobiotic conjugate requires Phase III proteins, which are ATP-dependent transporters with the ability to recognize and bind to GSH.

GSH plays a major role both in a plant's endogenous cellular processes as well as in the plant's defense responses. GSH is ubiquitous and abundant with roles spanning protein and nucleic acid synthesis, including modulation of enzyme activity and adaptation to environmental stress.[28] Environmental stressors are quite variable, ranging from temperature extremes to xenobiotic stress. GSH is found in two forms within a cell: an oxidized form, in which a disulfide bond is formed between two glutathione molecules (GSSG), and the reduced form (GSH). The ratio of the two forms is crucial to how a plant adapts to its stressor. A lack of free GSH can diminish a plant's ability to mount an appropriate response to a stressor. Understanding the GSH biosynthetic pathway and the mechanisms by which it is utilized by various enzymes will provide insight into xenobiotic detoxification by plants. Activation of GSH biosynthesis is based on the ability of proteins involved in photosynthesis to act as an intricate sensory system to respond to variations in redox potential caused by environmental stress.

Environmental stress also induces other enzymes that are part of the detoxification pathway. Glutathione reductase is an enzyme that reduces GSSG to GSH, thus increasing the concentration of free GSH. Glutathione peroxidases (GPX) are antioxidant enzymes. A plant, as a reaction to environmental stress, produces ROS to contain the stressor within the site where the stressor is introduced. GPXs are used to prevent oxidative damage by oxidizing two GSHs to form GSSG. GSTs also have peroxidase activity (although they are encoded by a different gene family) and their mode of action is conjugation of GSH to the oxidant.[29] The functional difference between the two types of peroxidases is based on the substrate acted upon. GPXs reduce ROS, while GST peroxidases conjugate electrophiles such as lipid peroxides that are the result of ROS.

9.6 INDUCTION OF GSTS IN *PHASEOLUS VULGARIS* AND *ZEA MAYS* BY CHLORTETRACYCLINE

The authors of this chapter performed similar experiments to those reported earlier by Batchelder[9] to examine the physiological basis of the observed differences in response between maize and pinto beans grown in antibiotic-treated soil. Ten-day-old maize and pinto beans were transplanted into soil pretreated with 20 mg kg^{-1} of chlortetracycline (CTC), with concentrations similar to those found in the

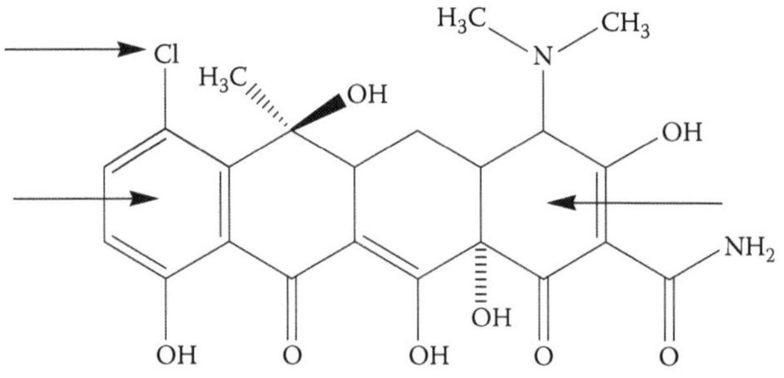

FIGURE 9.2 The chemical structure of chlortetracycline. The arrows depict potential sites of glutathione conjugation.

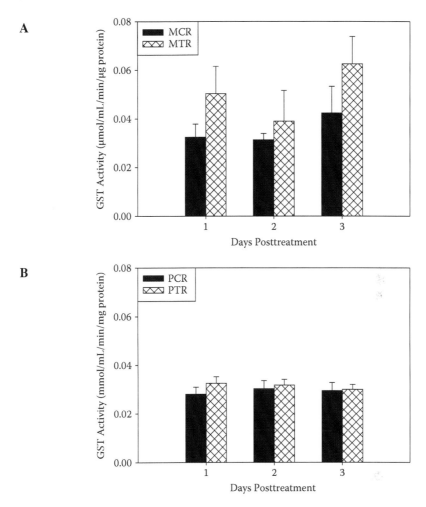

FIGURE 9.3 GST activity measured from total protein extracts at 1, 2, and 3 days after plants were treated with CTC. (A) Maize control (MCR) and CTC-treated (MTR) plants. (B) Pinto bean control (PCR) and CTC-treated (PTR) plants. Values represent the mean and standard deviation of six replicates. Asterisks denote statistically significant data ($p < 0.05$).

environment.[30] CTC is a good candidate for initial investigations into a plant's response for two reasons: it is widely used with high application rates in agriculture, and it contains both hard and soft nucleophiles (Figure 9.2). The plants were then harvested daily for 3 days and extracted for analysis of total proteins. To gain a general perspective of the response of the plants to the antibiotic, the total proteins were subjected to SDS-Polyacrylamide Gel Electrophoresis (SDS-PAGE). Interestingly, a distinct increase in bands at the range of 20 to 30 kDa from the maize samples grown in CTC-treated soil was observed, relative to the maize control (untreated) samples. This was not observed in CTC-treated pinto beans. The increase in the proteins banding at this size range was indicative of GST induction in the treated maize plants.

To verify that these induced proteins were in fact GSTs, enzyme activity assays were performed using the crude extracts from the plants. The assay used was based on the standard GST-catalyzed conjugation reaction of GSH to 1-chloro-2,4-dinitrobenzene (CDNB).[31] Indeed, GST activity in the CTC-treated maize samples was significantly higher relative to the control (untreated plants) on the first and third days of exposure (Figure 9.3A). The GST activities in the protein extracts from pinto beans showed no significant differences between treated and control plants in any of the days sampled (Figure 9.3B), consistent with the SDS-PAGE results.

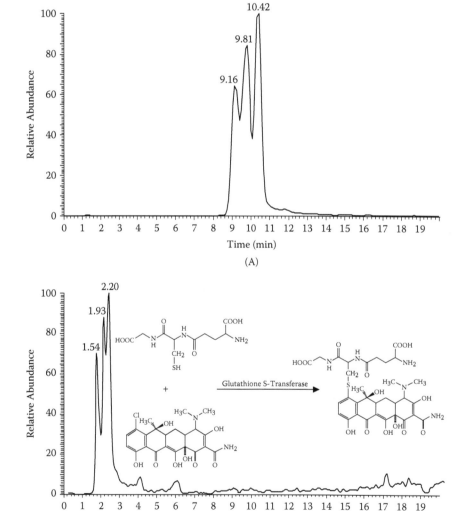

FIGURE 9.4 *(Continued)*

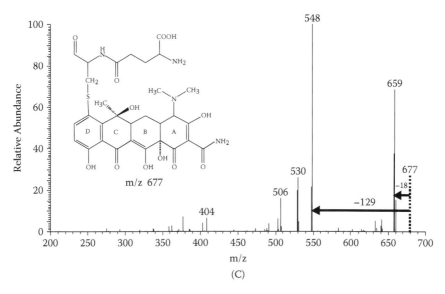

FIGURE 9.4 *In vitro* LC/MS/MS data for GST-mediated CTC-GSH conjugation. (A) Chromatogram of m/z 677 conjugate and hypothesized product of a GST-mediated CTC-GSH reaction (inset). (B) Chromatogram of CTC demonstrating the difference in polarity relative to the observed conjugate and the similarities in isomeric peaks. (C) chemical structure of the fragment ion m/z 677 and its MS/MS fragmentation spectrum.

9.7 MASS SPECTRAL CHARACTERIZATION OF ANTIBIOTIC-GSH CONJUGATES

To demonstrate the involvement of GSTs in CTC transformation, *in vitro* conjugation reactions were performed using affinity-purified GSTs from CTC-treated maize.[32] The enzyme reaction products were characterized using liquid chromatography/ion-trap mass spectrometry (LC/MS/MS). Since GSH may also conjugate with xenobiotics nonenzymatically, *in vitro* control reactions were conducted containing all reactants, excluding the GST enzyme (no plant extract added). Results from the LC/MS/MS analysis of the reaction products revealed peaks corresponding to dechlorinated CTC-GSH conjugate (Figure 9.4A). These peaks were characterized by the absence of a chlorine signature in the mass spectra and very short chromatographic retention time, indicating increased polarity relative to the unconjugated CTC (Figure 9.4B). MS/MS analysis revealed an ion with a m/z of 677, which corresponds to the GSH-CTC conjugate with the loss of glycine (MW 75 Da) (Figure 9.4C). Losses of m/z 18 (a water molecule) and m/z129 (glutamic acid) are characteristic fragmentation patterns for GSH. Another important feature of the chromatogram of the CTC-GSH conjugate was the existence of three isomeric peaks characteristic of CTC, which were maintained after conjugation. This is of interest because the products formed during nonenzymatic conjugation were two different conjugates (m/z 654 and m/z 695), each of which eluted as single peaks, with retention times very close to the

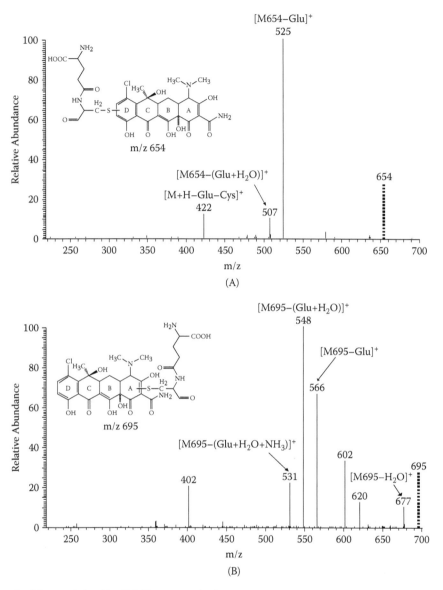

FIGURE 9.5 LC/ESI-MS/MS spectra of m/z 654 (A) and m/z 695 (B) formed in nonenzymatic *in vitro* reactions. Insets represent hypothesized position of GSH conjugation.

CTC standard. Furthermore, the mass spectra of these nonenzymatically formed conjugates indicated that the chlorine atom was retained (Figures 9.5A and 9.5B). This suggests that GSH was able to conjugate to CTC nonenzymatically but at sites other than the chlorine atom.

Enzymatic conjugation of GSH to CTC occurred when either maize or pinto bean GSTs were used to catalyze the reaction. However, while both control samples and CTC-treated samples produced CTC-GSH conjugates, treated maize GST samples produced twice as much product as the control maize GSTs. Pinto beans, on the other hand, showed no difference in product formation between the treated and control samples. In addition, the pinto bean samples only produced 2 to 3% of the conjugate as compared to the maize-treated GST samples. The nonenzymatic conjugation reactions produced about 1% of the maize-treated GST samples, and these reactions incubated six times longer. Therefore, these data suggest that CTC-induced toxicity in pinto beans is likely caused by the inability of GSTs to efficiently detoxify the antibiotic, allowing it to negatively affect growth. These findings also indicate that nonenzymatic conjugation is much too slow to provide any support for the GSTs.

9.8 OTHER ENVIRONMENTALLY IMPORTANT ANTIBIOTICS

The CTC-GSH conjugates observed in our study provide the first evidence that a specific antibiotic can be transformed by plant GSTs. It is important to understand the adaptability of this process more thoroughly by extending this research to additional antibiotics. Due to the ability of maize to respond well to high concentrations of CTC treatments, it was used in the subsequent experiments to determine plant responses to three different classes of antibiotics. In this experiment, maize was treated separately with 1 mg kg^{-1} each of tylosin, sulfadimethoxine, or erythromycin and then harvested as described in the previous section. GST activity assays of the crude extracts from these treated plants showed a significant increase in activity relative to the control on the third day posttreatment (Figure 9.6). Interestingly,

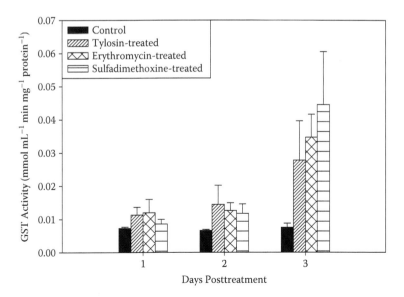

FIGURE 9.6 GST activities of maize plants treated with tylosin, erythromycin, and sulfadimethoxine. Maize plants were transplanted into separate soils treated with 1 μg kg^{-1} of each antibiotic. Plants were harvested and total proteins were extracted daily for 3 days after treatment. GST activity was determined using 1-chloro-2,4-dinitrobenzene (CDNB) as substrate.

sulfadimethoxine-treated plants showed the highest GST activity, which was unexpected considering the known phytotoxicity of sulfadimethoxine to maize.

In vitro conjugation reactions provided insight as to the results of the GST activity assays. It was found that tylosin was the only antibiotic that formed a GSH conjugate when catalyzed by maize GSTs; no conjugate was formed in the absence of GSTs. The inability of GSTs to catalyze the transformation of sulfadimethoxine via GSH conjugation may explain the reported toxicity of this antibiotic toward maize, despite the increased production of GSTs resulting from antibiotic exposure. Erythromycin, lacking nucleophilic sites, also did not form GSH conjugate even in the presence of high level of GSTs. It is possible that Phase I enzymes may act upon erythromycin prior to GSTs, *in vivo*. This is a likely scenario for the antibiotics lacking electrophilic atoms such as erythromycin, where a Phase I enzyme like Cytochrome P450 monooxygenase may add a functional group, *in vivo*, prior to GST-mediated GSH conjugation. This scenario, however, is currently only a hypothesis.

9.9 HYPOTHESIS FOR ANTIBIOTIC-INDUCED PHYTOTOXICITY

According to the findings presented here, it appears that GSTs play a significant role in the transformations and potential detoxification of some antibiotics. However, phytotoxicity in susceptible plants is still observed when GSTs are induced, but cannot detoxify the antibiotics via GSH conjugation, as in the case of sulfadimethoxine in maize. The data suggest that there are two different mechanisms for antibiotic-induced phytotoxicity. First, in the case of CTC toxicity in pinto beans, it appears that the plant GSTs do have an innate ability to bind and conjugate CTC to GSH; however, pinto beans do not appear to have the ability to induce expression of GSTs upon exposure to CTC.

On the other hand, while GST activity assays of sulfadimethoxine-treated maize show a significantly increased enzyme activity relative to the untreated controls, these enzymes are unable to catalyze GSH conjugation of sulfadimethoxine, which retains its phytotoxic effects. Thus, the plant defense system does appear to recognize the sulfadimethoxine as a xenobiotic but cannot detoxify it because the myriad GSTs available cannot bind the compound to effect its transformation. These two hypothetical mechanisms need further investigation and better understanding of the GST biochemical pathways, as well as clarification of mechanisms that trigger a response under antibiotic stress.

9.10 AREAS REQUIRING FURTHER RESEARCH

Residues of veterinary antibiotics that are unintentionally applied to soil through the land application of manure may have phytotoxic effects on susceptible agricultural crops, which could lead to losses in production. Whether the negative effects of antibiotics on crop production are significant enough to outweigh the benefits of using manure as fertilizer is yet to be determined. While plant uptake of antibiotics has been shown to be detrimental to some crops,[5,10,11] it is not yet known whether antibiotics bioaccumulate in edible plant tissues (tubers, leaves, fruit) and potentially expose consumers to low concentrations of antibiotics. For instance, uptake of other organic pollutants (such as trichloroethylene) have been quantified in the fruit of apple and pear trees by Chard and coworkers.[13] On the other hand, our research demonstrates the involvement of the glutathione conjugation pathway in dechlorination of chlortetracycline by maize. Therefore, it is possible that certain agricultural crops may be used for the natural remediation of antibiotic-contaminated soil. A recent study by Gujarathi and Linden[8] showed that exudates from hairy roots of sunflower cultures promote oxidation of oxytetracycline into products devoid of antibiotic activity. Taken together, these studies indicate that there may be several agricultural crops that can be used in phytoremediation of antibiotics from crop fields. However, basic knowledge on the types of plants capable of detoxifying antibiotics and the conditions that favor effective biodegradation of these contaminants in the environment is needed. With more information, one could develop ways to utilize agricultural crops in phytoremediation of antibiotics and devise crop rotation programs that would prevent the accumulation of persistent antibiotics in soil. Phytoremediation is an expanding area of research where field studies have shown to be effective in remediating sites contaminated with pollutants such as heavy metals, pesticides, and explosives.[33] The findings that GST detoxification is involved in the biotransformation of various antibiotics is encouraging; it suggests they may be removed by the same mechanisms as other known pollutants via glutathione conjugation.

Clearly, GST-mediated detoxification of xenobiotics is a beneficial mechanism used by plants in their defense against environmental stresses. However, questions have been raised about the ability of antibiotics to induce herbicide resistance in plants that they are intended to act upon. This is an area of research that could have significant impact on agriculture and needs to be studied more thoroughly. On the other hand, antibiotic-induced herbicide resistance could be beneficial in phytoremediation. For instance, if chlortetracycline can induce GST expression in maize, then perhaps the isozymes induced can enhance the plant's ability to detoxify the herbicide. Using *in vitro* assays similar to those described earlier, preliminary results comparing CTC-treated GSTs (i.e., GSTs isolated from CTC-treated plants) to nontreated control GSTs using three chloroacetanilide herbicides (alachlor, propachlor, metolachlor) as substrates showed that, in maize, CTC-treated GSTs have a reduced ability to conjugate these herbicides to GSH. While this research is still in its early stages, it emphasizes the need for a better understanding of the effects of applying antibiotic-laden manure to crop fields that will subsequently be treated with herbicides.

REFERENCES

1. McManus, P.S., Stockwell, V.O., Sundin, G.W., and Jones, A.L., Antibiotic use in plant agriculture, *Annual Review of Phytopathology,* 40, 443–465, 2002.
2. Capleton, A.C., Courage, C., Rumsby, P., Holmes, P., Stutt, E., Boxall, A.B.A., and Levy, L.S., Prioritising veterinary medicines according to their potential indirect human exposure and toxicity profile, *Toxicology Letters,* 163, 213–223, 2006.
3. Sedlak, D.L. and Pinkston, K.E., Factors affecting the concentrations of pharmaceuticals released to the aquatic environment, *Journal of Contemporary Water Research and Education,* 1, 56–65, 2001.

4. Khachatourians, G.G., Agricultural use of antibiotics and the evolution and transfer of antibiotic-resistant bacteria, *CMAJ*, 159, 1129–1136, 1998.

5. Kumar, K., Gupta, S.C., Baidoo, S.K., Chander, Y. and Rosen, C.J., Antibiotic uptake by plants from soil fertilized with animal manure, *Journal of Environmental Quality*, 34, 2082–2085, 2005.

6. Pomati, F., Castiglioni, S., Zuccato, E., Fanelli, R., Vigetti, D., Rossetti, C., and Cala-mari, D., Effects of a complex mixture of therapeutic drugs at environmental levels on human embryonic cells, *Environ. Sci. Technol.* 40, 2442–2447, 2006.

7. Forni, C., Cascone, A., Fiori, M., and Migliore, L., Sulphadimethoxine and azolla filic-uloides lam: a model for drug remediation, *Water Research*, 36, 3398–3403, 2002.

8. Gujarathi, N.P. and Linden, J.C., Oxytetracycline inactivation by putative reactive oxy-gen species released to nutrient medium of helianthus annuus hairy root cultures, *Bio-technology and Bioengineering*, 92, 393–402, 2005.

9. Batchelder, A.R., Chlortetracycline and oxytetracylcine effects on plant growth and development in soil systems, *Journal of Environmental Quality*, 11, 675–678, 1982.

10. Migliore, L., Civitareale, C., Cozzolino, S., Casoria, P., Brambilla, G., and Gaudio, L., Laboratory models to evaluate phytotoxicity of sulphadimethoxine on terrestrial plants, *Chemosphere*, 37, 2957–2961, 1998.

11. Migliore, L., Cozzolino, S., and Fiori, M., Phytotoxicity to and uptake of enrofloxacin in crop plants, *Chemosphere*, 52, 1233–1244, 2003.

12. Brain, R.A., Reitsma, T.S., Lissemore, L.I., Bestari, K.J., Sibley, P.K., and Solomon, K.R., Herbicidal effects of statin pharmaceuticals in lemna gibba, *Environ. Sci. Tech-nol.*, 40, 5116–5123, 2006.

13. Chard, B.K., Doucette, W.J., Chard, J.K., Bugbee, B. and Gorder, K., Trichloroethyl-ene uptake by apple and peach trees and transfer to fruit, *Environ. Sci. Technol.*, 40, 4788–4793, 2006.

14. Halling-Sorensen, B., Nors Nielsen, S., Lanzky, P.F., Ingerslev, F., Holten Lutzhoft, H.C. and Jorgensen, S.E., Occurrence, fate and effects of pharmaceutical substances in the environment—A review, *Chemosphere*, 36, 357–393, 1998.

15. Thiele-Bruhn, S., Pharmaceutical antibiotic compounds in soils—A review, *Journal of Plant Nutrition and Soil Science*, 166, 145–167, 2003.

16. Kulshrestha, P., Giese, R.F., and Aga, D.S., Investigating the molecular interactions of oxytetracycline in clay and organic matter: Insights on factors affecting its mobility in soil, *Environ. Sci. Technol.* 38, 4097–4105, 2004.

17. Tolls, J., Sorption of veterinary pharmaceuticals in soils: a review, *Environ. Sci. Tech-nol.*, 35, 3397–3406, 2001.

18. Gujarathi, N.P., Haney, B.J., Park, H.J., Wickramasinghe, S.R., and Linden, J.C., Hairy roots of helianthus annuus: a model system to study phytoremediation of tetracycline and oxytetracycline, *Biotechnol. Prog.*, 21, 775–780, 2005.

19. Coleman, J., Blake-Kalff, M., and Davies, E., Detoxification of xenobiotics by plants: chemical modification and vacuolar compartmentation, *Trends in Plant Science*, 2, 144–151, 1997.

20. Schroder, P. and Collins, C., Conjugating enzymes involved in xenobiotic metabolism of organic xenobiotics in plants, *International Journal of Phytoremediation*, 4, 247–265, 2002.

21. Mozer, T.J., Tiemeier, D.C., and Jaworski, E.G., Purification and characterization of corn glutathione s-transferase, *Biochemistry*, 22, 1068–1072, 1983.

22. Hatton, P.J., Dixon, D., Cole, D.J., and Edwards, R., Glutathione transferase activities and herbicide selectivity in maize and associated weed species, *Pesticide Science*, 46, 267–275, 1996.

23. Walton, J.D. and Casida, J.E., Specific binding of a dichloroacetamide herbicide safener in maize at a site that also binds thiocarbamate and chloroacetanilide herbicides, *Plant Physiol.*, 109, 213–219, 1995.
24. Sheehan, D., Meade, G., Foley, V.A., and Dowd, C.A., Structure, function and evolution of glutathione transferases: implications for classification of non-mammalian members of an ancient enzyme superfamily, *Biochemical Journal*, 360, 1–16, 2001.
25. Fujisawa, T., Kurosawa, M., and Katagi, T., Uptake and transformation of pesticide metabolites by duckweed (lemna gibba), *Journal of Agricultural and Food Chemistry*, 54, 6286–6293, 2006.
26. Dixon, D.P., Cummins, I., Cole, D.J., and Edwards, R., Glutathione-mediated detoxification systems in plants, *Current Opinion in Plant Biology*, 1, 258–266, 1998.
27. Aga, D.S., Thurman, E.M., Yockel, M.E., Zimmerman, L.R., and Williams, T.D., Formation and transport of the sulfonic acid metabolites of alachlor and metolachlor in soil, *Environ. Sci. Technol.* 30, 592–597, 1996.
28. Edwards, R., Dixon, D.P., and Walbot, V., Plant glutathione s-transferases: enzymes with multiple functions in sickness and in health., *Trends in Plant Science*, 5, 193–198, 2000.
29. May, M., Vernoux, T., Leaver, C., Van Montagu, M., and Inze, D., Glutathione homeostasis in plants: implications for environmental sensing and plant development, *Journal of Experimental Botany* 49, 649–667, 1998.
30. Aga, D.S., O'Connor, S., Ensley, S., Payero, J.O., Snow, D., and Tarkalson, D., Determination of the persistence of tetracycline antibiotics and their degradates in manure-amended soil using enzyme-linked immunosorbent assay and liquid chromatography-mass spectrometry, *J. Agric. Food Chem.*, 53, 7165–7171, 2005.
31. Habig, W.H., Pabst, M.J., and Jakoby, W.B., Glutathione s-transferases. the first enzymatic step in mercapturic acid formation, *J. Biol. Chem.*, 249, 7130–7139, 1974.
32. Farkas, M.H., Berry, J.O., and Aga, D.S., Chlortetracyline detoxification in maize via induction of glutathione s-transferases after antibiotic exposure, *Environ. Sci. Technol.*, 41, 1450–1456, 2007.
33. Flathman, P.E. and Lanza, G.R., Phytoremediation: current views on an emerging green technology, *Soil and Sediment Contamination* (formerly *Journal of Soil Contamination*), 7, 415–432, 1998.

Part III

Treatment of Pharmaceuticals in Drinking Water and Wastewater

10 Drugs in Drinking Water
Treatment Options

Howard S. Weinberg,
Vanessa J. Pereira, and Zhengqi Ye

Contents

10.1 INTRODUCTION

The high usage of drugs throughout the world, their partial metabolism after ingestion, and inconsistencies in the way they are disposed make their presence in the
aquatic environment inevitable. Their sources in natural waters are not limited to
excretion of parent compounds and their metabolites by individuals and pets but also
include disposal of unused medications to sewage systems, underground leakage
from sewage system infrastructures, release of treated or untreated hospital wastes,
disposal by pharmacies and physicians, and humanitarian drug surplus to domestic
sewage systems. Transmission routes include release to private septic fields; treated
effluent from domestic sewage treatment plants discharged to surface waters; overflow of untreated sewage from storm events and system failures directly to surface
waters; transfer of sewage solids to land; release from agriculture; dung from medi-
cated domestic animals and confined animal feeding operations; direct release via
washing, bathing, or swimming; discharges from industrial manufacturing and
clandestine drug laboratories, as well as illicit drug usage; leaching from defective
landfills; and release from aquaculture.[1] After release into the environment, most
pharmaceutically active compounds (PhACs) are eventually transported to the aqueous domain and are expected to be only partially degraded and transformed into
other products by phototransformative, physicochemical, and biological degradation
reactions. The environmental fate of only a fraction of these compounds has been
evaluated in laboratory studies, and only recently has their occurrence in drinking

water been considered.[2] Their presence at low levels presents analytical challenges, and the environmental impact and public health effects of long-term, low-level exposure and combinatory effects of these compounds requires further study. Particular concern should be raised over those compounds that resist wastewater and drinking water treatment. Treatment processes that are expected to efficiently remove PhACs from water include adsorption using activated carbon, ozonation, ultraviolet (UV), and advanced oxidation processes. The concentrations investigated in laboratory-controlled evaluations of treatment options are often higher than those expected in the aquatic environment. This is a standard procedure in laboratory-scale studies for determination of rate constants and other fundamental process parameters minimizing interference and analytical constraints, since working at these levels allows the analyte reduction to be followed over at least an order of magnitude without involving extensive extractions or preparation procedures.

Although the lifetime ingestion of drinking water may result in consumer exposure to PhACs at an order of more or less than a single therapeutic dose, little is known about long-term, low-level exposure to humans or the potential synergisms that may arise from exposure to multiple compounds. Consequently, it is prudent to consider the options available to prevent PhACs from reaching drinking water.

10.2 OCCURRENCE IN THE AQUATIC ENVIRONMENT

The occurrence of different classes of PhACs, such as analgesics and antiinflammatories, antibiotics, antiepileptics, beta-blockers, blood lipid regulators, contrast media, oral contraceptives, cytostatic, and bronchodilator drugs, has been reported in sewage, surface, ground and drinking water.[3,4,5-13] Maximum occurrence levels for some of these compounds reported in different countries are presented in Table 10.1, which also presents estimates of quantities of PhACs sold for use in human medicine in Germany in 1997, where scrip data are more widely accessible than in the United States, together with secondary wastewater treatment plant (WWTP) removal efficiencies obtained by collecting composite raw influent and final effluent samples over a period of 6 days.[3] The reported WWTP removal efficiencies were highly variable, and during other sampling events conducted at the same plant lower removal efficiencies were observed. These results make clear the need to investigate further the fate and potential remediation options for those PhACs that were found to resist wastewater treatment and that were found in drinking water, such as clofibric acid, iopromide, carbamazepine, diclofenac, and ibuprofen.

10.3 DRINKING WATER TREATMENT

A conventional surface water treatment process that consists of coagulation, flocculation, sedimentation, filtration, and disinfection is often employed in drinking water treatment facilities. Ozone and UV can be used as oxidants and disinfectants, but chlorine and chloramines are most often employed for final disinfection in the United States so that a persistent residual is maintained in the distribution system.

TABLE 10.1

Pharmaceuticals Sold in Germany, Wastewater Removal Efficiency in Germany, and Maximum Concentrations Reported by Several Authors in Different Countries (References Given in Parentheses)

Compound	Therapeutic Class	PhACs Sold in Germany (Tons)	WWTP Removal Efficiency (%)	Maximum Concentrations Reported			
				WWTP Effluent (µg/L)	Surface Water (µg/L)	Groundwater (µg/L)	Drinking Water (µg/L)
Acetaminophen	Analgesics/ nonsteroidal antiinflammatories		>99 (3)	6.0 (6)	10 (7)		
Diclofenac		75 (3)	69 (3)	2.5 (6)	1.2 (6)		0.006 (3)
Ibuprofen		180 (3)	90 (3)	85 (4)	2.7 (4)		0.003 (3)
Ketoprofen				0.38 (6)	0.12 (6)		
Naproxen			66 (3)	3.5 (11)	0.4 (11)		
Oxytetracycline	Antibiotics				0.34 (7)		
Tetracycline					1.0 (5)		
Ciprofloxacin				0.132 (10)	0.07 (9)	0.018 (9)	
Carbamazepine	Antiepileptic	80 (3)	~0 (3)	6.3 (6)	1.1 (6)		0.258 (12)
Metoprolol	Beta-blockers	52 (3)	67 (3)	2.2 (6)	2.2 (6)	1.1 (4)	
Clofibric acid	Antilipemic		51 (3)	1.6 (13)	0.55 (13)	4.0 (4)	0.270 (4)
Iohexol	Contrast media			7.0 (8)	0.5 (8)		
Iopromide		130 (3)		20 (8)	4.0 (8)		
17α-ethinylestradiol	Oral contraceptives	0.050 (3)		0.003 (4)	0.831 (7)		0.086 (3)
Ifosfamide	Cytostatic			2.9 (6)			
Salbutamol	Bronchodilator		>90 (3)		0.035 (6)		

10.3.1 Pretreatment

The potential for the removal of PhACs from drinking water by different treatment processes has been reviewed.[14] Neither coagulation, which is expected to remove only hydrophobic compounds associated with particulate or colloidal material with high organic carbon content, nor flocculation would be efficient tools for removing most of these compounds from water. This was confirmed in a bench-scale coagulation/flocculation/sedimentation study on antibiotic removal with alum and iron salts.[15] Under simulated drinking water treatment conditions with ng/L initial concentrations and 68 mg/L alum coagulant dose at pH 6 ~ 8, the removal rates of sulfamethoxazole and trimethoprim were below 20%, while erythromycin-H_2O could be removed by up to 33%.[16] Erythromycin is much more hydrophobic than sulfamethoxazole and trimethoprim (log Kow = 3.06, 0.89, and 0.91, respectively) and is therefore more likely to partition onto solids and have higher removal rates. Diclofenac, carbamazepine, bezafibrate, and clofibric acid were also poorly removed by ferric chloride precipitation.[17]

10.3.2 Filtration

Adsorption using activated carbon could play an important role in the removal of PhACs, but competition with more polar or larger compounds, including natural organic matter (NOM), has a major impact. For example, even though the addition of 10 to 20 mg/L of powdered activated carbon (PAC) efficiently removed seven antibiotics from distilled water (50 to greater than 99% removal), when the same experiment was conducted in river water the removal decreased by 10 to 20%.[14] The percent removal of sulfonamides, trimethoprim, and carbadox in a filtered (0.45 μm) surface water sample with dissolved organic content of 10.7 mg/L ranged from 49 to 73% and 65 to 100% at PAC dosages of 10 and 20 mg/L, respectively.[15] In another laboratory-controlled batch study at an initial antibiotic concentration of 30 to 150 ng/L, sulfamethoxazole, trimethoprim, and erythromycin-H_2O were removed through PAC adsorption by 21%, 93%, and 65%, respectively, in a natural water of DOC 3.5 mg/L with 4 mg/L PAC dose and contact time of 4 h.[16] These findings show that even though the NOM in surface water may compete with the antibiotics for some of the adsorption sites on PAC, this process might still be somewhat effective as a treatment tool.

Membrane filtration processes are used for water treatment and various industrial applications when production and distribution of water with high chemical and microbiological quality is required. Processes such as reverse osmosis, nanofiltration, and ultrafiltration were found to efficiently remove many PhACs from water.[18,19] However, a major disadvantage of using any of these filtration processes is that the removal of PhACs is accompanied by production of a rejection concentrate that will be much more concentrated than the feed water with respect to suspended and dissolved constituents and will consequently require additional treatment and disposal.[20]

10.3.3 CHLORINE-BASED DISINFECTION

Chlorine disinfectants, such as free chlorine and chloramines, are widely used in drinking water disinfection in the United States. Free aqueous chlorine (HOCl/OCl⁻) can be formed by dissolving chlorine gas or hypochlorite into water, while chloramines can be formed by reaction of free chlorine with ammonia. Free chlorine, as a strong oxidant, is reactive toward many organic pollutants and produces chlorination byproducts. Chloramines, as relatively weaker oxidants, are expected to react much more slowly with organics.[21]

Aliphatic amines react rapidly with HOCl to produce *N*-chloramines, and direct correlations were observed between degree of nucleophilicity of amines and reaction rate with chlorine.[22] *N*-chloro compounds with a hydrogen atom on the carbon α- to the amine could undergo elimination reactions to form an imide, which subsequently hydrolyzes, resulting in bond cleavage between the nitrogen and carbon atoms and removal of the α-carbon side-chain.[23] Aromatic amines tend to form ring-substituted rather than *N*-chlorinated products. Chlorination of phenol proceeds via a typical electrophilic substitution pathway. The phenolate anion has a higher electron density and, hence, reacts quite rapidly with HOCl. Among antibiotics, sulfonamides contain an aromatic amine group that is susceptible to free chlorine attack, while the aliphatic amine groups in the structures of fluoroquinolones, tetracyclines, and macrolides are likely to react with free chlorine to form *N*-chloroamines that can further decompose.

The kinetics and mechanisms of sulfamethoxazole, trimethoprim, and three fluoroquinolone antibiotics (ciprofloxacin, enrofloxacin, and flumequine) in reaction with free chlorine and chloramines have been studied, albeit at a far lower disinfectant-to-analyte ratio (~10) than would be typical with full-scale water treatment (~6000).[24–26] All these antibiotics react rapidly with free chlorine and at slower rates with preformed chloramines, except for flumequine, lacking the characteristic piperizine ring in its structure, which exhibits no apparent reactivity toward chlorine oxidants. Sulfamethoxazole yields an *N*-chlorinated adduct, which rearranges to a ring chlorination product or leads to rupture of the sulfonamide moiety to form the major product *N*-chloro-p-benzoquinoneimine. Reaction of trimethoprim appears to occur primarily on the molecule's trimethoxybenzyl moiety at pH <5, while at pH ≥5 an *N*-chlorinated intermediate is generated, which may react further or rearrange to a number of stable substitution products. Ciprofloxacin reacts very rapidly to form a chloramine intermediate that spontaneously decays in water by piperazine fragmentation. Enrofloxacin reacts relatively slowly to form a chlorammonium intermediate that can catalytically halogenate the parent in aqueous solution. The incomplete oxidation of fluoroquinolones may not completely eliminate the biological effect of these compounds.[26] However, the substantial structural modification resulting from reaction of sulfamethoxazole with free chlorine may lead to a significant reduction of that parent molecule's antimicrobial activities.[25] Nevertheless, sulfonamides were demonstrated to be readily removed from drinking water at near neutral pH although barely affected by monochloramine.[27] The antimicrobial activity of trimethoprim might not be significantly reduced via chlorination due to the formation of primarily stable and multiple-substituted products.[24] The reaction kinetics of carbadox with chlorine are highly pH dependent, with the apparent second-order rate constants ranging from 51.8 to 3.15×10^4 M^{-1} s^{-1} at pH 4 to 11.[28] Carbadox was completely removed to below detection levels by both free chlorine (0.1 mg/L) and monochloramine (1 mg/L) within 1 min of contact time in both deionized and surface water.[27] However, chlorination did not appear to remove the antibacterial activity of the parent compound, as the identified chlorination byproducts of carbadox retain their biologically active functional groups. Although dechlorination agents were not used in many of these studies, they are employed when samples are collected for disinfection byproduct analysis, but only after they have been evaluated to determine if they do not affect the stability of the analytes the samples are being collected for. Such an approach must also be used before field samples are collected for the analysis of PhACs.

Chlorine dioxide (ClO_2) is an alternative to chlorine for disinfection, and it is a highly selective oxidant for specific functional groups of organic compounds.[29] For example, sulfamethoxazole and roxithromycin were reactive to ClO_2 with second-order rate constants (pH 7, 20°C) of 6.7×10^3 M^{-1} s^{-1} and 2.2×10^2 M^{-1} s^{-1}, respectively.[30]

Organic pollutants can associate with dissolved NOM in the aquatic phase via the same mechanism as their sorption to particulate natural organic matter. For example, the sorption coefficient ($K_{d, DOM}$) values of tetracyclines on Aldrich humic acid were 2060 and 1430 L/Kg at pH 4.6 and 6.1, respectively,[31] which is comparable to the K_d value of tetracycline. Association of antibiotics with dissolved NOM may facilitate their transport in the aquatic environment along with the dissolved NOM.

10.3.4 OZONE AND ADVANCED OXIDATION TREATMENT

Ozone (O_3), with its high standard oxidation potential, is expected to oxidize organic compounds more quickly than chlorine or chlorine dioxide. Ozonation is used in drinking water treatment plants to achieve disinfection and oxidation for purposes such as color, taste and odor control, control of iron and manganese, destabilization of colloidal material to aid flocculation, oxidation of disinfection byproduct (DBP) precursors, and elimination of organic compounds.[32] Ozone is a very selective oxidant that will react with double bonds, activated aromatic compounds, and deprotonated amines, whereas hydroxyl (OH) radicals generated when ozone is employed in advanced oxidation mode react with most water constituents with nearly diffusion controlled rates.[33] Diclofenac was efficiently degraded in a semibatch reactor in distilled and river water at an ozone dose of 1 mg/L but not ibuprofen or clofibric acid ($C_o = 2\mu g/L$ and reaction time = 10 min).[34] On the other hand, greater than 70% removal of each in a pilot-scale plant was achieved using 2.5mg/L ozone.[2] These three compounds were effectively degraded by advanced oxidation processes (AOPs) using two O_3 to H_2O_2 ratios (3.7:1.4 and 5.0:1.8 mg/mg).[34]

Batch experiments have also been conducted to determine the degradation rate constants of several pharmaceuticals (bezafibrate, carbamazepine, diazepam, diclofenac, 17α-ethinylestradiol, ibuprofen, sulfamethoxazole, and roxithromycin) with ozone and OH radicals.[35] Carbamazepine, diclofenac, 17α-ethinylestradiol, sulfamethoxazole, and roxithromycin were completely degraded during ozonation, and the rates obtained in pure water solutions could efficiently be applied to predict these compounds' behavior in natural waters (bank filtrate, well, and lake waters) with different dissolved organic carbon content and alkalinity. A rapid reaction of ozone with the double bond in carbamazepine has been reported[36] as has the formation of byproducts containing quinazoline-based functional groups that can be further oxidized by reaction with OH radicals. Moreover, in a full-scale ozonation plant, removal of sulfamethoxazole and carbamazepine, at occurrence levels of 9.7 and 2.4 ng/L in the source water, to below detection limit (<1 ng/L) was observed.[2] The actual mechanism of removal, however, remains unclear. In a pilot-scale study, samples of coagulated/settled/filtered (dual media) water illustrated little change in the levels of carbamazepine in the plant's source water. However, when ozone was introduced prior to coagulation, 66 to 96% reduction was observed, although it could not be determined if this was the result of enhanced coagulation rather than ozone alone.[37]

10.3.5 ULTRAVIOLET (UV) AND ADVANCED OXIDATION TREATMENT

Even though few studies of degradation of PhACs by UV light treatment process exist, in combination with ozone or hydrogen peroxide this process may effectively transform the compounds. UV radiation is widely used for drinking water disinfection in Europe. In the United States, this technology is currently gaining importance, since its use can reduce the chlorine dose applied for final disinfection, therefore, decreasing the levels of DBPs formed.[38] AOPs using UV in place of O_3 can also be used for DBP precursor removal and are attractive due to lower cost and lower potential for producing alternative chemical byproducts.

Degradation of organic compounds can also be obtained using direct photolysis and AOPs. For a compound to be photolabile it needs to have the capacity to absorb light. As a consequence of that light absorption, it will undergo transformation. It can also undergo degradation by receiving energy from other excited species (sensitized photolysis) or by chemical reactions involving very reactive and short-lived species such as hydroxyl-radicals, peroxy-radicals, or singlet oxygen.[39]

UV radiation can be generated using low pressure (LP) lamps that emit monochromatic light at 254 nm or medium pressure (MP) lamps that emit a broadband ranging from 205 to above 500 nm.[40] MP lamps were found to achieve a more effective degradation of bisphenol A, ethinyl estradiol, and estradiol as compared to direct photolysis using LP lamps.[41]

The kinetic degradation constant of carbamazepine and reaction intermediates formed using LP UV/H_2O_2 have been studied.[42] Even though direct photolysis in the absence of H_2O_2 leads to negligible degradation, an effective removal of carbamazepine can be obtained, and pathways were suggested to describe the degradation to acradine, a potentially mutagenic and carcinogenic byproduct.

The effectiveness of ozonation and LP UV/H_2O_2 processes were compared to test the degradation of paracetamol and diclofenac and identify the main byproducts formed.[43,44] Both processes proved to be effective in inducing the degradation of both xenobiotics and achieved degrees of mineralization of approximately 30 and 40% for ozonation and H_2O_2 photolysis, respectively.

LP and MP ultraviolet systems were evaluated in batch-sale laboratory reactors[45,46] to investigate the UV photolysis and UV/H_2O_2 oxidation of PhACs that were found to occur in the aquatic environment and belong to different therapeutic classes. The chemicals investigated were carbamazepine (antiepileptic agent), clofibric acid (metabolite of the lipid regulator clofibrate), iohexol (x-ray contrast agent), ciprofloxacin (antibiotic), naproxen, and ketoprofen (both analgesics). Fundamental direct and indirect photolysis parameters obtained in laboratory-grade water were reported and used to model the UV photolysis and UV/H_2O_2 oxidation of the pharmaceuticals in a surface water using LP and MP lamps. MP-UV photolysis and MP-UV/H_2O_2 oxidation modeling predicted the experimental results very well. The LP-UV model predicted the experimental UV photolysis removals well but underestimated the LP-UV/H_2O_2 oxidation results. Overall, MP lamps proved to be more efficient at maximizing the degradation of the selected group of compounds by both UV photolysis and UV/H_2O_2 oxidation in the bench-scale experiments conducted. The UV fluences required to achieve 50% removal of the selected pharmaceuticals from surface and laboratory-grade water ranged from 34 to 3466 mJ/cm^2 using MP-UV photolysis, 39 to 23105 mJ/cm^2 using LP-UV photolysis, 91 to 257 mJ/cm^2 using MP-UV/H_2O_2 oxidation, and 108 to 257 mJ/cm^2 using LP-UV/H_2O_2 oxidation.

It should be emphasized that the irradiance measurement in a batch reactor is much less complex than in a full-scale UV reactor with multiple light sources. Future studies should validate these results in pilot and full-scale facilities and evaluate whether the use of high UV fluences and AOPs that could degrade a wide variety of organic compounds are economically feasible and competitive when compared to the use of other treatment processes (such as ozonation and membranes). The comparison should take into consideration the possibility of byproduct formation during photolysis, ozonation, and AOPs as well as how to deal with the membrane rejection concentrate.

Photocatalysis is an AOP that has proven to be efficient for application in water disinfection and degradation of pollutants. It relies on the formation of strongly oxidative hydroxy radicals that inactivate microorganisms and degrade resilient organic micropollutants relatively nonselectively and may be carried out in the presence of a semiconductor (heterogeneous photocatalysis) or in the presence of chemical oxidants such as iron and hydrogen peroxide (photo-Fenton).

Among the heterogeneous catalysts widely tested, titanium dioxide (TiO_2) appears to be one of the most promising materials in promoting a good level of disinfection and efficient destruction of chemical compounds. Its advantages include chemical inertness, photostability, absence of toxicity, and low cost, and it has therefore been considered for a wide range of applications.[47] To be catalytically active, titanium dioxide requires irradiation with a source in a wavelength range lower than 390 nm that will induce the photoexcitation of an electron, since it has an energy band gap of about 3.2 eV.

After finding that photocatalysis in controlled laboratory-scale experiments appeared to reduce persistent substances such as NOM, carbamazepine, clofibric acid, iomeprol, and iopromide (even if they are present in a complex matrix),[48] the process was evaluated in combination with microfiltration at the pilot scale to test the degradation of some of these compounds in a model solution without the presence of NOM.[49] High photocatalytic degradation of carbamazepine and clofibric acid was accompanied by elimination of the model solution's dissolved organic carbon showing that the xenobiotics were mineralized to some extent. On the other hand, the photocatalytic degradation of iomeprol was accompanied by formation of degradation products and intermediates. The combination of photocatalysis with cross-flow microfiltration allowed the efficient separation and reuse of the TiO_2 particles.

10.4   CONCLUSION

Despite the low levels of PhACs expected in the environment, their constant infusion can cause them to become more persistent and, therefore, even if the half-lives of these compounds are short, long-term exposure effects and combinatory effects need to be addressed. The ecotoxicological potential of 10 prescription drugs has been evaluated, and even though for most of the substances toxicities were moderate, tests with combinations of various pharmaceuticals revealed stronger effects than expected from the effects measured individually.[50] The potential for indirect human exposure to pharmaceuticals from drinking water supplies was studied, and the margin between potential indirect daily exposure via drinking water and daily therapeutic dose was higher by at least three orders of magnitude or more.[51] Despite these findings, concerns are raised about long-term, low-level human exposure to pharmaceutical products, their metabolites, and degradation compounds via drinking water. Research has also shown that the presence of antibiotics in the aquatic environment poses a potential threat to ecosystem function and human health.[10,52] Minor side effects from prescribed drugs are common, and even though they are usually outweighed by the health benefits of the medication, they can possibly have adverse effects in routine unintended exposure.

This chapter has attempted to consider what does and does not work in remediating the presence of drugs in drinking water. There is no question that subtherapeutic doses of these compounds are finding their way into the surface and groundwaters that ultimately become consumers' drinking water and that, for now, the levels found in that finished drinking water are most often close to the analytical limits of detection. Nevertheless, there are insufficient occurrence data for us to conclude that all conventional drinking water treatment plants are generating a finished product that is "drug free." If the surface water source is impacted by an upstream wastewater discharge from a major population center, there is a strong likelihood that a small downstream conventional drinking water plant will receive elevated levels of pharmaceutically active compounds that will survive to some degree into the finished drinking water. One study even suggests that subsequent chlorination of such water could generate an even more toxic end product so that a switch to chloramination, already favored for reducing disinfection byproduct formation, might be preferable. Also, the introduction of advanced treatments such as AOPs or photolysis that target reactive centers in the chemical contaminant offer some degree of remediation that, when coupled with adsorption, appear to offer a fair degree of protection to the consumer.

Further studies should focus on evaluating the environmental and human impact of these compounds to determine to what extent they should be removed from drinking water. In addition, economic viability studies of using higher UV fluences than those typically employed for drinking water treatment should be conducted. Furthermore, the fate of the target analytes during treatment should also be studied to determine if the pharmaceuticals are being mineralized or formation of photolysis products occurs. Last, if photolysis byproduct formation occurs, their toxicity and treatability should also be addressed.

REFERENCES

1. Daughton, C.G. In *Water: Science and Issues*; Dasch, J., Ed.; Macmillan Reference USA: New York, 2003; Vol. 1, pp 158–164.
2. Snyder, S.A. et al. Occurrence of EDCs and pharmaceuticals in US drinking waters. Paper presented at the American Water Works Association Water Quality Technology Conference, Quebec City, November, 2005.
3. Ternes, T. In *Pharmaceuticals and Personal Care Products in the Environment: Scientific and Regulatory Issues*; Daughton, C.G. and Jones-Lepp, T.L., Eds.; American Chemical Society: Washington, DC, 2001; pp 39–54.
4. Heberer, T. Occurrence, fate, and removal of pharmaceutical residues in the aquatic environment: A review of recent research data, *Toxicol. Lett.* 2002, 131, 5–17.
5. Halling-Sørensen, B., Nors Nielson, S., Lanzky, P.F., Ingerslev, F., Holten Lutzhoft, H.C., and Jorgensen, S.E. Occurrence, fate and effects of pharmaceutical substances in the environment—a review, *Chemosphere* 1998, 36, 357–393.
6. Daughton, C.G. and Ternes, T.A. Pharmaceuticals and personal care products in the environment: agents of subtle change?, *Environ. Health Persp.* 1999, 107, 907–938.
7. Kolpin, D.W., Furlong, E.T., Meyer, M.T., Thurman, E.M., Zaugg, S.D., Barber, L.B., and Buxton, H.T. Pharmaceuticals, hormones, and other organic wastewater contaminants in US streams, 1999–2000: a national reconnaissance, *Environ. Sci. Technol.* 2002, 36, 1202–1211.
8. Putschew, A., Schittko, S., and Jekel, M. Quantification of triiodinated benzene derivatives and X-ray contrast media in water samples by liquid chromatography-electrospray tandem mass spectrometry, *J. Chromatogr. A.* 2001, 930, 127–134.
9. Ye, Z., Weinberg, H.S., and Meyer, M.T. Occurence of antibiotics in drinking water, *4th International Conference on Pharmaceuticals and Endocrine Disrupting Chemicals in Water*, Minneapolis, MN, 2004.
10. Costanzo, S.D., Murby, J., and Bates, J. Ecosystem response to antibiotics entering the aquatic environment, *Mar. Pollut. Bull.* 2005, 51, 218–223.
11. Öllers, S., Singer, H.P., Fässler, P., and Müller, S.R. Simultaneous quantification of neutral and acidic pharmaceuticals and pesticides at the low-ng/l level in surface and waste water, *J. Chromatogr. A.* 2001, 911, 225–234.
12. Stackelberg, P.E., Furlong, E.T., Meyer, M.T., Zaugg, S.D., Henderson, A.K., and Reissman, D.B. Persistence of pharmaceutical compounds and other organic wastewater contaminants in a conventional drinking-water-treatment plant, *Sci. Total Environ.* 2004, 329, 99–113.
13. Ternes, T.A. Occurrence of drugs in German sewage treatment plants and rivers, *Water Res.* 1998, 32, 3245–3260.
14. Snyder, S.A., Westerhoff, P., Yoon, Y., and Sedlak, D.L. Pharmaceuticals, personal care products, and endocrine disruptors in water: implications for the water industry, *Environ. Eng. Sci.* 2003, 20(5), 449–469.
15. Adams, C., Wang, Y., Loftin, K., and Meyer, M. Removal of antibiotics from surface and distilled water in conventional water treatment processes, *J. Environ. Eng.* 2002, 128(3), 253–260.

16. Westerhoff, P., Yoon, Y., Snyder, S., and Wert, E. Fate of endocrine-disruptor, pharmaceutical, and personal care product chemicals during simulated drinking water treatment processes, *Environ. Sci. Technol.* 2005, 39(17), 6649–6663.

17. Sacher, F., Haist-Gulde, B., Brauch, H.-J., Zullei-Seibert, N., Preuss, G., Meisenheimer, M., Welsch, H., and Ternes, T.A. Behavior of selected pharmaceuticals during drinking-water treatment, *Book of Abstracts, 219th ACS National Meeting, San Francisco, CA, March 26–30,* 2000, ENVR-045.

18. Kim, S.D., Cho, J., Kim, I.S., Vanderford, B.J., and Snyder, S.A. Occurrence and removal of pharmaceuticals and endocrine disruptors in South Korean surface, drinking, and waste waters, *Water Res.* 2007, 41(5), 1013–1021.

19. Yoon, Y., Westerhoff, P., Snyder, S.A., and Wert, E.C. Nanofiltration and ultrafiltration of endocrine disrupting compounds, pharmaceuticals and personal care products, *J. Membrane Sci.* 2006, 270(1–2), 88.

20. USEPA, *Membrane Filtration Guidance Manual,* U.S. Environmental Protection Agency, Cincinnati, OH, 2005.

21. Rice, R.G. and Gomez-Taylor, M. Occurrence of by-products of strong oxidants reacting with drinking water contaminants—Scope of the problem, *Environ. Health Perspect.* 1986, 69, 31–44.

22. Abia, L., Armesto, X.L., Canle, L.M., Garcia, M.V., and Santaballa, J.A. Oxidation of aliphatic amines by aqueous chlorine, *Tetrahedron* 1998, 54, 521–530.

23. Armesto, X.L., Canle, M.L., Garcia, M.V., and Santaballa, J.A. Aqueous chemistry of N-halo-compounds, *Chem. Soc. Rev.* 1998, 27, 453–460.

24. Dodd, M.C. Chemical oxidation of aquatic antibiotic microcontaminants by free and combined chlorine. MS thesis, Georgia Institute of Technology, 2003.

25. Dodd, M.C. and Huang, C.-H. Transformation of the antibacterial agent sulfamethoxazole in reactions with chlorine: kinetics, mechanisms, and pathways, *Environ. Sci. Technol.* 2004, 38, 5607–5615.

26. Dodd, M.C., Shah, A.D., Von Gunten, U., and Huang, C.-H. Interactions of fluoroquinolone antibacterial agents with aqueous chlorine: reaction kinetics, mechanisms, and transformation pathways, *Environ. Sci. Technol.* 2005, 39(18), 7065–7076.

27. Chamberlain, E. and Adams, C. Oxidation of sulfonamides, macrolides, and carbadox with free chlorine and monochloramine, *Water Res.* 2006, 40, 2517–2526.

28. Shah, A.D., Kim, J.-H., and Huang, C.-H. Reaction kinetics and transformation of carbadox and structurally related compounds with aqueous chlorine, *Environ. Sci. Technol.* in press.

29. Hoigné, J. and Bader, H. Kinetics of reactions of chlorine dioxide (OClO) in water – I Rate constants for inorganic and organic compounds, *Water Res.* 1994, 28, 45–55.

30. Huber, M.M., Korhonen, S., Ternes, T.A., and Von Gunten, U. Oxidation of pharmaceuticals during water treatment with chlorine dioxide, *Water Res.* 2005, 39, 3607–3617.

31. Sithole, B.B. and Guy, R.D. Models for oxytetracycline in aquatic environments. II. Interaction with humic substances, *Water Air Soil Pollut.* 1987, 32, 315–321.

32. Haas, C.N., Ed. *Water Quality and Treatment*; 4th edition; McGraw-Hill: New York, 1990.

33. von Gunten, U. Ozonation of drinking water: part I. oxidation kinetics and product formation, *Water Res.* 2003, 37, 1443–1467.

34. Zwiener, C. and Frimmel, F.H. Oxidative treatment of pharmaceuticals in water, *Water Res.* 2000, 34, 1881–1885.

35. Huber, M.M., Canonica, S., Park, G.-Y., and von Gunten, U. Oxidation of pharmaceuticals during ozonation and advanced oxidation processes, *Environ. Sci. Technol.* 2003, 37, 1016–1024.

36. McDowell, D.C., Huber, M.M., Wagner, M., von Gunten, U., and Ternes, T.A. Ozonation of carbamazepine in drinking water: identification and kinetic study of major oxidation products, *Environ. Sci. Technol.* 2005, 39(20), 8014–8022.
37. Hua, W., Bennett, E.R., and Letcher, R.J. Ozone treatment and the depletion of detectable pharmaceuticals and atrazine herbicide in drinking water sourced from the Upper Detroit River, Ontario, Canada, *Water Res.* 2006, 40, 2259–2266.
38. Sharpless, C.M. and Linden, K.L. UV photolysis of nitrate: effects of natural organic matter and dissolved inorganic carbon and implications for UV water disinfection, *Environ. Sci. Technol.* 2001, 35, 2949–2955.
39. Schwarzenbach, R.P., Gschwend, P.M., and Imboden, D.M. *Environmental Organic Chemistry*; Wiley-Interscience: New York, 1993.
40. Sharpless, C.M. and Linden, K.L. Experimental and model comparisons of low- and medium-pressure Hg lamps for the direct and H_2O_2 assisted UV photodegradation of N-nitrosodimethylamine in simulated drinking water, *Environ. Sci. Technol.* 2003, 37, 1933–1940.
41. Rosenfeldt, E. and Linden, K.G. Degradation of endocrine disrupting chemicals bisphenol A, ethinyl estradiol, and estradiol during UV photolysis and advanced oxidation processes, *Environ. Sci. Technol.* 2004, 38(20), 5476–5483.
42. Vogna, D., Marotta, R., Andreozzi, R., Napolitano, A. and d'Ischia, M. Kinetic and chemical assessment of the UV/H_2O_2 treatment of antiepileptic drug carbamazepine, *Chemosphere* 2004, 54(4), 497–505.
43. Andreozzi, R., Caprio, V., Marotta, R., and Vogna, D. Paracetamol oxidation from aqueous solutions by means of ozonation and H_2O_2/UV system, *Water Res.* 2003, 37(5), 993–1004.
44. Vogna, D., Marotta, R., Napolitano, A., Andreozzi, R. and d'Ischia, M. Advanced oxidation of the pharmaceutical drug diclofenac with UV/H2O2 and ozone, *Water Res.* 2004, 38(2), 414–422.
45. Pereira, V.J., Weinberg, H.S., Linden, K.G., and Singer, P.C. UV degradation kinetics and modeling of pharmaceutical compounds in laboratory grade and surface water via direct and indirect photolysis at 254 nm, *Environ. Sci. Technol.* 2007, 41(5) 1682–1688.
46. Pereira, V.J., Linden, K.G., and Weinberg, H.S. Evaluation of UV irradiation for photolytic and oxidative degradation of pharmaceutical compounds in water. *Water Res.* (in press), available online June 2007.
47. Fujishima, A., Rao, T.N., and Tryk, D.A. Titanium dioxide photocatalysis, *J. Photochem. Photobiol. C: Photochem. Rev.* 2000, 1, 1–21.
48. Doll, T.E. and Frimmel, F.H. Photocatalytic degradation of carbamazepine, clofibric acid and iomeprol with P25 and Hombikat UV100 in the presence of natural organic matter (NOM) and other organic water constituents, *Water Res.* 2005, 39(2–3), 403–411.
49. Doll, T.E. and Frimmel, F.H. Cross-flow microfiltration with periodical back-washing for photocatalytic degradation of pharmaceutical and diagnostic residues-evaluation of the long-term stability of the photocatalytic activity of TiO_2, *Water Res.* 2005, 39(5), 847–854.
50. Cleuvers, M. Aquatic ecotoxicity of pharmaceuticals including the assessment of combination effects, *Toxicol. Lett.* 2003, 142, 185–194.
51. Webb, S., Ternes, T., Gibert, M., and Olejniczak, K. Indirect human exposure to pharmaceuticals via drinking water, *Toxicol. Lett.* 2003, 142, 157–167.
52. Schwartz, T., Kohnen, W., Jansen, B., and Obst, U. Detection of antibiotic-resistant bacteria and their resistance genes in wastewater, surface water, and drinking water biofilms, *FEMS Microbiol. Ecol.* 2003, 43, 325–335.

11 Removal of Endocrine Disruptors and Pharmaceuticals during Water Treatment

Shane A. Snyder, Hongxia Lei, and Eric C. Wert

Contents

11.1   INTRODUCTION

Over the past decade a great amount of interest has arisen regarding the occurrence and fate of trace organic contaminants in the aquatic environment. Of particular concern are human hormones and pharmaceuticals, many of which are ubiquitous contaminants in conventional municipal wastewater treatment plant effluents when measured with ng/L detection limits. As analytical procedures and bioassay techniques become more readily available and increasingly sensitive, new contaminants will be discovered. The presence or absence of any chemical in a wastewater effluent is essentially a function of analytical detection capability. This poses a unique challenge for drinking water treatment plants intent on the removal of organic contaminants, as complete removal is merely a reflection of reporting limits. The project described in this chapter was designed to investigate the attenuation of a group of structurally diverse emerging contaminants by conventional and advanced water treatment processes.

These data represent a portion of the findings from a study sponsored by the American Water Works Association Research Foundation (AwwaRF) entitled "Removal of EDCs and Pharmaceuticals in Drinking and Reuse Processes" (Project #2758). This project sought to determine the treatment efficacy of various processes at bench, pilot, and full scale for the removal of emerging contaminants by monitoring the concentration decrease of the parent compounds. Reaction byproducts and the corresponding structural changes were beyond the scope of this study; however, oxidative processes generally do not result in appreciable mineralization, and reaction byproducts are expected. This chapter focuses on findings related to oxidation, magnetic ion-exchange, and activated carbon. For the sake of brevity, experiments using Colorado River water (CRW) will be the highlight of this chapter while noting AwwaRF Project 2758 investigated several natural waters.

This study shows that the majority of emerging contaminants can be readily oxidized using ozone or ultraviolet (UV)-advanced oxidation. Free chlorination effectively removed more target compounds than chloramination. Magnetic ion-exchange (MIEX®) provided minimal contaminant removal; however, contaminants that were negatively charged at ambient pH were well removed. Activated carbon, both in powdered and granular forms, was effective for contaminant adsorption. Carbon type, contact time, and dose or regeneration are influential parameters in removal efficacy by activated carbon. Although not discussed in this chapter, tight membrane filtration (reverse osmosis and nanofiltration) was more effective than loose membrane filtration (ultrafiltration and microfiltration). No single treatment process was capable of removing all contaminants consistently to below the analytical method reporting limit. Moreover, each treatment process provided advantages and disadvantages that will be discussed in this chapter. A multibarrier approach would provide the most comprehensive removal strategy for the treatment of organic contaminants.

11.2 BACKGROUND

In 1965, Stumm-Zollinger and Fair of Harvard University published the first known report indicating that steroid hormones are not completely eliminated by wastewater treatment.[1] In an article published in 1970, Tabak and Bunch investigated the fate of human hormones during wastewater treatment and stated "since they (hormones) are physiologically active in very small amounts, it is important to determine to what extent the steroids are biodegraded."[2] As early as the 1940s, scientists were aware that certain chemicals had the ability to mimic endogenous estrogens and androgens.[3,4] In 1977 researchers from the University of Kansas published the first known report specifically addressing the discharge of pharmaceuticals from a wastewater treatment plant.[5] Despite these early findings, the issue of steroids and pharmaceuticals in wastewater outfalls did not gain significant attention until the 1990s, when the occurrence of natural and synthetic steroid hormones in wastewater was linked to reproductive impacts in fish living downstream of outfalls.[6–8]

Since the initial link between trace contaminants (sub-µg/L) in wastewater effluents and ecological impacts in receiving waters, many studies have focused on the occurrence of these contaminants.[9–17] As a result, pharmaceuticals and steroid hormones have been detected in many water bodies around the world.[16,18,19] One major contributor of such widespread contamination is municipal wastewater discharge, which impacts surface water quality by contaminating receiving water bodies with chemicals not completely removed by current wastewater treatment processes. Indirect potable water reuse, either planned or unplanned, can occur when wastewater treatment plant discharge comprises a significant portion of the receiving stream's total flow. In some cases, effluent-dominated surface waters are used as source waters for drinking water treatment facilities. Global water sustainability depends in part upon effective reuse of water. In particular the reuse of municipal wastewater is critical for irrigation and augmentation of potable water supplies. However, public perception and concern regarding trace hormones and pharmaceuticals has generated resistance to reuse projects. It is necessary to obtain accurate information on the elimination of these contaminants from wastewater, the impact of wastewater discharge on surface water or groundwater drinking water supplies, and the removal efficiency of the remaining contaminants by conventional and advanced drinking water treatment processes.

A significant number of articles have investigated the fate of trace hormones and pharmaceuticals through water treatment processes.[13,20–33] The ability of a particular treatment process to remove organic contaminants depends mostly on the structure and concentration of the contaminant. In addition, the operational parameters of the process (e.g., oxidant dose and contact time) will also determine the degree of attenuation of a particular contaminant.

Tixier et al. studied the concentration variation of six pharmaceuticals in wastewater effluent and river waters.[34] The concentration profiles of these contaminants in the water column of a lake in Switzerland were measured over 3 months. Phototransformation, adsorption, and biodegradation were identified as the main elimination processes for these contaminants in the lake water. Loraine and Pettigrove reported the occurrence of pharmaceuticals and personal-care products in raw drinking water at four water treatment plants impacted by sewage treatment plant effluent.[35] Some compounds were detected in finished drinking water, demonstrating that the treatment processes employed were not capable of complete removal of all trace contaminants. The concentrations in raw water showed high seasonal variations during low and high flow conditions in one of its two source waters. Temperature change can also result in seasonal variation of pharmaceuticals, as observed in Finland at a drinking water plant that was impacted by wastewater.[36]

The treatment efficiency of pharmaceuticals was evaluated in a few full-scale treatment facilities. Results obtained from one with two-stage coagulation, followed by granular activated carbon (GAC) filtration and chlorine disinfection, have confirmed previous findings from laboratory investigations,[36] where pharmaceuticals were poorly removed by coagulation, while GAC was very efficient in removing these contaminants.[25,32,37] Another team investigated a conventional full-scale drinking water treatment plant, consisting of coagulation, filtration, and disinfection for the treatment efficiency of organic contaminants. Pharmaceuticals and other wastewater-related organic contaminants were monitored during low flow conditions.[38] Forty of the monitored compounds were detected in raw or source water and several were detected in finished water, suggesting some contaminants were persistent and demonstrated the inefficiency of commonly employed treatment processes.

To date, the number and type of organic contaminants investigated are still limited considering the large pool of undetected chemicals. The sparse information on the occurrence and removal efficiency of many yet-to-be investigated organic contaminants stimulated the work presented in this chapter, which was performed at both bench and pilot scales, and subsequently evaluated by full-scale investigations.

11.3 SELECTION OF TARGET COMPOUNDS

Target compounds were selected from various classifications of organic contaminants. Classes included pharmaceuticals, hormones, personal-care products, suspected endocrine disruptors, and other model compounds. Pharmaceuticals were selected to encompass several therapeutic classifications including: antibiotics, analgesics, birth control, psychoactive drugs, and cholesterol-lowering medications. Target compounds included seven hormones that represent both estrogens and androgens. Several suspected endocrine disruptors, including atrazine and triclosan, were also included. These compounds present a diverse group of chemical structures, such that the fate of future contaminants can be predicted based upon their structure. Four primary criteria were considered during the selection process.

The first criterion was likelihood of occurrence in the environment. Previous publications from peer-reviewed journals and government reports were considered. Pharmaceuticals have been identified in surface waters, groundwaters, and wastewaters in the United States, Europe, Asia, and other continents around the world.[35,39–44] Among these, the nationwide occurrence surveys conducted by U.S. Geological Survey (USGS) draw the most attention, for which pharmaceuticals comprise a significant portion of the contaminants investigated.[16,45] These reports were considered during the initial selection process to warrant the inclusion (or rejection) of various candidate contaminants.

The second criterion was toxicological relevance and public interest. Pharmaceuticals were selected based on concerns over long-term, low-dose exposure, which have been implicated as a possibility in increased antibiotic resistance and allergic reactions.[46–48] Steroids were considered because they are biologically active at low concentrations and frequently included in many studies due to serious concern of their estrogenic effect to aquatic species.

The third criterion was structural diversity and treatability. This is reflected by the wide range of molecular weight (MW), water solubility (S), octanol/water partitioning coefficient (log K_{ow}), and acidic/basic properties (Table 11.1). The removal of each compound is intrinsically linked to its chemical structure and the particular treatment process. For instance, phenolic compounds are amenable to oxidation treatment and hydrophobic compounds can readily adsorb to GAC. An herbicide, a pesticide, two fragrances, and a flame retardant were included to increase structural diversity. This breadth of chemical structure ensures a reasonable span of treatability for an objective evaluation of each unit process.

The fourth and final criterion was analytical capability. The limiting factors are the availability of analytical standards and method performance for specific compounds with analytical equipment available. Compounds for which purified standards were not commercially available were not included. Significant quantities of compounds were required in order to conduct the spiked batch and dynamic

TABLE 11.1
Physicochemical Properties of Selected Pharmaceuticals*

Compounds	Classes	CAS	MW	S (mg/L)	Log K_{OW}	pKa
Acetaminophen	Analgesic	103-90-2	151.2	1.40E+4	0.46	9.38
Androstenedione	Hormone	63-05-8	286.4	57.8	2.75	na
Atrazine	Herbicide	1912-24-9	215.7	34.7	2.61	1.7
Caffeine	Psychoactive	58-08-2	194.2	2.16E+4	−0.07	10.4
Carbamazepine	Psychoactive	298-46-4	236.3	18[50]	2.45	13.9[51]
DEET**	Insect repellant	134-62-3	191.3	9.9[52]	2.18	0.7 (est)
Diazepam	Psychoactive	439-14-5	284.7	50	2.82	3.4
Diclofenac	Analgesic	15307-86-5	296.2	2.37	4.51	4.15
Dilantin	Psychoactive	57-41-0	252.3	32	2.47	8.33
Erythromycin	Antimicrobial	114-07-8	733.9	1.44 (est)	3.06	8.88
Estriol	Hormone	50-27-1	288.4	441 (est)	2.45	9.85 (est)
Estradiol	Hormone	50-28-2	272.4	3.6	4.01	10.4[53]
Estrone	Hormone	53-16-7	270.4	30	3.13	10.4[53]
Ethynyl estradiol	Hormone	57-63-6	296.4	11.3	3.67	10.4[54]
Fluoxetine	Psychoactive	54910-89-3	309.3	60.3 (est)	4.05	10.3 (est)
Galaxolide	Fragrance	1222-05-5	258.4	1.75[55]	5.9 [55]	na
Gemfibrozil	Antilipidemic	25812-30-0	250.3	19 (est)	4.33 (est)	4.42
Hydrocodone	Analgesic	125-29-1	299.4	6870 (est)	2.16 (est)	8.35 (est)
Ibuprofen	Analgesic	15687-27-1	206.3	21	3.97	4.91
Iopromide	x-ray contrast agent	73334-07-3	791.1	23.8 (est)	−2.05	10.2 (est)
Meprobamate	Psychoactive	57-53-4	218.3	4700	0.7	10.9 (est)
Metolachlor	Pesticide	51218-45-2	283.8	530	3.13	na
Musk ketone	Fragrance	81-14-1	294.3	0.46,[56] 1.9[57]	4.3[56]	na
Naproxen	Analgesic	22204-53-1	230.3	15.9	3.18	4.15
Pentoxifylline	Vasodilator	6493-05-6	278.3	7.70E+4	0.29	1.49 (est)
Progesterone	Hormone	57-83-0	314.5	8.81	3.87	na
Sulfamethoxazole	Antimicrobial	723-46-6	253.3	610	0.89	5.5[58]
TCEP***	Flame retardant	115-96-8	285.5	7000	1.44	na
Testosterone	Hormone	58-22-0	288.4	23.4	3.32	na
Triclosan	Antimicrobial	3380-34-5	289.5	10	4.76	7.9[59]
Trimethoprim	Antimicrobial	738-70-5	290.3	400	0.91	7.12

*Unless indicated, all are experimental values from Environmental Science Database SRC PhysProp
** Chemical name: N,N-diethyl-meta-toluamide
*** Chemical name: Tri(chloroethyl)phosphate

treatment studies. Some compounds could not be considered due to a lack of, or the cost of, available standards. Since all compounds were intended to be extracted by a single solid-phase extraction (SPE) method, some analytes could not be accommodated with the operational parameters considered.[49]

Table 11.1 provides the list of target compounds discussed in this chapter following the aforementioned selection criteria. Also summarized in this table are the compound classification and molecular properties pertinent to water treatment processes. These particular properties provide information on molecular size, hydrophobicity, and molecular charge under ambient pH.

11.4 ANALYTICAL METHODS

One-liter samples were collected in silanized amber glass bottles for bench-, pilot-, and full-scale investigations. Samples were preserved by adjusting the pH to 2 for microbial activity control and ascorbic acid was used to quench disinfectant residual.[17] The pH adjustment was found important to prevent the biodegradation of certain steroids and the pharmaceuticals trimethoprim, acetaminophen, and fluoxetine.[17] Grab samples were taken by personnel from each utility with general water qualities monitored simultaneously. For each plant, at least one site was selected for duplicate samples (usually the raw water) for quality control. Additionally, travel blanks, laboratory blanks, and matrix spikes were prepared with each sampling event. The samples were shipped back to the laboratory in an ice-packed cooler by overnight express shipping.

All samples were extracted in batches of six samples using automated-SPE (ASPE) utilizing 500-mg hydrophilic-lipophilic balance (HLB) cartridges from Waters Corp. (Millford, Massachusetts). Cartridges were sequentially preconditioned by methyl tertbutyl ether (MTBE), methanol, and reagent water. The absorbed analytes were then eluted with 10/90 (v/v) methanol/MTBE solution and pure methanol, followed by concentration to a volume of 1 mL with nitrogen.

Pharmaceuticals, hormones, and atrazine were analyzed by liquid chromatography with tandem mass spectrometric detection (LC-MS/MS) using electrospray ionization (ESI) in both positive and negative modes and atmospheric pressure chemical ionization (APCI) in positive mode following a method published previously.[17]

The remaining chemicals were analyzed by gas chromatography with tandem mass spectrometric detection (GC-MS/MS).[49] The same SPE procedure was used; however, a portion of the resulting extract was removed and reextracted using dichloromethane in order to remove compounds that interfere with GC-MS/MS analysis.[49]

Method performance in terms of instrument detection limits (IDLs) and spiked recoveries remain unchanged from previously published work.[17,49] The spike recoveries of most target compounds were generally above 70% for LC-MS/MS and above 60% for GC-MS/MS analysis with the exception of acetaminophen and galaxolide (41% and 30%, respectively). A conservative method reporting limit (MRL) of 1 ng/L for most LC-MS/MS compounds and 10 ng/L for GC-MS/MS compounds was established for this method. While concentrations are reported without adjustment for the recovery, all treatment data are shown as percent removal through a treatment process. The percent removal is not impacted by analytical recovery, since the recovery was constant in both the "raw" and "finished" water of each process.

11.5 BENCH-SCALE EVALUATIONS

11.5.1 BENCH-SCALE EXPERIMENTAL PROCEDURES

Evaluation of contaminant oxidation by chlorine, chloramine, ozone, and UV, and adsorption by GAC and MIEX was performed using Colorado River water (CRW) spiked with 100 to 300 ng/L of each of the target compounds. The spiking concentration was selected such that removal to the MRL would provide approximately 2-log removal (99%). Experimental conditions for evaluation are summarized in Table 11.2. The CRW quality, as sampled from the Southern Nevada Water Authority intakes in Lake Mead, is relatively stable with average values provided in Table 11.3.

Chlorination experiments were performed using liquid sodium hypochlorite (NaOCl, Fisher Scientific, Pittsburgh, Pennsylvania). Stock solutions of chlorine were initially prepared in deionized (DI) water at 1200 mg/L. Chlorine was dosed directly into 1-L bottles containing the source water. Chlorine doses were determined based upon the chlorine demand in order to achieve a residual goal of approximately 0.5 mg/L after 24 hours. Free chlorine residuals were measured by the N,N-diethyl-p-phehylenediamine (DPD) Method using a Hach DR4000 spectrophotometer (U.S. Environmental Protecton Agency [USEPA]-approved Hach Method #8021, Hach Company, Loveland, Colorado). Experiments were conducted at ambient water pH and at a suppressed pH of 5.5. After the 24-hour contact time, residual chlorine was quenched with 50 mg/L of ascorbic acid. Analytical surrogates sensitive to chlorine were added just prior to solid-phase extraction; therefore, analytical recovery of these surrogates provided definitive information on whether or not the chlorine residual was quenched.

Chloramine experiments were performed by first adding ammonia to a 1-L sample of raw water followed by sodium hypochlorite addition. Stock solutions of each chemical were created from 29% ammonium hydroxide (J.T. Baker, Phillipsburg, New Jersey) and 5% sodium hypochlorite (NaOCl, Fisher Scientific). A chlorine: ammonia ratio of 4:1 was targeted because this is commonly used in drinking water treatment. This sequence of chemical addition was selected over preformed mono-chloramine solution to closely simulate actual water treatment plant conditions. The

---

TABLE 11.2
Experimental Matrix for Bench-Scale Studies

Tests	Dosage	Contact Time
Ozone	2.5 mg/L	5 min
Free chlorine	3, 3.5 mg/L	24 hr
UV	40 mJ/Cm^2	
Chloramine	2, 3 mg/L	24 hr
MIEX	5, 15, 20 mL/L	10 min
GAC	n/a	7.6 min EBCT*

*EBCT: Empty bed contact time.

TABLE 11.3

Average Water Quality of Colorado River Water

Water Quality Parameter	Average	Minimum	Maximum
pH	7.79	7.68	7.99
Temperature (°C)	21	20	22
Turbidity (NTU)	0.73	0.35	1.5
UV254 (1/cm)	0.029	0.025	0.037
Alkalinity (mg/L)	133	131	134
Bromide (mg/L)	0.08	0.05*	0.10
Chloride (mg/L)	84	75	94
Hardness (mg/L)	288	260	310
Sulfate (mg/L)	246	230	250
SUVA	1.2	—	—
TDS (mg/L)	619	602	642
TOC (mg/L)	3.23	2.9	3.4

* Below method reporting level.

experiments were conducted at room temperatures (approximately 20°C) at ambient pH. The chloramine residual was quenched with ascorbic acid using a ratio of 20:1 ascorbic acid:chlorine.

Ozone experiments were conducted by injecting a high concentration of dissolved ozone into a 1-L sample bottle containing the source water. Dissolved ozone stock solution was prepared by dissolving a high concentration of gaseous ozone into DI water at 2°C. Dissolved ozone stock solution concentrations and dissolved ozone residuals were measured spectrophotometrically using the indigo method (Standard Methods 4500-O_3). The ozone dose was determined using an ozone demand/decay curve generated for the CRW.[30]

MIEX is a process that utilizes magnetic polymer microspheres as an ion exchange resin. The resin is composed of a polymer and inorganic magnetic materials such as Fe_3O_4, Fe_2O_3, nickel, and cobalt so that the resin exhibits the characteristics of both ionic exchange and magnetism in order to achieve rapid and improved separation for regeneration and recycle. MIEX resin (Orica Watercare, Inc., Watkins, Colorado) from an on-site pilot plant provided the resin for the jar testing. Gravity separation of the resin is very efficient because of the "magnetically" enhanced agglomeration of individual resin beads that yields resin settling against the rapidly rising water in the settler. Since the MIEX resin used for bench-scale testing was cycled several times through the pilot system, the binding capacity was likely decreased, and was more representative of a resin that would be in service as compared to virgin material. MIEX concentrations of 5, 15, and 20 mL/L were evaluated with a 10-minute contact time. These experiments mimicked conditions used during previous pilot testing.[60]

Bench-scale testing was conducted using rapid small-scale column tests (RSSCTs) to predict GAC performance. A detailed description of these tests was published previously.[31] The RSSCT design used constant diffusivity similitude, where the ratio of empty bed contact time is proportional to the square of the ratio of the grain diameters.[61] Constant diffusivity similitude was selected because Crittenden and others have found this to be more appropriate than proportional diffusivity for "smaller" molecular weight discrete molecules that do not have a high charge.[61] Proportional diffusivity has been found most appropriate for natural organic matter with an average molecular weight of 1000 Daltons or higher, which also has a considerable charge, and for highly charged inorganic species such as perchlorate.[62] Moreover, constant diffusivity is the more "conservative" similitude to use (i.e., if RSSCT exhibits that an organic molecule will be removed within a certain bed life when using constant diffusivity, then the predicted bed life at full scale will be the same or longer). Tests were performed with the lignite-based GAC Hydrodarco 4000 (HD4000) manufactured by Norit. In accordance with constant diffusivity similitude for the target compounds, a column with 0.307-cm^3 empty bed volume was utilized with the test carbons crushed and wet-sieved to 75 to 90-μm particle size (170 × 200 US mesh). With this design, RSSCTs simulated a full-scale column that operates at a 7.6-minute empty bed contact time (EBCT). CRW was spiked with 100 to 200 ng/L of the target compounds. The test column was maintained at room temperature between 20 and 25°C.

UV bench-scale experiments were conducted using a collimated beam system equipped with medium pressure lamp (Calgon Carbon Corporation, Pittsburgh, Pennsylvania). Aliquots of spiked CRW (500 mL) were irradiated in a 600-mL glass beaker on a magnetic stir plate centered with respect to the light beam. Samples were irradiated for predetermined periods of time corresponding to selected UV fluences (UV dose). Irradiance measurements were collected with a 1700 International Light Research Radiometer with SED40 detector, calibrated at each wavelength within a 200- to 340-nm range. UV fluences were calculated using a spreadsheet based on the incident irradiance, sample geometry, and water absorption spectrum, as described elsewhere.[63] During advanced oxidation experiments, hydrogen peroxide was added 1 minute prior to exposure, and residuals were quenched with 0.2 mg/L bovine catalase.

11.5.2 Results from Bench-Scale Studies

The comparisons for the removal of target analytes by UV, free chlorine, and ozone are presented in Figure 11.1 through Figure 11.5, according to compound classifications. Overall, UV is not able to provide significant removal to most target analytes under a common disinfection dose of 40 mJ/cm^2. Free chlorine disinfection is significantly more efficient than UV disinfection for contaminant removal, while ozone disinfection can oxidize nearly all target compounds investigated. It should be

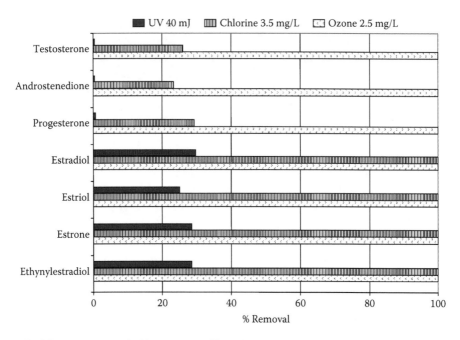

FIGURE 11.1 Removal of hormones by UV, chlorine, and ozone.

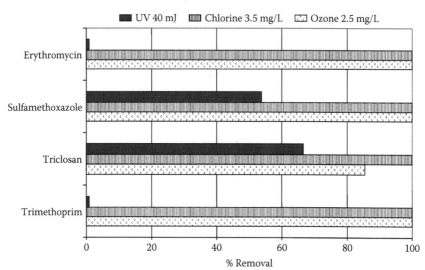

FIGURE 11.2 Removal of antimicrobials by UV, chlorine, and ozone.

noted that no process was capable of removing all target compounds to less than the method reporting limits.

Two principal mechanisms for contaminant removal by UV irradiation are possible. First, UV can directly cleave bonds in organic molecules by direct photolysis of the target molecule. Second, UV reacts with inorganic constituents in water to form highly reactive intermediates, with the formation of hydroxyl radicals (HO•)

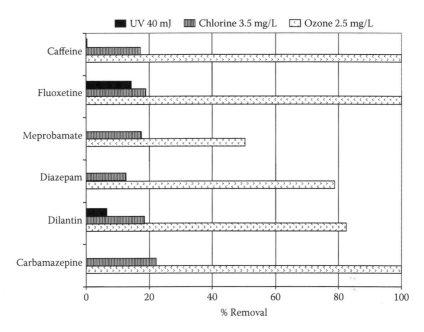

FIGURE 11.3 Removal of psychoactive compounds by UV, chlorine, and ozone.

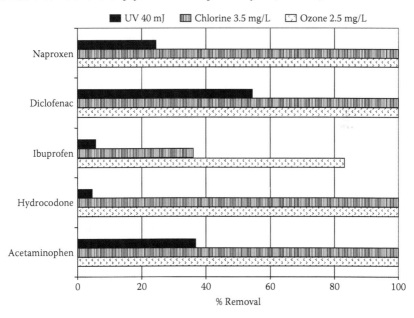

FIGURE 11.4 Removal of analgesic pharmaceuticals by UV, chlorine, and ozone.

contributing to the oxidation of organic compounds. In this case, the total energy delivered by UV lamp is relatively low and explains the minimal removal observed at disinfection dosages. This finding is consistent with a study conducted by Rosenfeldt and Linden who examined the degradation of 17α-ethynyl estradiol and 17β-estradiol

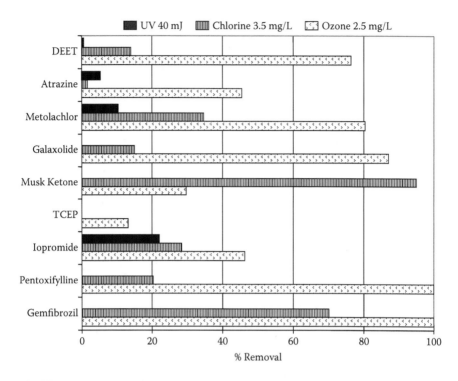

FIGURE 11.5 Removal of model compounds by UV, chlorine, and ozone.

via UV and UV/hydrogen peroxide processes using either low- or medium-pressure arc lamp collimated beam reactors.[64] Low degradation yields were reported for both compounds, ranging from 2 to 5%, or from 15 to 22%, at a UV dose of 1000 mJ/cm^2 delivered by a low- or a medium-pressure lamp, respectively. The chemical structure of the target analyte controls whether it can be oxidized by UV light. The conjugated aromatic structure in diclofenac and triclosan causes these two compounds to exhibit relatively higher removal.

Chlorine mostly reacts via electrophilic substitution and addition. Functional groups are critical in determining a compound's reactivity with chlorine. Chlorine can also act as an oxidant, by reacting selectively with electron rich bonds of organic chemicals (e.g., $C = C$ bonds in aromatic ring). Electron donating substituents in organic molecules tend to increase reactivity, while electron withdrawing groups decrease reactivity.[25,33,65,66] All four antimicrobials and four of the five analgesics showed complete removal by free chlorine due to their electron donating substituents (Figure 11.2 and Figure 11.4). Phenolic steroids (e.g., estradiol, estrone, estriol, and ethynyl estradiol) were readily oxidized by chlorine, while ketone steroids (e.g., testosterone and progesterone) were not effectively oxidized (Figure 11.1). The rapid reactions of phenolic compounds with free chlorine are mainly through the electrophilic attack of hypochlorous acid to the deprotonated phenolate anion.[67,68] Psychoactive compounds were removed by less than 20%, while the removal of model compounds ranged from no removal to over 90% consistent with their structure diversity intentionally included for objective evaluation.

Free chlorine is present in the form of either hypochlorite ion or hypochlorous acid (HOCl) with a pK_a of 7.5. So HOCl dominates at pH < 7.5 and OCl⁻ constitutes a higher percentage at pH > 7.5. HOCl is generally considered a stronger oxidizing and substituting agent than OCl⁻.[69] The oxidative power of aqueous chlorine highly depends on pH. Results summarized in Figure 11.6 confirmed this point. The removal capability of free chlorine almost doubled when pH decreased from 8.2 to 5.5.

Chloramine in the form of monochloramine is commonly used for the distribution system in drinking water supplies to maintain chlorine residual. It is much more stable than free chlorine and exhibits less tendency to form chlorinated disinfection byproducts (DBPs). Bench-scale testing results indicated that chloramine is less potent in removing target contaminants. Most target compounds were less than 10% removed by 2 or 3 mg/L of chloramine after 24 hours of contact time (Figure 11.7). Only triclosan, diclofenac, and hydrocodone were removed by 30 to 90%, all of which were susceptible to both free chlorine and ozone oxidation. This suggests that chloramine is not a viable option when considering the removal of trace amounts of pharmaceuticals.

Ozone is a strong oxidant and disinfectant that decays within minutes after its addition to water. Ozone reacts with organic contaminants either through the direct reaction with molecular ozone or through oxidation by free radicals, including the hydroxyl radical (HO•). During ozonation tests, most of the target compounds investigated showed over 80% removal at typical drinking water dosages and were removed within 5 minutes of contact time (Figure 11.1 through Figure 11.5). As expected, electron-donating groups enhance the reactivity of aromatic compounds toward ozone, while electron-withdrawing groups inhibit the reactivity. As a result,

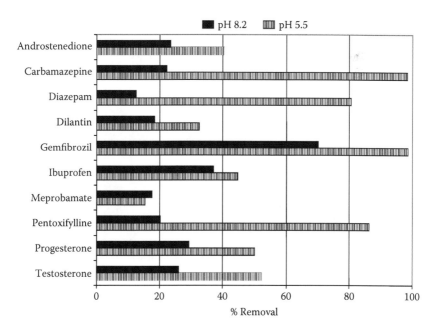

FIGURE 11.6   Impact of pH on oxidation by chlorine (3.5 mg/L).

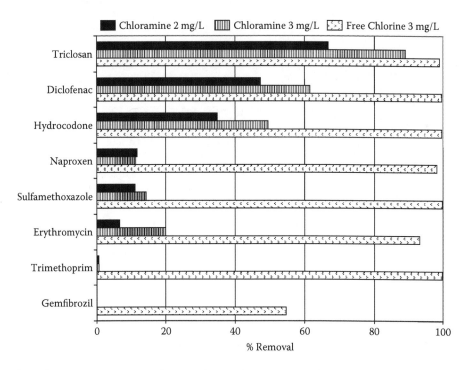

FIGURE 11.7 Comparison of chloramine and free chlorine oxidation.

all target compounds with phenolic structures, such as acetaminophen, and several hormones were removed to below analytical detection limits (Figure 11.8). A number of compounds have also exhibited complete removals (carbamazepine, gemibrozil, hydrocodone, diclofenac, erythromycin, naproxen, sulfamethoxazole, and trimethoprim) due to their increased affinity to electrophiles (ozone in this case) from highly conjugated systems and/or multiple electron-donating substituents. The concentrations of musk ketone and tris(2-chloroethyl) phosphate (TCEP) were only decreased by 12% and 30%, respectively. Androstenedione, progesterone, and testosterone were oxidized less efficiently than any of the estrogen compounds due to ketone functional groups on these hormones, which decreases the reactivity of ozone with the adjacent carbons.

During MIEX bench testing most compounds showed less than 20% removal when using a contact time of 10 minutes. The remaining compounds showed a dose response related to the MIEX concentration (Figure 11.9). Only triclosan and diclofenac were removed by greater than 50%, both of which were negatively charged at test pH and favored by the anionic exchange. When the contact time was increased to 20 minutes, the same group of target compounds again showed less than 20% removal (data not shown). The remaining compounds showed slightly higher removal at the increased contact time.[60]

Activated carbon (AC) removes organic contaminants via adsorption. The activation process for the manufacture of AC creates highly porous materials with a distribution of pore sizes and surface areas that promotes hydrophobic interaction with

Phenolic Structure

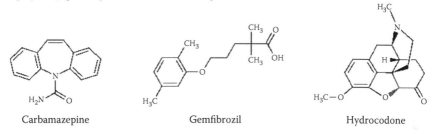

| Acetaminophen | Estradiol | Estrone |

Highly Conjugated System/Electron Donating Substituent

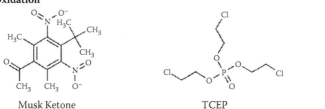

| Carbamazepine | Gemfibrozil | Hydrocodone |

Resistant to Oxidation

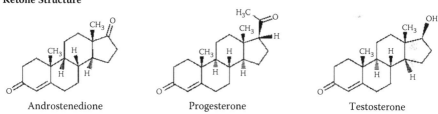

| Musk Ketone | TCEP |

Ketone Structure

| Androstenedione | Progesterone | Testosterone |

FIGURE 11.8 Chemical structures of some endocrine disruptors and pharmaceuticals.

organic contaminants. Those contaminants with low aqueous solubility, and a size conducive to fitting within the pore structure, are most readily adsorbed. In drinking water treatment, the concentration and type of natural organic matter (NOM) in the source water will compete with organic contaminants for surface adsorption sites, reducing the effectiveness of the AC.[70–72] Figure 11.10 demonstrates that GAC is highly effective; however, water-soluble contaminants can break through the GAC more rapidly than strongly bound hydrophobic contaminants. The removal of neutral compounds was correlated with their octanol-water partition coefficients (K_{ow}). This is logical since adsorption of neutral compounds on AC is dominated by hydrophobic interactions, for which K_{ow} is a good indicator. However, K_{ow} was

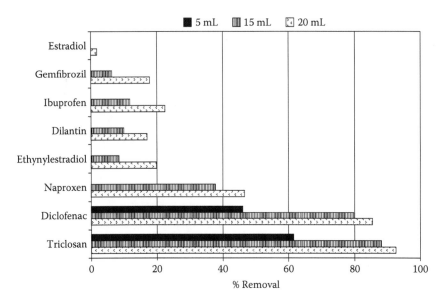

FIGURE 11.9 Removal of selected pharmaceuticals based on different amounts of **MIEX.**

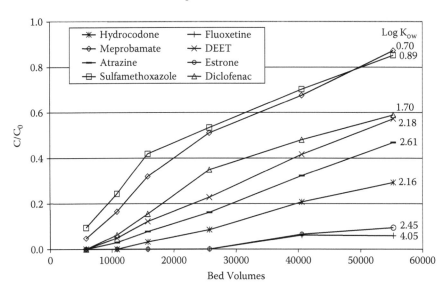

FIGURE 11.10 Removal using HD4000 granular activated carbon.

generally not a good indicator for deprotonated acids or protonated bases, because most reported K_{ow} values are for the neutral form of the molecules. For these species, the octanol–water distribution coefficient (D_{ow}) should be more appropriate to represent their hydrophobicity and relate to their removal tendency. D_{ow} is the ratio of the sum of the concentrations of all species of the compound in octanol to the sum of the concentrations of all species of the compound in water, and its value highly depends on the extent of ionization. Since the experimental pKa values of these compounds are not readily available it is difficult to conclude definitively about the relevance of removal to D_{ow}.

11.6 PILOT-SCALE EVALUATIONS

11.6.1 Pilot-Scale Experimental Procedures

Ozone and UV with and without hydrogen peroxide were evaluated using flow-through pilot equipment. Ozone pilot experiments were conducted at the Southern Nevada Water Authority using CRW. UV pilot experiments were conducted using dechlorinated tap water at the Trojan UV Technologies research facility in London, Ontario, Canada. Only LC-MS/MS target compounds were evaluated during pilot-scale experiments; therefore, galaxolide, musk ketone, and metolachlor were not considered.

Ozone pilot systems with flow rates of 1.0 L/min and 23 L/min were supplied with CRW. These pilot systems have been described in detail previously.[30] The bench-top pilot plant (BTPP) with a flow rate of 1.0 L/min was used to conduct multiple ozone and ozone/hydrogen peroxide oxidation experiments. The BTPP ozone contactor consisted of 12 glass chambers each providing 2 minutes of contact time. The BTPP testing was performed using two 170-L batches of CRW spiked with target compounds. The larger ozone pilot plant, with a flow rate of 23 L/min, was used to evaluate ozone oxidation without hydrogen peroxide addition. A syringe pump was used to introduce the target compounds into the process stream. Two static mixers followed the contaminant spike to provide homogenization. The pilot ozone contactor consisted of 12 cells to provide approximately 24 minutes of contact time at the design flow rate of 23 L/min.

The UV pilot contained a medium-pressure UV reactor operated through an electronic ballast such that the lamp power level could be varied from 30 to 100%, static mixers, an electronic flowmeter, ports for spiking hydrogen peroxide and other solutions, sampling ports, and metering peristaltic pumps. The pilot setup contained two 41,600-L tanks. Dechlorinated drinking water (30,000 L) was spiked with 100 to 200 ng/L of the target compounds in a large feed tank, and recirculated for 1 hour at 2840 L/min to ensure thorough mixing. UV energies delivered to the water were varied by adjusting the flow rate through the UV reactor. UV fluences required for contaminant oxidation were delivered using four lamps at 60% power level and flow rates from 416 to 1400 L/min. While multiple UV experiments were conducted with and without hydrogen peroxide, only results from experiments using hydrogen peroxide (AOP) are shown here.

MIEX pilot plant testing was conducted using CRW spiked with target compounds. A MIEX concentration of 20 mL/L was selected as the optimum from bench-scale results. Samples were collected after each contactor chamber having total elapsed contact times of 10 and 20 minutes.

11.6.2 Results from Pilot-Scale Studies

Results from ozone BTPP experiments are presented in Figure 11.11 and Figure 11.12. Ozone was found to be highly effective for the removal of most target compounds even at relatively low ozone doses. Figure 11.11 and Figure 11.12 show only compounds for which removals were incomplete; therefore, compounds not shown were removed to less than the MRL. Compounds are presented in order of removal efficacy, with TCEP the most difficult to oxidize. Only TCEP, atrazine, meprobamate, and iopromide had removal efficacies of less than 50% at the lowest ozone dose applied (Figure 11.11). The addition of 0.25 mg/L hydrogen peroxide with an ozone dose of 1.3 mg/L resulted in a significant (>10%) decrease in removal of meprobamate, iopromide, N,N-diethyl-m-toluamide (DEET), dilantin, and ibuprofen (Figure 11.11). Conversely the addition of hydrogen peroxide increased the removal of androstenedione, progesterone, testosterone, caffeine, and pentoxifylline. At an increased ozone dose of 2.7 mg/L, removal of target compounds increased. However, the addition of 0.5 mg/L hydrogen peroxide showed a similar trend as with the lower dose, yet the degree of impact was less dramatic (Figure 11.12). Results from the 23-L/min pilot were remarkably similar to those obtained using the BTPP (Figure 11.13).

Target compounds not removed to less than the MRL with the lowest medium-pressure UV-AOP dose are shown in Table 11.4. Acetaminophen, diclofenac, estradiol, estrone, ethynyl estadiol, hydrocodone, naproxen, and triclosan were all removed to less than the MRL using even the lowest UV/hydrogen peroxide dose (216 mJ/cm^2 to 4.6 mg/L). Carbamazepine, TCEP, meprobamate, and erythromycin were

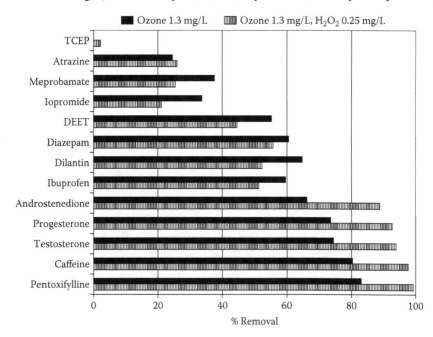

FIGURE 11.11 Bench-top ozone pilot plant removal—low dose.

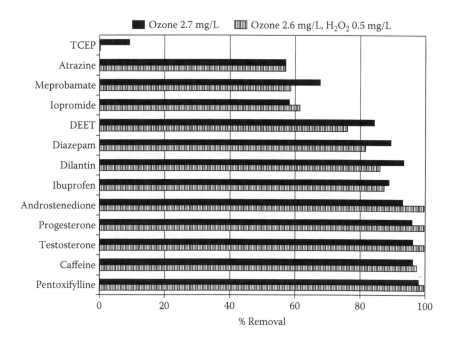

FIGURE 11.12   Bench-top ozone pilot plant removal—high dose.

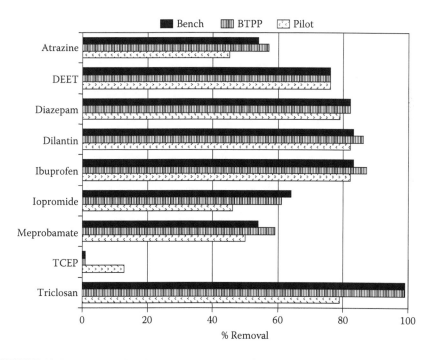

FIGURE 11.13   Impact of scaling at an ozone dose of 2.5 mg/L obtained with the 23-L/min pilot system.

the only target compounds with less than 50% reduction at the minimum UV-AOP dose applied (Table 11.4). A constant UV dose of approximately 370 mJ/cm^2 was applied with hydrogen peroxide doses of 5.8 and 7.6 mg/L. Interestingly, the increase in hydrogen peroxide at a constant UV dose did not offer a large increase in contaminant removal. Increasing the UV dose to approximately 540 mJ/cm^2 with a hydrogen peroxide dose of 7.5 mg/L showed a modest increase in contaminant oxidation. These data demonstrate diminishing returns with large increases in UV and hydrogen peroxide doses, suggesting that lower doses may provide nearly equivalent contaminant reduction with less energy and peroxide cost. Contaminant removal by UV-AOP and ozone-AOP appear to be highly correlated, as would be expected as these processes both induce the formation of the less selective, yet highly potent, hydroxyl radical.

Results consistent with bench-scale testing were obtained during MIEX pilot plant testing. Triclosan and diclofenac were the only two compounds removed by over 50% with both 10 and 20 minutes of contact time. Slightly higher removals were observed for progesterone and estrone during pilot-scale experiments. However, the general trend in which the compounds were removed was similar to the bench-scale testing results.

TABLE 11.4
Medium-Pressure UV-AOP Pilot

UV Dose (mJ/cm^2)	216	366	379	537
Peroxide Dose (mg/L)	4.6	5.8	7.6	7.5
	Percent Removal (%)			
Androstenedione	81	83	89	96
Atrazine	61	61	67	80
Caffeine	66	68	76	89
Carbamazepine	16	49	67	>88
DEET	64	67	78	89
Diazepam	73	74	81	93
Dilantin	84	86	91	97
Erythromycin	0	35	19	64
Fluoxetine	92	93	96	>98
Gemfibrozil	76	78	85	95
Ibuprofen	73	74	83	94
Iopromide	76	80	79	91
Meprobamate	48	45	58	75
Pentoxifylline	68	68	78	90
Progesterone	84	86	91	98
Sulfamethoxazole	95	97	97	>99
TCEP	10	0	8	16
Testosterone	83	85	90	97
Trimethoprim	76	77	85	94

11.7 FULL-SCALE EVALUATIONS

Samples of raw and finished drinking water were collected from 20 drinking water treatment plants from geographically diverse locations across the United States. Treatment plants were selected based upon known wastewater influence in the source water. Therefore, these utilities have the highest potential to contain EDC/PPCP compounds, making them good candidates for evaluating efficiencies of full-scale treatment process. Summary results from drinking water EDC/PPCP monitoring are presented in Table 11.5 and Table 11.6. These two tables show the number of detections above the MRL (detects), percent frequency of detection (% freq), minimum (min), maximum (max), median, and average (ave). It is important to note that the min, max, median, and ave values are calculated only from detectable values, thus values less than the MRL are not factored into the ave reported. Diazapem, diclofenac, estradiol, ethynyl estradiol, fluoxetine, pentoxifylline, and testosterone were not detected in any of the raw or finished water samples evaluated.

TABLE 11.5
Concentrations of Target Compounds in Source Waters (n = 20)

Compound	Detects	% Freq.	Min (ng/L)	Max (ng/L)	Median (ng/L)	Ave (ng/L)
DEET	20	100	2.8	28	6.9	9.0
Carbamazepine	18	90	1.2	39	3.1	6.2
Dilantin	18	90	1.1	13	3.2	3.5
Atrazine	17	85	1.3	571	28	101
Sulfamethoxazole	17	85	1.2	44	8.1	14
Ibuprofen	16	80	1.1	24	4.2	6.1
Meprobamate	16	80	1.4	16	5.9	7.0
TCEP	15	75	2	66	13	18
Caffeine	14	70	9.1	87	27	34
Iopromide	14	70	2.2	46	7.6	12
Gemfibrozil	13	65	1.2	11	4.8	5.2
Naproxen	10	50	1.1	16	2.2	5.7
Erythromycin	8	40	1	3.5	2.2	2.2
Acetaminophen	7	35	1.1	9.5	1.6	2.7
Metalochlor	7	35	11	174	15	71
Triclosan	6	30	1	30	1.9	6.4
Galaxolide	3	15	25	30	28	28
Musk Ketone	3	15	14	17	16	16
Trimethoprim	3	15	1	2.3	2.2	1.8
Estrone	2	10	1	1.4	1.2	1.2
Androstenedione	1	5	1.9	1.9	1.9	1.9
Hydrocodone	1	5	1.9	1.9	1.9	1.9
Progesterone	1	5	1.1	1.1	1.1	1.1

TABLE 11.6

Emerging Contaminants in U.S. Drinking Water (n = 20)

Compound	Hits	% Freq	Min (ng/L)	Max (ng/L)	Median (ng/L)	Ave (ng/L)
DEET	18	90	2.1	30	5.1	8.2
Atrazine	15	75	1.4	430	29	74
Meprobamate	15	75	1.6	13	3.8	6.1
Dilantin	14	70	1.1	6.7	2.3	2.7
Ibuprofen	13	65	1	32	3.8	7.9
Iopromide	13	65	1.1	31	6.5	8.5
Caffeine	12	60	2.6	83	23	25
Carbamazepine	11	55	1.1	5.7	2.8	2.8
TCEP	7	35	3	19	5.5	10.1
Gemfibrozil	5	25	1.3	6.5	4.2	3.9
Metalochlor	4	20	14	160	86	86
Estrone	2	10	1.1	2.3	1.7	1.7
Progesterone	2	10	1.1	1.1	1.1	1.1
Erythromycin	1	5	1.3	1.3	1.3	1.3
Musk Ketone	1	5	17	17	17	17
Naproxen	1	5	8	8	8	8.0
Sulfamethoxazole	1	5	20	20	20	20
Triclosan	1	5	43	43	43	43
Trimethoprim	1	5	1.3	1.3	1.3	1.3

In both raw and finished drinking water samples, N,N-diethyl-m-toluamide (DEET) had highest frequency of occurrence (100 and 90%, respectively). The U.S. Environmental Protection Agency (EPA) has estimated that approximately one third of the U.S. population uses DEET-containing products (http://www.epa.gov/pesticides/ factsheets/chemicals/deet.htm). The pharmaceuticals carbamazepine and dilantin were detected in 90% of raw water samples with maximum concentrations of 39 and 13 ng/L, respectively. In finished drinking water, carbamazepine and dilantin were less frequently detected (55 and 70%, respectively) with maximum concentrations of 5.7 and 6.7 ng/L, respectively. These data show that drinking water treatment will significantly reduce the concentrations of these pharmaceuticals. Atrazine was detected at the greatest concentration in drinking waters. Atrazine was detected in 85% of raw waters and 75% of finished waters, with maximum concentrations of 571 and 430 ng/L, respectively. Drinking water treatment processes evaluated here were largely ineffective for the removal of atrazine, with the exception of activated carbon. The antibiotic sulfamethoxazole occurred in 85% of raw water samples collected, but was only detected in one finished drinking water sample. This facility had 40 ng/L of sulfamethoxazole in the raw water and 20 ng/L in the finished water, thus 50% of this antibiotic was removed through the plant. The poor removal at this facility can be explained by the use of chloramines for primary disinfection, which was shown to be much less reactive with sulfamethoxazole than free chlorine (Figure 11.6). The pharmaceuticals ibuprofen and meprobamate and the x-ray contrast media iopromide are compounds that were found difficult to remove by conventional water treatment processes. These compounds were detected in over 65% of the raw and finished drinking waters evaluated, with only small amounts removed during treatment. The antimicrobial triclosan was detected in 30% of the raw water samples at low concentrations ranging from 1 to 30 ng/L, while it was detected in only one finished water sample at 45 ng/L. The facility that contained the highest level of triclosan in the raw water also contained the only detection of triclosan in the finished water. This utility was also the only location with sulfamethoxazole detected in the finished water. Both triclosan and sulfamethoxazole have rapid removal when in contact with chlorine or chloramines; therefore, the finished water sample clearly had a low amount of contact time with chloramine before the oxidant was quenched. The synthetic fragrances musk ketone and galaxolide were detected in only 2 of the 20 raw water samples analyzed, and only musk ketone was detected in one finished water at 17 ng/L.

While steroid hormones seem to garner the greatest amount of concern regarding public health from regulatory agencies, these compounds were rarely detected in U.S. drinking waters. Estrone was detected in two raw water samples at 1.0 and 1.4 ng/L and in two finished waters at 1.1 and 2.3 ng/L. These values are near the MRL of 1.0 ng/L. Progesterone and androstenedione were each detected in one raw water, each at less than 2 ng/L, while progesterone was detected twice at 1.1 ng/L in finished water.

Removal efficacy of individual unit processes also was evaluated. In general, coagulation, flocculation, and filtration had very little impact on target compound removal at full scale (data not shown). However, disinfection played a major role in contaminant removal. Table 11.7 shows the removal of compounds by chlorine disinfection at full scale. In general, predicted removal at bench scale agreed well with observations at full-scale utilities. Only compounds detected at 2× the MRL are shown in Table 11.7 in order to determine percent removal. Compounds that appear to deviate from the bench-scale predictions are generally those with occurrence near the 2× MRL threshold. Removal by UV disinfection, ozone disinfection, and GAC at full-scale utilities are shown in Table 11.8 through Table 11.10. As predicted at bench scale, UV disinfection was largely inefficient, while ozone disinfection was highly efficient for trace contaminant removal. Moreover, full-scale GAC proved to be effective for contaminant removal as long as the GAC was replaced or regenerated on a regular basis.[31] Conversely, activated carbon has a limited service life for contaminant adsorption and will become completely ineffective for contaminant removal.[31] Powdered activated carbon is also a viable option for organic contaminant removal, and results specific to these target compounds have been published.[33] The results shown here depict the overall removal through the process and do not account for differences in dose, contact time, or water quality. Interestingly, the trends from bench-scale predictions are in good agreement with full-scale observations even with the variability in operational process parameters. Although not shown here, full-scale treatment plants using reverse osmosis membranes effectively removed all target analytes, while membrane filtration using ultrafiltration and microfiltration was generally ineffective for contaminant rejection.[31]

TABLE 11.7
Full-Scale Removal Using Free Chlorine

		Observed Percent Removal				Bench-Scale
	# Utilities	>80	50–80	20–50	<20	Predicted
Acetaminophen	1	1				>80
Atrazine	10			1	9	<20
Caffeine	8		2		6	<20
Carbamazepine	8	1	1	3	3	<20
DEET	12			1	11	<20
Dilantin	8			3	5	<20
Erythromycin	3	2	1			>80
Estrone	1				1	>80
Galaxolide	2				2	20–50
Gemfibrozil	5	2	1	1	1	50–80
Ibuprofen	8			2	6	<20
Iopromide	8			5	2	<20
Meprobamate	8				8	<20
Metolachlor	4				4	<20
Naproxen	4	4				>80
Sulfamethoxazole	7	6	1			>80
TCEP	5				4	<20
Triclosan	1	1				>80
Trimethoprim	2	2				>80

Summarily, these results show that removal of trace contaminants in a full-scale drinking water treatment plant will largely be a function of the primary disinfectant, the secondary disinfectant, and contact time. Carbon adsorption and membranes are also viable removal tools, but are far less common in drinking water treatment applications.

11.8 CONCLUSION

The use of oxidants for the attenuation of trace contaminants offers an economically viable option for water treatment systems. It is important to note that even the strongest oxidation techniques investigated do not result in complete mineralization (oxidation to carbon dioxide and water) at the doses investigated. Therefore, oxidation byproducts will be produced and should be considered. However, the concentration of natural organic matter will be orders of magnitude greater in concentration as compared to pharmaceuticals, endocrine disruptors, and other trace contaminants. The oxidation byproducts of organic carbon should be considered collectively, as there is a far greater probability of forming toxic byproducts from mg/L of NOM as compared to ng/L of trace contaminants. Both ozone and UV-AOP processes tend to form less aromatic and more biodegradable oxidation products.[32,73–75] More research is needed to determine the fate of these byproducts in subsequent adsorption and

TABLE 11.8
Full-Scale Removal Using UV Disinfection
(Approximately 40 mJ/cm^2)

	Facility A % Rem	Facility B % Rem	Facility C % Rem	Bench-Scale % Rem
Acetaminophen	>44	NA	NA	50–20
Atrazine	4	<1	<1	<20
Caffeine	42	<1	<1	<20
Carbamazepine	>17	NA	<1	<20
DEET	22	19	<1	<20
Dilantin	15	NA	<1	<20
Erythromycin	<1	NA	<1	<20
Galaxolide	9	>23	8	<20
Gemfibrozil	69	NA	<1	<20
Ibuprofen	<1	<1	<1	<20
Iopromide	<1	NA	<1	<20
Meprobamate	<1	<1	<1	<20
Sulfamethoxazole	>83	NA	<1	80–50
TCEP	5	<1	<1	<20

NA = Not applicable; % Rem = %Removal

TABLE 11.9
Full-Scale Removal with GAC

Compound	Facility A % Rem	Facility B % Rem	Facility C % Rem
Atrazine	>99	5.9	NA
Caffeine	>41	36	<1
Carbamazepine	>54	NA	<1
DEET	>44	38	10
Dilantin	>44	NA	29
Estradiol	NA	>84	NA
Erythromycin	>44	NA	NA
Galaxolide	NA	>9	NA
Gemfibrozil	>16	NA	NA
Ibuprofen	>9	>58	52
Iopromide	>69	<1	14
Meprobamate	>16	NA	16
Metolachlor	>91	NA	NA
Sulfamethoxazole	>83	NA	NA
TCEP	NA	NA	40

NA = Not applicable; % Rem = %Removal

TABLE 11.10
Full-Scale Ozone

Facility	A	B	C	D	BTPP
Compound	% Rem	% Rem	% Rem	% Rem	% Rem
Atrazine	>47	>28	NA	38	20–50
Caffeine	NA	NA	>72	95	>80
Carbamazepine	>79	>71	>93	>70	>80
DEET	>87	48	48	NA	50–80
Dilantin	57	52	56	74	20–50
Erythromycin	NA	NA	>60	NA	>80
Estrone	NA	>28	NA	NA	>80
Galaxolide	NA	NA	55	NA	>80
Gemfibrozil	>50	NA	>79	>58	>80
Ibuprofen	>41	NA	76	NA	50–80
Iopromide	50	NA	25	NA	20–50
Meprobamate	25	28	28	41	20–50
Naproxen	NA	NA	>91	NA	>80
Sulfamethoxazole	>91	>90	>90	>91	>80
Triclosan	NA	NA	>68	>66	>80

NA = Not Applicable; BTPP O3 = 1.25 mg/L, Contact Time = 24 min

biodegradation processes. It is likely that many of the byproducts from ozone and UV-AOP will be biodegradable.

Without question, trace pharmaceuticals, endocrine disruptors, and other organic contaminants will be detectable in source waters using sensitive analytical methodologies. This report contributes to the further understanding of the fate of several emerging contaminants during common disinfection processes. Additionally, this chapter provides the first known data on the removal efficacy of magnetic ion-exchange media for emerging contaminant removal. The data provided here also show that activated carbon is capable of adsorbing organic contaminants; however, more water-soluble contaminants can breach the carbon relatively quickly. A multibarrier approach that combines oxidation and adsorption or rejection (membranes) will be highly effective for contaminant removal. Regardless, some trace contaminants will still be detectable even after the most advanced treatment processes. It is critical that toxicological relevance be determined in order that treatment goals and analytical method reporting limits can be established that are protective of public health.

ACKNOWLEDGMENTS

The authors would like to thank the American Water Works Association Research Foundation (AwwaRF Project #2758) for providing the resources to undertake this project. The authors would particularly like to thank Kim Linton from AwwaRF, who provided guidance, encouragement, and support in all aspects of this research.

The authors are grateful for the kind assistance and support from their colleagues in the Water Quality Research & Development Division of the Southern Nevada Water Authority, particularly Linda Parker, David Rexing, Rebecca Trenholm, Brett Vanderford, Ron Zegers, and Janie Zeigler-Holady. The authors would like to acknowledge and thank the co-principal investigators for this project, Paul Wester-hoff from Arizona State University and Yeomin Yoon from CH2M Hill. The authors also wish to thank Alex Mofidi and Connie Lee from the Metropolitan Water District of Southern California, who conducted the collimated beam UV work presented in this chapter. The authors also thank Adam Redding and Fred Cannon from Penn State University, who conducted the RSSCT test for GAC adsorption. The authors also thank Trojan Technologies for conducting the UV-AOP pilot tests presented in this chapter. In particular, the authors wish a special thanks to Mihaela Stefan for her insight and guidance on the UV-AOP research related to this project. The authors thank the Black & Veatch Corporation for providing assistance for MIEX testing, in particular Jessica Edwards-Brandt.

REFERENCES

1. Stumm-Zollinger, E. and Fair, G.M., Biodegradation of steroid hormones. *J. Water Poll. Contr. Fed.* 1965, 37, 1506–1510.
2. Tabak, H.H. and Bunch, R.L., Steroid hormones as water pollutants. I. Metabolism of natural and synthetic ovulation-inhibiting hormones by microorganisms of activated sludge and primary settled sewage. *Dev. Ind. Microbiol.* 1970, 11, 367–376.
3. Schueler, F.W., Sex-hormonal action and chemical constitution. *Science* 1946, 103, 221–223.
4. Sluczewski, A. and Roth, P., Effects of androgenic and estrogenic compounds on the experimental metamorphoses of amphibians. *Gynecol. and Obstet.* 1948, 47, 164–176.
5. Hignite, C. and Azarnoff, D.L., Drugs and drug metabolites as environmental contaminants: chlorophenoxyisobutyrate and salicylic acid in sewage water effluent. *Life Sci.* 1977, 20, 337–341.
6. Purdom, C.E., Hardiman, P.A., Bye, V.J., Eno, N.C., Tyler, C.R., and Sumpter, J.P., Estrogenic effects of effluents from sewage treatment works. *Chem. Ecol.* 1994, 8, 275–285.
7. Desbrow, C., Routledge, E.J., Brighty, G.C., Sumpter, J.P., and Waldock, M., Identification of estrogenic chemicals in STW effluent. 1. Chemical fractionation and in vitro biological screening. *Environ. Sci. Technol.* 1998, 32, 1549–1558.
8. Routledge, E.J., Sheahan, D., Desbrow, C., Brighty, G.C., Waldock, M., and Sumpter, J.P., Identification of estrogenic chemicals in STW effluent. 2. *In vivo* responses in trout and roach. *Environ. Toxicol. Chem.* 1998, 32, 1559–1565.
9. Halling-Sorensen, B., Nielsen, S.N., Lanzky, P.F., Ingerslev, F., Lutzhoft, H.C.H., and Jorgensen, S.E., Occurrence, fate and effects of pharmaceutical substances in the environment—A review. *Chemosphere* 1998, 36, 357–393.
10. Ternes, T.A., Hirsch, R., Mueller, J., and Haberer, K., Methods for the determination of neutral drugs as well as betablockers and β_2-sympathomimetics in aqueous matrices using GC/MS and LC/MS/MS. *Fresenius' J. Anal. Chem.* 1998, 362, 329–340.
11. Daughton, C.G. and Ternes, T.A., Pharmaceuticals and personal care products in the environment: Agents of subtle change? *Environ. Health Perspectives* 1999, 107, 907–938.

12. Snyder, S.A., Keith, T.L., Verbrugge, D.A., Snyder, E.M., Gross, T.S., Kannan, K., and Giesy, J.P., Analytical methods for detection of selected estrogenic compounds in aqueous mixtures. *Environ. Sci. Technol.* 1999, 33, 2814–2820.

13. Metcalfe, C.D., Koenig, B., Ternes, T.A., and Hirsch, R., Drugs in sewage treatment plant effluents in Canada. In Proceedings for ACS National Meeting, 2000, vol 40, 100–102.

14. Ternes, T.A. and Hirsch, R., Occurrence and behavior of x-ray contrast media in sewage facilities and the aquatic environment. *Environ. Sci. Technol.* 2000, 34, 2741–2748.

15. Snyder, S.A., Kelly, K.L., Grange, A.H., Sovocool, G.W., Snyder, E.M., and Giesy, J.P., Pharmaceuticals and personal care products in the waters of Lake Mead, Nevada. In *Pharmaceuticals and Personal Care Products in the Environment: Scientific and Regulatory Issues*, Daughton, C.G.; Jones-Lepp, T.L., Eds., American Chemical Society: Washington, D.C., 2001; Symposium Series 791, pp 116–140.

16. Kolpin, D.W., Furlong, E.T., Meyer, M.T., Thurman, E.M., Zaugg, S.D., Barber, L.B., and Buxton, H.T., Pharmaceuticals, hormones, and other organic waste contaminants in U.S. Streams, 1999–2000: A national reconnaissance. *Environ. Sci. Technol.* 2002, 36, 1202–1211.

17. Vanderford, B.J., Pearson, R.A., Rexing, D.J., and Snyder, S.A., Analysis of endocrine disruptors, pharmaceuticals, and personal care products in water using liquid chromatography/tandem mass spectrometry. *Anal. Chem.* 2003, 75, 6265–6274.

18. Cargouet, M., Perdiz, D., Mouatassim-Souali, A., Tamisier-Karolak, S., and Levi, Y., Assessment of river contamination by estrogenic compounds in Paris area (France). *Sci. Tot. Environ.* 2004, 324, 55–66.

19. Petrovic, M., Eljarrat, E., Lopez de Alda, M.J., and Barcelo, D., Endocrine disrupting compounds and other emerging contaminants in the environment: a survey on new monitoring strategies and occurrence data. *Anal. Bioanal. Chem.* 2004, 378, 549–562.

20. Ternes, T.A., Kreckel, P., and Mueller, J., Behavior and occurrence of estrogens in municipal sewage treatment plants—II. Aerobic batch experiments with activated sludge. *Sci. Total Environ.* 1999, 225, 91–99.

21. Ternes, T.A., Stumpf, M., Mueller, J., Haberer, K., Wilken, R.D., and Servos, M., Behavior and occurrence of estrogens in municipal sewage treatment plants—I. Investigations in Germany, Canada and Brazil. *Sci. Total Environ.* 1999, 225, 81–90.

22. D'Ascenzo, G., Di Corcia, A., Gentili, A., Mancini, R., Mastropasqua, R., Nazzari, M., and Samperi, R., Fate of natural estrogen conjugates in municipal sewage transport and treatment facilities. *Sci. Total Environ.* 2003, 302, 199–209.

23. Huber, M.M., Korhonen, S., Ternes, T.A., and von Gunten, U., Oxidation of pharmaceuticals during water treatment with chlorine dioxide. *Water Res.* 2005, 39, 3607–3617.

24. Zwiener, C. and Frimmel, F.H., Oxidative treatment of pharmaceuticals in water. *Water Res.* 2000, 34, 1881–1885.

25. Adams, C., Wang, Y., Loftin, K., and Meyer, M., Removal of antibiotics from surface and distilled water in conventional water treatment processes. *J. Environ. Eng.—ASCE* 2002, 128, 253–260.

26. Drewes, J.E., Heberer, T., and Reddersen, K., Fate of pharmaceuticals during indirect potable reuse. *Water Sci. Technol.* 2002, 46, 73–80.

27. Snyder, S.A., Westerhoff, P., Yoon, Y., and Sedlak, D.L., Pharmaceuticals, personal care products, and endocrine disruptors in water: Implications for the water industry. *Environ. Sci. Technol.* 2003, 20, 449–469.

28. Huber, M.M., Göbel, A., Joss, A., Hermann, N., Löffler, D., McArdell, C.S., Reid, A., Siegrist, H., Ternes, T.A., and von Gunten, U., Oxidation of pharmaceuticals during ozonation of municipal wastewater effluents: a pilot study. *Environ. Sci. Technol.* 2005, 39, 4290–4299.

29. Yoon, Y., Westerhoff, P., Snyder, S.A., Wert, E.C., and Yoon, J., Removal of endocrine disrupting compounds and pharmaceuticals by nanofiltration and ultrafiltration membranes. *Desalination* 2006, 202, 16–23.
30. Snyder, S.A., Wert, E.C., Rexing, D.J., Zegers, R.E., and Drury, D.D., Ozone oxidation of endocrine disruptors and pharmaceuticals in surface water and wastewater. *Ozone-Sci. Eng.* 2006, 28, 445–460.
31. Snyder, S.A., Adham, S., Redding, A.M., Cannon, F.S., DeCarolis, J., Oppenheimer, J., Wert, E.C., and Yoon, Y., Role of membranes and activated carbon in the removal of endocrine disruptors and pharmaceuticals. *Desalination* 2006, 202, 156–181.
32. Ternes, T.A., Meisenheimer, M., Mcdowell, D., Sacher, F., Brauch, H.-J., Haist-Gulde, B., Preuss, G., Wilme, U., and Zulei-Seibert, N., Removal of pharmaceuticals during drinking water treatment. *Environ. Sci. Technol.* 2002, 36, 3855–3863.
33. Westerhoff, P., Yoon, Y., Snyder, S., and Wert, E., Fate of endocrine-disruptor, pharmaceutical, and personal care product chemicals during simulated drinking water treatment processes. *Environ. Sci. Technol.* 2005, 39, 6649–6663.
34. Tixier, C., Singer, H.P., Oellers, S., and Mueller, S.R., Occurrence and fate of carbamazepine, clofibric acid, diclofenac, ibuprofen, ketoprofen, and naproxen in surface waters. *Environ. Sci. Technol.* 2003, 37, 1061–1068.
35. Loraine, G.A. and Pettigrove, M.E., Seasonal variations in concentrations of pharmaceuticals and personal care products in drinking water and reclaimed wastewater in southern California. *Environ. Sci. Technol.* 2006, 40, 687–695.
36. Vieno, N.M., Tuhkanen, T., and Kronberg, L., Seasonal variation in the occurrence of pharmaceuticals in effluents from a sewage treatment plant and in the recipient water. *Environ. Sci. Technol.* 2005, 39, 8220–8226.
37. Vieno, N., Tuhkanen, T., and Kronberg, L., Removal of pharmaceuticals in drinking water treatment: effect of chemical coagulation. *Environ. Tech.* 2006, 27, 183–192.
38. Stackelberg, P.E., Furlong, E.T., Meyer, M.T., Zaugg, S.D., Henderson, A.K., and Reissman, D.B., Persistence of pharmaceutical compounds and other organic wastewater contaminants in a conventional drinking-water-treatment plant. *Sci. Total Environ.* 2004, 329, 99–113.
39. Zuccato, E., Castiglioni, S., Fanelli, R., Reitano, G., Bagnati, R., Chiabrando, C., Pomati, F., Rossetti, C., and Calamari, D., Pharmaceuticals in the environment in Italy: causes, occurrence, effects and control. *Environ. Sci. Pollut. R.* 2006, 13, 15–21.
40. Castiglioni, S., Bagnati, R., Fanelli, R., Pomati, F., Calamari, D., and Zuccato, E., Removal of pharmaceuticals in sewage treatment plants in Italy. *Environ. Sci. Technol.* 2006, 40, 357–363.
41. Zuehlke, S., Duennbier, U., and Heberer, T., Determination of polar drug residues in sewage and surface water applying liquid chromatography-tandem mass spectrometry. *Anal. Chem.* 2004, 76, 6548–6554.
42. Moldovan, Z., Occurrences of pharmaceutical and personal care products as micropollutants in rivers from Romania. *Chemosphere* 2006, 64, 1808–1817.
43. Yasojima, M., Nakada, N., Komori, K., Suzuki, Y., and Tanaka, H., Occurrence of levofloxacin, clarithromycin and azithromycin in wastewater treatment plant in Japan. *Water Sci. Technol.* 2006, 53, 227–233.
44. Nakada, N., Yasojima, M., Okayasu, Y., Komori, K., Tanaka, H., and Suzuki, Y., Fate of oestrogenic compounds and identification of oestrogenicity in a wastewater treatment process. *Water Sci. Technol.* 2006, 53, 51–63.
45. Kolpin, D.W., Skopec, M., Meyer, M.T., Furlong, E.T., and Zaugg, S.D., Urban contribution of pharmaceuticals and other organic wastewater contaminants to streams during differing flow conditions. *Sci. Total Environ.* 2004, 328, 119–130.

46. Hunt, J.W., Anderson, B.S., Phillips, B.M., Tjeerdema, R.S., Puckett, H.M., and de Vlaming, V., Patterns of aquatic toxicity in an agriculturally dominated coastal watershed in California. *Agr. Ecosyst. Environ.* 1999, 75, 75–91.

47. Schwarzbauer, J., Heim, S., Brinker, S., and Littke, R., Occurrence and alteration of organic contaminants in seepage and leakage water from a waste deposit landfill. *Water Res.* 2002, 36, 2275–2287.

48. Turnidge, J., Antibiotic use in animals-prejudices, perceptions and realities. *J. Antimicrob. Chemoth.* 2004, 53, 26–27.

49. Trenholm, R.A., Vanderford, B.J., Holady, J.C., Rexing, D.J., and Snyder, S.A., Broad range analysis of endocrine disruptors and pharmaceuticals using gas chromatography and liquid chromatography tandem mass spectroscopy. *Chemosphere* 2006, 65, 1990–1998.

50. Carballa, M., Omil, F., and Lema, J.M., Removal of cosmetic ingredients and pharmaceuticals in sewage primary treatment. *Water Res.* 2005, 39, 4790–4796.

51. Jones, O.A.H., Voulvoulis, N., and Lester, J.N., Aquatic environmental assessment of the top 25 English prescription pharmaceuticals. *Water Res.* 2002, 36, 5013–5022.

52. Qiu, H.C., Jun, H.W., and McCall, J.W., Pharmacokinetics, formulation, and safety of insect repellent N,N-diethyl-3-methylbenzamide (DEET): a review. *J. Am. Mosquito Contr.* 1998, 14, 12–27.

53. Deborde, M., Rabouan, S., Duguet, J.-P., and Legube, B., Kinetics of aqueous ozone-induced oxidation of some endocrine disruptors. *Environ. Sci. Technol.* 2005, 39, 6086–6092.

54. Huber, M.M., Canonica, S., Park, G.-Y., and von Gunten, U., Oxidation of pharmaceuticals during ozonation and advanced oxidation processes. *Environ. Sci. Technol.* 2003, 37, 1016–1024.

55. Noaksson, E., Gustavsson, B., Linderoth, M., Zebuhr, Y., Broman, D., and Balk, L., Gonad development and plasma steroid profiles by HRGC/HRMS during one reproductive cycle in reference and leachate-exposed female perch (Perca fluviatilis). *Toxicol. Appl. Pharm.* 2004, 195, 247–261.

56. EC, Risk assessment musk ketone. European Union Risk Assessment Report. Final Draft, June 2003.

57. Simonich, S.L., Begley, W.M., Debaere, G., and Eckhoff, W.S., Trace analysis of fragrance materials in wastewater and treated wastewater. *Environ. Sci. Technol.* 2000, 34, 959–965.

58. Dodd, M.C., Buffle, M.O., and von Gunten, U., Oxidation of antibacterial molecules by aqueous ozone: Moiety-specific reaction kinetics and application to ozone-based wastewater treatment. *Environ. Sci. Technol.* 2006, 40, 1969–1977.

59. Loftsson, T. and Hreinsdottir, D., Determination of aqueous solubility by heating and equilibration: a technical note. *AAPS Pharm. Sci. Tech.* 2006, 7, E1–E4.

60. Snyder, S.A., Wert, E.C., Edwards, J., Budd, G., Long, B., and Rexing, D., Magnetic ion exchange (MIEX) for the removal of endocrine disrupting chemicals and pharmaceuticals. In Proceedings of Water Quality and Technology Conference, AWWA: San Antonia, TX, 2004.

61. Crittenden, J.C., Sanongraj, S., Bulloch, J.L., Hand, D.W., Rogers, T.N., Speth, T.F., and Ulmer, M., Correlation of aqueous-phase adsorption isotherms. *Environ. Sci. Technol.* 1999, 33, 2926–2933.

62. Na, C., Cannon, F.S., and Hagerup, B., Perchlorate removal via iron-reloaded GAC and borohydride regeneration. *J. Am. Water Works Assoc.* 2002, 94, 90–102.

63. Bolton, J.R., and Linden, K.G., Standardization of methods for fluence (UV dose) determination in bench-scale experiments. *J. Environ. Eng.—ASCE* 2003, 129, 209–216.

64. Rosenfeldt, E. and Linden, K.G., Degradation of endocrine disrupting chemicals bisphenol A, ethinyl estradiol, and estradiol during UV photolysis and advanced oxidation processes. *Environ. Sci. Technol.* 2004, 38, 5476–5483.
65. Deborde, M., Rabouan, S., Gallad, H., and Legube, B., Aqueous chlorination kinetics of some endocrine disruptors. *Environ. Sci. Technol.* 2004, 38, 5577–5583.
66. Lee, B.-C., Kamata, M., Akatsuka, Y., Takeda, M., Ohno, K., Kamei, T., and Magara, Y., Effects of chlorine on the decrease of estrogenic chemicals. *Water Res.* 2004, 38, 733–739.
67. Faust, S.D. and Hunter, J.V., *Principles and Applications of Water Chemistry.* John Wiley and Sons: New York, 1967.
68. Gallard, H. and von Gunten, U., Chlorination of phenols: Kinetics and formation of chloroform. *Environ. Sci. Technol.* 2002, 36, 884–890.
69. Snyder, S.A., Leising, J., Westerhoff, P., Yoon, Y., Mash, H., and Vanderford, B.J., Biological attenuation of EDCs and PPCPs: Implications for water reuse. *Ground Water Monit. R.* 2004, 24, 108–118.
70. Li, Q.L., Snoeyink, V.L., Marinas, B.J., and Campos, C., Pore blockage effect of NOM on atrazine adsorption kinetics of PAC: the roles of PAC pore size distribution and NOM molecular weight. *Water Res.* 2003, 37, 4863–4872.
71. Knappe, D.R.U., Matsui, Y., Snoeyink, V.L., Roche, P., Prados, M.J., and Bourbigot, M.-M., Predicting the capacity of powder activated carbon for trace organic compounds in natural waters. *Environ. Sci. Tech.* 1998, 32, 1694–1698.
72. Zhang, Y. and Zhou, J.L., Removal of estrone and 17β-estradiol from water by adsorption. *Water Res.* 2005, 39, 3991–4003.
73. Westerhoff, P., Debroux, J., Aiken, G., and Amy, G., Ozone-induced changes in natural organic matter (NOM) structure. *Ozone-Sci. Eng.* 1999, 21, 551–570.
74. Bose, P., Bezbarua, B.K., and Reckhow, D.A., Effect of ozonation on some physical and chemical properties of aquatic natural organic-matter. *Ozone-Sci. Eng.* 1994, 16, 89–112.
75. Vuorio, E., Vahala, R., Rintala, J., and Laukkanen, R., The evaluation of drinking water treatment performed with HPSEC. *Environ. Int.* 1998, 24, 617–623.

12 Reaction and Transformation of Antibacterial Agents with Aqueous Chlorine under Relevant Water Treatment Conditions

Ching-Hua Huang,
Michael C. Dodd, and Amisha D. Shah

Contents

12.1 INTRODUCTION

Each year large quantities of antibacterial agents (referred to as antibacterials hereafter) are used to treat diseases and infections in humans and animals. Applications of antibacterials in human medicine can ultimately lead to significant discharges of such compounds into surface waters, via excretion of the unmetabolized parent compounds into municipal sewage systems and subsequent passage through municipal wastewater treatment facilities.[1] In addition, antibacterials are utilized for a number of agricultural applications, including use as growth promoters and feed efficiency enhancers for livestock[2] and in aquaculture and fruit orchards.[3] It has been estimated that nearly 50% of the total antibacterial usage in the United States was for agriculture.[4] Significant proportions of the administered antibacterials can be excreted from the dosed animals with little metabolic transformation.[1,5] A nationwide reconnaissance study published by the U.S. Geological Survey (USGS) in 2002 reported the presence of a wide variety of chemicals including many pharmaceuticals and personal-care products in U.S. streams.[6] Other similar findings regarding the ubiquity of pharmaceuticals in the aquatic environment have also been reported in the United States and other parts of the world.[1,7–19] Among the pharmaceuticals, several widely applied human-use and veterinary antibacterial classes, such as fluoroquinolone, sulfonamide, tetracycline, macrolide, and so forth, have been repeatedly detected in concentrations ranging from low ng/L to low µg/L.[6–19]

The presence of antibacterial residues in natural surface waters and wastewater effluents merits concern for a number of reasons. First, the continuous exposure of wastewater-borne or environmental bacterial communities to mixtures of antibacterial residues may promote induction or dissemination of low-level resistant bacterial phenotypes, which have significant indirect implications for human health.[5,20,21] Second, studies have shown that, once present in bacterial populations, numerous resistance phenotypes are stable over many bacterial generations, even in the absence of selective pressure from the antibacterial compounds themselves.[22–25] Furthermore, some antibacterials and their metabolites are reported to exhibit carcinogenic or genotoxic effects, which may be of direct significance to human health.[26–28] To properly evaluate the risks posed by antibacterial micropollutants, and to ensure provision of safe potable water supplies, the behavior of antibacterials during relevant water treatment processes should be critically evaluated. Chlorination is an important treatment process that is likely to affect the fate of numerous antibacterials, on account of its common use in water and wastewater treatment for disinfection purposes, and because many antibacterial compounds contain electron-rich functional groups that are susceptible to reaction with electrophilic chlorine.

This chapter summarizes the authors' recent contribution to developing a fundamental understanding of the interactions of four structural classes (quinoxaline N,N'-dioxide,[29] fluoroquinolone,[30] sulfonamide,[31] and pyrimidine[32]) of antibacterials with aqueous chlorine under relevant water treatment conditions. In contrast to the previous publications in which each structural class was dealt with separately, this chapter discusses these four structural classes simultaneously, highlighting similarity and difference in their interactions with aqueous chlorine. The investigations were undertaken to elucidate the chemical reactivity, reaction kinetics, products, and pathways by which antibacterials are transformed by free chlorine. Reaction kinetics were determined over a wide pH range and evaluated by a second-order kinetic model that incorporated the acid-base speciation of each reactant (i.e., oxidant and antibacterial). Various structurally related compounds that resemble the hypothesized reactive and nonreactive moieties of the target antibacterials were examined to probe reactive functional groups. Results obtained from kinetic experiments were supplemented by product identification analyses by liquid chromatography/mass spectrometry (LC/MS), gas chromatography/mass spectrometry (GC/MS), nuclear magnetic resonance spectroscopy (^{1}H-NMR), and other techniques to facilitate identification of reaction pathways and mechanisms. Additional experiments were conducted in real municipal wastewater and surface water matrices to assess the field-applicability of observations obtained for reagent water systems in the laboratory.

12.2 BACKGROUND

12.2.1 ANTIBACTERIAL AGENTS OF INVESTIGATION

Representative antibacterial agents from four structural classes—quinoxaline N,N'-dioxide, fluoroquinolone, sulfonamide, and pyrimidine—were investigated (see Table 12.1 and Table 12.2 for the structures of investigated antibacterials and associated model compounds). Carbadox (CDX) and olaquindox (QDX) represent the quinoxaline N,N'-dioxide group of veterinary antibacterial agents, which are widely used in swine production for promoting growth and preventing dysentery and bacterial enteritis. In recent years CDX and its major metabolite desoxycarbadox (DCDX) have been shown to exhibit carcinogenic and genotoxic effects.[26–28] Such concerns led the European Union to ban the use of CDX in animal feeds in 1999[33] and Health Canada to issue a ban on CDX sales in 2001 after reports of misuse and accidental contamination.[28] Ciprofloxacin (CIP) and enrofloxacin (ENR) belong to the fluoroquinolone structural class, a group of synthetic, broad-spectrum antibacterial agents that interfere with bacterial DNA replication,[5] and are used in a multitude of human and veterinary applications.[2] CIP is one of the most frequently prescribed human-use fluoroquinolones in North America and Europe,[1] whereas ENR was popular for disease prevention and control in the U.S. poultry production industry until recently.[34] Sulfamethoxazole (SMX) is one of the most popular sulfonamide antibacterials used to treat diseases and infections in humans.[5] Sulfonamides, often referred to as sulfa drugs, are synthetic antibacterials widely used in human and veterinary medicines and as growth promoters in feeds for livestock.[5] SMX is commonly prescribed in tandem with the synthetic pyrimidine antibacterial trimethoprim (TMP) under the name cotrimoxazole.[35] These two antibacterials function synergistically as inhibitors of bacterial folic acid synthesis.[5]

Recent studies have reported frequent detection of fluoroquinolones, sulfonamides, and TMP in the aquatic environment. Reported concentrations of various fluoroquinolones range from ~1 to 125 µg/L in untreated hospital sewage,[10,11] ~70 to 500 ng/L in secondary wastewater effluents,[7,14–16] to ~10 to 120 ng/L in surface waters.[6,14,15] SMX is the most frequently detected sulfonamide antibacterial at concentrations of 70 to 150 ng/L in surface waters[6,12–14] and 200 to 2000 ng/L in secondary wastewater effluents.[7,12–14,17] Occurrence of TMP was reported at several-hundred ng/L in secondary municipal wastewater effluents,[18,19] and at concentrations from approximately 10 to several-hundred ng/L in surface waters,[12,17] particularly those receiving substantial discharges of treated wastewater.[8] Concentrations of CDX in surface waters and wastewaters were reported in two studies to be below the detection limits of 0.1 µg/L and 5 ng/L, respectively.[6,17]

TABLE 12.1
Structures and Apparent Second-Order Rate Constants (k_{app}) for Reactions of CIP, ENR, SMX, CDX, and Related Model Compounds with FAC at pH 7–7.2 and 25°C

Compound	k_{app} (M⁻¹s⁻¹)	Compound	k_{app} (M⁻¹s⁻¹)
CDX 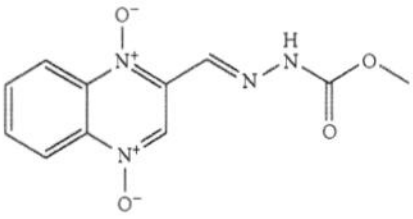	1.3×10^4	CIP 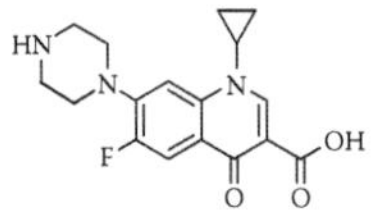	6.5×10^5
Desoxy-CDX 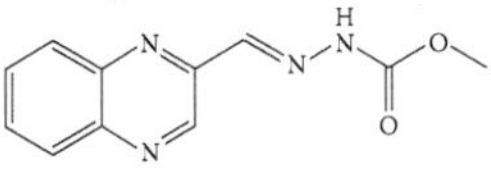	2.3×10^3	ENR 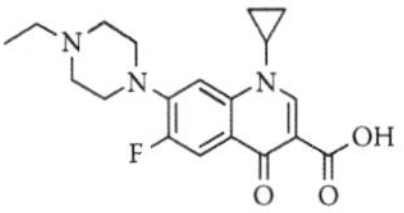	5.3×10^2
ODX 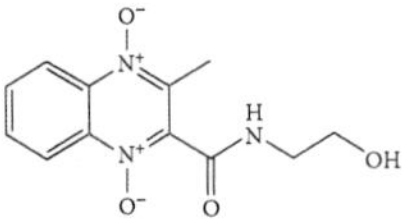	~ 0	FLU 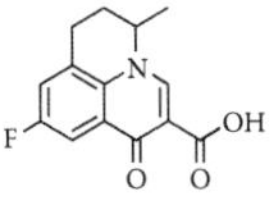	~ 0
QDX 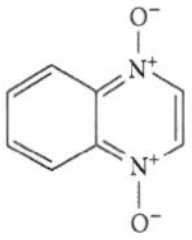	~ 0	SMX 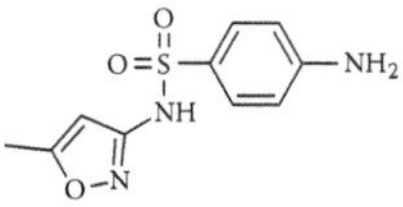	1.5×10^3
QXO 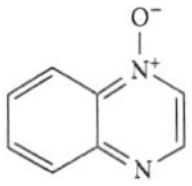	~ 0	MMIB 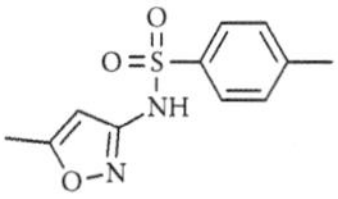	~ 0
QNO 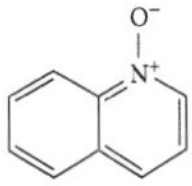	~ 0	APMS 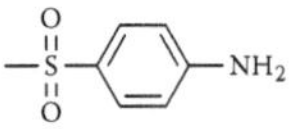	1.1×10^2
		DMI 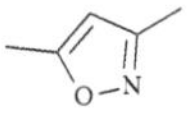	~ 0

QDX = Quindoxin
QXO = Quinoxaline N-oxide
QNO = Quinoline N-oxide
FLU = Flumequine
MMIB = 4-Methyl-N-(5-methyl-isoxazol-3-yl)-benzenesulfonamide
APMS = 4-aminophenyl methyl sulfone
DMI = 3,5-dimethylisoxazole

TABLE 12.2

Structures and Apparent Second-Order Rate Constants (k_{app}) for Reactions of Trimethoprin (TMP) and Related Model Compounds with Free Available Chlorine (FAC) at 25°C

Compound	k_{app} (M^{-1}s^{-1})
TMP	14.2 (pH 4)
	48.1 (pH 7)
	0.78 (pH 9)
TMT	59.8 (pH 4)
	3.22 (pH 7)
	0.24 (pH 9)
DAMP	0.46 (pH 4)
	23.9 (pH 7)
	0.77 (pH 9)

TMT = 3,4,5-trimethoxytoluene
DAMP = 2,4-diamino-5-methylpyrimidine

As shown in Figure 12.1, each of these antibacterials contains acidic or basic functional groups in their structures that undergo proton exchange in aqueous systems. CDX and DCDX each contain a hydrazone side-chain, in which an N-H group can deprotonate with an estimated pK_a of 9.6 for CDX (by Strock et al. using Chemaxon).[36] As illustrated by CIP, fluoroquinolones exhibit pH-dependent speciation in cationic, neutral, zwitterionic, or anionic forms. Because the neutral and zwitterionic microspecies are often difficult to distinguish by simple potentiometric titration techniques, macroscopic constants K_{a1} and K_{a2} are often used to describe the equilibrium between the cationic and neutral/zwitterionic forms and the equilibrium between the neutral/zwitterionic and anionic forms, respectively. Although the macroscopic constant is not for a particular functional group, literature has linked K_{a1} to the carboxylate group and K_{a2} to the piperazinyl N4 atom of fluoroquinolones because of similar pK_a values to those of monofunctional analogs.[37] Although not shown, ENR's pH speciation pattern is similar to that of CIP with reported pK_{a1} and pK_{a2} values at 6.1 and 7.7, respectively.[38] Sulfonamides contain two moieties connected by way of the characteristic sulfonamide linkage (-NH-S(O$_2$)-); the aniline moiety in *para*-connection to the sulfonyl S is common among almost all sulfonamides, and a variety of different structures may be connected to the sulfonamide N.[5] Sulfonamides exhibit two acid dissociation constants: one involving protonation

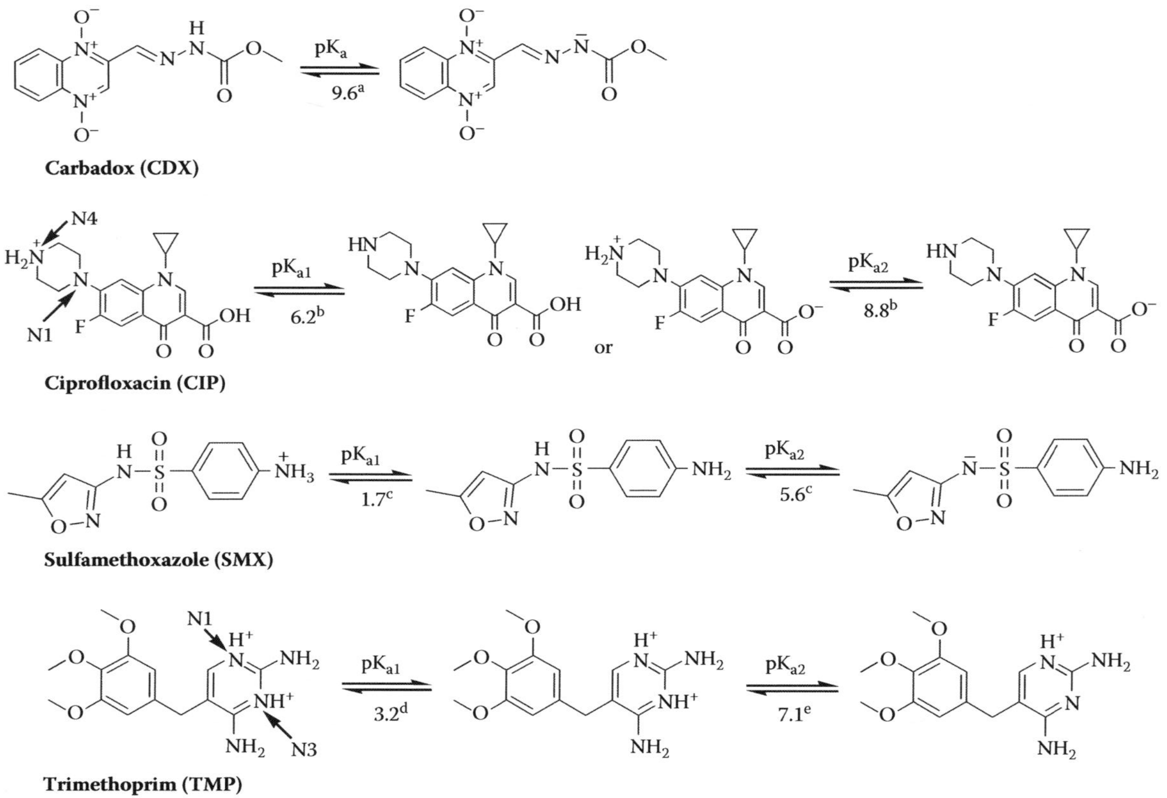

FIGURE 12.1 Structures and pH speciation of representative antibacterial agents. (a) Reference 36; (b) Reference 37; (c) Reference 39; (d) Reference 40; and (e) Reference 41.

of the aniline N and the other corresponding to deprotonation of the sulfonamide N.[39] TMP can undergo protonation at the heterocyclic N1 and N3 nitrogen atoms contained within its 2,4-diamino-5-methylpyrimidinyl moiety, leading to positively charged species at circumneutral to lower pHs.[40,41] As will be discussed in this chapter, variations in acid-base speciation of these antibacterial agents under environmentally relevant conditions strongly affect their reactivity with aqueous chlorine.

12.2.2 CHEMICAL OXIDATION BY AQUEOUS CHLORINE

Aqueous chlorine (HOCl / OCl⁻) is an important drinking water disinfectant and is used in both drinking water and wastewater treatment to achieve chemical oxidation of undesirable taste-, odor-, and color-causing compounds and reduced inorganic species.[42,43] Aqueous chlorine is typically present either as hypochlorous acid (HOCl) or its dissociated form, hypochlorite ion (OCl⁻), at pH > 5 (Equation 12.1), and may form molecular $Cl_{2(aq)}$ at very low pH or high Cl⁻ concentrations (Equation 12.2) . The combination of these aqueous chlorine species (generally only HOCl + OCl⁻ under typical water treatment conditions) is referred to as free available chlorine (FAC) hereafter in this chapter.

$$HOCl \rightleftharpoons OCl^- + H^+ \qquad pK_a = 7.4 - 7.5 \text{ (at } 25°C)^{44,45} \qquad (12.1)$$

$$Cl_2 + H_2O \rightleftharpoons HOCl + Cl^- + H^+ \quad K = 5.1\times 10^{-4} \text{ M}^2 \text{ (at } 25°C)^{46} \qquad (12.2)$$

In recent decades the electrophilic character of aqueous chlorine has drawn substantial attention due to reactions with natural organic matter (NOM), leading to the formation of harmful chlorinated disinfection byproducts (DBPs) (e.g., trihalomethanes [THMs] and haloacetic acids [HAAs]).[42] Substrates such as NOM are readily oxidized, since they consist of organic molecules with electron-rich sites that are susceptible toward electrophilic attack. The reaction mechanisms of aqueous chlorine with NOM are complex and may involve oxidation with oxygen transfer and substitutions or additions that lead to chlorinated byproducts.[42,47] The above reactions may then be followed by a number of nonoxidation processes, such as elimination, hydrolysis, and rearrangement reactions,[42] further complicating the range of byproducts generated.

Synthetic organic compounds such as antibacterials can be considered targets for transformation by aqueous chlorine. Since many antibacterials possess structural moieties and functional groups that are electron rich, such as activated aromatic rings and amines (as seen in Figure 12.1), chemical transformation of these compounds during chlorine treatment is likely. The rate and extent of such reactions, as well as the byproducts formed, will be highly dependent on the antibacterials' chemical properties, the applied chlorine dose, and water conditions such as pH, temperature, and concentrations and types of dissolved organic or inorganic species.

12.2.3 PRIOR WORK ON THE REACTION OF PHARMACEUTICALS WITH CHLORINE

Prior studies indicate that a number of pharmaceuticals are highly susceptible toward chlorine oxidation and are readily transformed under various drinking water and wastewater conditions. Thus far, studies have either assessed particular groups of pharmaceuticals that exhibit biological activity (e.g., endocrine disruptors,[48,49] β-blockers,[50] analgesics,[50] and antibiotics[51,52]), or have focused on the detailed reactivity of individual compounds (e.g., 17 β-estradiol,[53] acetaminophen,[54] naproxen,[55] caffeine,[56] and triclosan[57]). In many of these studies, laboratory-scale experiments were conducted by assessing the removal of spiked pharmaceuticals by aqueous chlorine in synthetic water or waters taken from water treatment plants or natural rivers. One particular study examined a large number of pharmaceuticals in which degradation varied greatly (<10 to >90%) after an initial chlorine dose of 2.8 to 6.75 mg/L as Cl_2, a 24-h contact time, and a solution pH of 5.5.[58] Concentrations of pharmaceuticals in full-scale treatment plants before and after chlorination were also monitored in order to evaluate the removal of these compounds by chlorine.[7] In many of these studies, reaction kinetics were examined in synthetic waters to determine whether the selected compounds were likely to be completely depleted during contact times typical of drinking water and wastewater treatment. In limited cases, byproduct analyses were conducted to assist in determining whether reaction products could potentially retain the biological activity of the corresponding parent compounds.[53]

Compared to the broad range of pharmaceuticals detected in surface waters, drinking water supplies, and wastewaters, the number of pharmaceutical compounds that have been investigated in depth regarding the mechanisms and products of their transformation by chlorine is still quite limited. A fundamental understanding of the reactions of pharmaceuticals with chlorine is critical because it enables identification of reactive functional groups/structural moieties and creates the basis for predicting the fate of other emerging contaminants on the structural basis. For example, many of the target pharmaceutical compounds contain aromatic functional groups with electron-donating substituents (e.g., substituted phenols and aromatic ethers)[48,50,53,54,57] that are known to react readily with chlorine.[59] In a study addressing chlorination of natural hormones (17β-estradiol, estrone, estriol, and progesterone) and one synthetic hormone (17α-ethinylestradiol), all molecules with a phenolic group were rapidly oxidized ($t_{1/2} = 6$ to 8 min at pH 7, $[chlorine]_0 = 1$ mg/L as Cl_2), whereas progesterone, which lacks a phenolic group, remained unchanged over 30 min in the presence of excess chlorine.[49] Another study addressing chlorination of analgesics found that all such compounds containing aromatic ether substituents were reactive toward excess chlorine, whereas those lacking such substituents (e.g., ibuprofen and ketoprofen) did not show any significant losses over 5 days.[50] Amine-containing compounds such as several β-blockers have also been shown to be reactive toward chlorine.[51]

In this chapter recent contributions by the authors toward elucidating the kinetics and transformation pathways of four structural classes of antibacterials (quinoxaline N,N'-dioxide, fluoroquinolone, sulfonamide, and pyrimidine) in reactions with free chlorine are discussed.[29–32] Significantly, these studies on the reactions of antibacterials with free chlorine are among the first conducted in such detail.

12.3 MATERIALS AND METHODS

12.3.1 CHEMICAL REAGENTS

The forms and commercial sources of the target antibacterials and structurally related model compounds were described previsouly.[29–32] DCDX, quindoxin (QDX), and quinoxaline *N*-oxide (QXO) were synthesized by methods described previously.[60] All commercial chemical standards were of >97% purity and used without further purification. NaOCl was obtained from Fisher Scientific at ~7% by weight. All other reagents used (e.g., buffers, colorimetric agents, reductants, solvents, etc. from Fisher Scientific or Aldrich) were of at least reagent-grade purity. Stock solutions of antibacterials and model compounds were prepared at 25 to 100 mg/L in reagent water (from a Barnstead or a Millipore water purification system) with 10 to 50% (v/v) methanol. FAC stocks at 0.1 to 1 g/L as Cl_2 were prepared by dilution of 7% NaOCl solutions and standardized iodometrically[61] or spectrophotometrically.[46]

12.3.2 SURFACE WATER AND WASTEWATER SAMPLES

Secondary wastewater effluent samples (collected after activated sludge processes and prior to disinfection) were obtained from pilot-scale or full-scale domestic wastewater treatment plants in Atlanta and Zurich. Surface water samples were collected from the Chattahoochee River in Atlanta—near the intake of a regional drinking water treatment plant, and from Lake Zurich in Switzerland—at the intake of one of Zurich's drinking water treatment plants. Samples were filtered through 0.45-µm filters, stored at 4 to 6°C, and used within 2 days of collection. Important characteristics of these samples were determined by standard methods or provided by the facilities where samples were taken, as summarized in Table 12.3.

12.3.3 REACTION SETUP AND MONITORING

Batch Reactions: For slower reactions, batch kinetic experiments were conducted in 25-mL amber glass vials at pH 3–11 under continuous stirring at 25°C. Reactions were buffered using 10 to 50 mM acetate (pH 4 to 5.5), phosphate (pH 6 to 8), or

TABLE 12.3
Surface Water (SW) and Wastewater (WW) Sample Characteristics

Sampling Site	pH	Alkalinity (mg/L, as $CaCO_3$)	NH_3 (mg NH_3-N/L)	DOC* (mg/L as C)
Atlanta WW	7.3	120	< 0.12[a]	14.0
Zurich WW 1	8.3	143	0.013	5.6
Zurich WW 2	8.1	500	1.5	12.3
Atlanta SW 1	6.6	13	< 0.12[a]	1.3
Atlanta SW 2	7.2	n.a.	n.a.	2.8
Lake Zurich SW 1	8.4	126	0.0068	1.6
Lake Zurich SW2	8.1	135	0.019	1.7

[a] Minimum detectable concentration was ~ 120 µg/L NH3-N (~ 7 × 10-6 mol/L)

n.a. = not available

*DOC = dissolved organic carbon

tetraborate (pH 9 to 11). The initial concentration of test compounds was typically 1 to 10 μM for kinetic studies. Reactions were initiated by adding excess (10×) molar amounts of FAC compared to the target compound. For SMX, 4-aminophenyl methyl sulfone (APMS), TMP, and 2,4-diamino-5-methylpyrimidine (DAMP), sample aliquots were periodically taken and quenched by a "soft" quenching technique—using NH_3 as a reductant—to minimize potential reversal conversion of reaction intermediates (e.g., N-chlorinated SMX) back to the parent compounds.[31] For the other compounds, sample aliquots were periodically taken and quenched with excess $Na_2S_2O_3$. Pseudo-first-order rate constants, k_{obs}, were calculated from linear ($0.95 > r^2 > 1$) plots of ln([antibacterial]) vs. time. Experiments to monitor the reaction rates of antibacterials with chlorine in real water samples were conducted by similar procedures used to measure rate constants in reagent water matrix. In these cases, FAC was added at concentrations at least tenfold greater than the corresponding antibacterial concentrations.

Competition Kinetics: The reactions of CDX and DCDX with FAC at pH > 5 and the reaction of CIP with FAC at pH 4, 5, 6, 10, and 11 were too fast to monitor by batch techniques. Instead, two competition kinetics methods were utilized for these measurements. In the method utilized for CDX and DCDX, the antibacterial and a selected competitor (with known k_{app}) were added at equimolar concentrations to batch reactors at various pHs. Varying substoichiometric amounts of free chlorine were then added. Sample aliquots were taken after each reaction was completed and analyzed for the concentrations of remaining antibacterial and the competitor. Plotting the data according to the following linear relationship allows determination of the second-order rate constant of the antibacterial:

$$\ln\left(\frac{[\text{competitor}]_{T,0}}{[\text{competitor}]_{T,t}}\right) = \frac{k_{app}^{\,competitor}}{k_{app}^{\,antibacterial}} \times \ln\left(\frac{[\text{antibacterial}]_{T,0}}{[\text{antibacterial}]_{T,t}}\right) \qquad (12.3a)$$

where $[\text{competitor}]_{T,t}$ and $[\text{antibacterial}]_{T,t}$ represent the remaining reactant concentrations after a specific substoichiometric dose of FAC has been added. 4,6-Dichlororesorcinol, with known k_{app} values reported in Rebenne et al.,[62] was selected as the competitor for CDX. Once the rate constant for CDX was determined at each pH, CDX was used as the competitor for DCDX.

In the method utilized for CIP, the antibacterial and the competitor 4,6-dichlororesorcinol were added to vials at varying molar ratios of competitor to antibacterial over a range of pH values. A fixed substoichiometric dose of free chlorine was then added under rapid mixing to each vial. After complete consumption of free chlorine, 1-mL sample aliquots were transferred to amber, borosilicate high-performance liquid chromatography (HPLC) vials, quenched with $Na_2S_2O_3$ to prevent reactions of the competitor with a N-chlorinated intermediate formed upon reaction of CIP with free chlorine, and stabilized with ~ 0.1 M H_3PO_4 prior to analysis by HPLC with fluorescence detection. Apparent second-order rate constants for the reaction of CIP with FAC at different pH were determined by monitoring yields of a product (presumably 2,4,6-trichloresorcinol) formed upon chlorination of the competitor, 2,4-dichloresorcinol, and plotting measured product yields according to Equation 12.3b:

$$\frac{[\text{product}]_{\text{absence}}}{[\text{product}]_{\text{presence}}} - 1 = \frac{k_{app}^{competitor}[\text{competitor}]_{T,0}}{k_{app}^{antibacterial}[\text{antibacterial}]_{T,0}} \qquad (12.3b)$$

where [product]$_{\text{absence}}$ represents product yield in the absence of CIP [product]$_{\text{presence}}$ represents product yields in the presence of varying CIP concentrations, and [competitor]$_{T,0}$ and [antibacterial]$_{T,0}$ represent reactant concentrations before addition of FAC.

Continuous-Flow Method: FAC decay was measured under pseudo-first-order conditions in the presence of a large excess of CIP, at pH ranging from 5–11, by a continuous-flow, quenched-reaction method described previously.[30] Pseudo-first-order rate constants, k_{obs}, were calculated from linear ($0.97 > r^2 > 1$) plots of ln([FAC]) vs. time.

Reactant and Reaction Product Analyses: Parent compound loss was monitored by reverse-phase HPLC with ultraviolet (UV), fluorescence, or mass spectrometry detection. Reaction intermediates and products were analyzed using reverse-phase HPLC/electrospray MS techniques. Details of the HPLC and MS instrumental conditions were described previously.[29–32] Residual oxidant concentrations were measured at the conclusion of each kinetic experiment by *N,N*-diethyl-*p*-phenylenediamine (DPD) colorimetry or DPD-ferrous ammonium sulfate (FAS) titrimetry.[61]

12.4 RESULTS AND DISCUSSION

12.4.1 REACTION KINETICS AND MODELING

For all of the antibacterials examined, reactions were found to be first order with respect to the antibacterial and FAC and so can be described by a general second-order rate expression:

$$\frac{d[\text{Antibacterial}]_T}{dt} = -k_{obs}[\text{Antibacterial}]_T = -k_{app}[\text{FAC}]_T[\text{Antibacterial}]_T \qquad (12.4)$$

where k_{obs} (in s^{-1}) is the observed pseudo-first-order rate constant, T represents the sum of all acid-base species for a given reactant, and k_{app} (in M^{-1}s^{-1}) is the pH-dependent apparent second-order rate constant for the overall reaction, which can be calculated from $k_{app} = \dfrac{k_{obs}}{[\text{FAC}]_T}$.

Kinetic experiments revealed a marked dependence of k_{app} values on pH (Figure 12.2). The variation in k_{app} from pH 4 to 11 can be attributed to the varying importance of specific reactions among the individual acid-base species of FAC and antibacterials. The acid-base speciation of FAC (i.e., Equation 12.1) and antibacterials

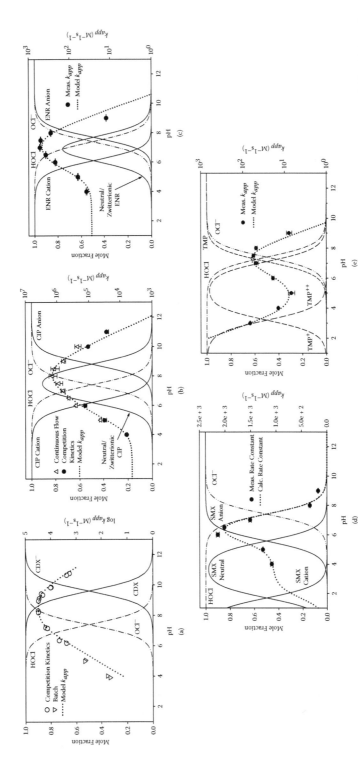

FIGURE 12.2 Effect of pH on the apparent second-order rate constants for the reactions of (a) CDX, (b) CIP, (c) ENR, (d) SMX, and (e) TMP. (Adapted from Reference 29 through Reference 32.)

(i.e., Figure 12.1) can be modeled according to mass balance relationships and known acid-base equilibria:

$$[FAC]_T = [HOCl] + [OCl^-] = \sum_{i=1,2} \alpha_i [FAC]_T \tag{12.5}$$

$$[Antibacterial]_T = \sum_{j=1,2 \text{ or } j=1,2,3} \beta_j [Antibacterial]_T \tag{12.6}$$

where $\alpha_1 = \dfrac{[H^+]}{[H^+] + K_{a,HOCl}}$ for HOCl and $\alpha_2 = \dfrac{K_{a,HOCl}}{[H^+] + K_{a,HOCl}}$ for OCl$^-$, $\beta_1 = \dfrac{[H^+]}{[H^+] + K_a}$ and

$\beta_2 = \dfrac{K_a}{[H^+] + K_a}$ for monoprotic antibacterial CDX, and $\beta_1 = \dfrac{[H^+]^2}{[H^+]^2 + K_{a1}[H^+] + K_{a1}K_{a2}}$,

$\beta_1 = \dfrac{[H^+]^2}{[H^+]^2 + K_{a1}[H^+] + K_{a1}K_{a2}}$, $\qquad \beta_2 = \dfrac{K_{a1}[H^+]}{[H^+]^2 + K_{a1}[H^+] + K_{a1}K_{a2}}$ and

$\beta_3 = \dfrac{K_{a1}K_{a2}}{[H^+]^2 + K_{a1}[H^+] + K_{a1}K_{a2}}$ for diprotic antibacterials CIP, ENR, SMX and TMP.

Incorporating Equation 12.5 and Equation 12.6 into Equation 12.4 can derive Equation 12.7,

$$k_{app} = \sum_{\substack{i=1,2 \\ j=1,2 \text{ or } j=1,2,3}} k_{ij} \alpha_i \beta_j \tag{12.7}$$

where k_{ij} represents the specific second-order rate constants for reactions of each oxidant species i with each antibacterial species j. Note that for fluoroquinolones, the zwitterionic and neutral species are combined as a single "effective" monoprotonated species (according to macroscopic pK_{a1} and pK_{a2} values) because reliable equilibrium constants are not available for the microspeciation of CIP and ENR.

As shown in Figure 12.2, k_{app} decreases sharply at pH greater than or equal to 7 for all of the compounds. The decrease in k_{app} can be attributed to deprotonation of HOCl to yield OCl$^-$, which is generally a much weaker electrophile than HOCl.[63] This indicates that reactions among OCl$^-$ and various antibacterial species are relatively unimportant and can be omitted from Equation 12.7 (as shown by Equation 12.8 through Equation 12.13 in Table 12.4). Further simplification of the kinetic equation can be conducted for the reactions of CDX and SMX with HOCl, respectively. For CDX, the reaction with HOCl appears to be dominated by the deprotonated CDX species (CDX$^-$), and the reaction of neutral CDX with HOCl is negligible (i.e., Equation 12.8) because: (i) the experimental k_{app} values can only be fitted if the neutral CDX species' specific rate constant (i.e., k_{11} is manually assigned a very small value or omitted, and (ii) the k_{app} exhibits a well-defined bell-shaped pH profile (Figure 12.2a) and reaches a maximum at the pH near the average of the pK_as of HOCl and CDX. For SMX, the reaction of HOCl with the cationic form of SMX (SMX$^+$) is neglected (i.e., Equation 12.11) on the basis of two assumptions: (i) the abundance of SMX$^+$ is rather low within the pH ranges studied (pK$_a$ for the aromatic amine is 1.7, which is 2.3 pH units lower than the lowest pH value studied), and (ii) protonation of the aniline's primary amino group should at the very least retard (if not prevent) the reaction between HOCl and SMX by coordinating the lone-pair electrons associated with this nitrogen.

For TMP, reactivity trends above pH 5 were similar to those observed for each of the other three antibacterial classes. However, in this case, k_{app} increases substantially with increasing acidity at pH below 5. This trend indicates that factors other than acid-base speciation of TMP and HOCl are playing a role in governing the kinetics of this reaction, as protonation of TMP should yield less nucleophilic species, which should in turn be less reactive toward electrophilic chlorine. One possible explanation for these observations is an acid-catalyzed reaction between HOCl and TMP's 3,4,5-trimethoxybenzyl moiety (i.e., $\text{TMP} + \text{HOCl} + \text{H}^+ \xrightarrow{k_{H^+}} \text{products}$), in parallel to trends previously observed for various phenols and methoxybenzenes.[50,57,59,62]

Accordingly, an acid catalysis term ($k_{H^+}[\text{H}^+]$) can be added to Equation 12.7 to yield Equation 12.12 in Table 12.4. Alternatively, the increase in reaction rate at pH

TABLE 12.4
Kinetic Models for the Reaction of Antibacterial Agents with FAC

Compound	Kinetic Model for the Apparent Second-Order Rate Constant k_{app} (M^{-1}s^{-1})	
Carbadox (CDX)	$k_{app,CDX} = k_{12}\alpha_1\beta_2$	(12.8)
Ciprofloxacin (CIP)	$k_{app,CIP} = k_{11}\alpha_1\beta_1 + k_{12}\alpha_1\beta_2 + k_{13}\alpha_1\beta_3$	(12.9)
Enrofloxacin (ENR)	$k_{app,ENR} = k_{11}\alpha_1\beta_1 + k_{12}\alpha_1\beta_2 + k_{13}\alpha_1\beta_3$	(12.10)
Sulfamethoxazole (SMX)	$k_{app,SMX} = k_{12}\alpha_1\beta_2 + k_{13}\alpha_1\beta_3$	(12.11)
Trimethoprim (TMP)	$k_{app,TMP} = k_{H^+}[\text{H}^+] + k_{11}\alpha_1\beta_1 + k_{12}\alpha_1\beta_2 + k_{13}\alpha_1\beta_3$	(12.12)
	$k_{app,TMP} = k_{Cl_{2(aq)}}\alpha_1\dfrac{[\text{Cl}^-][\text{H}^+]}{K_{Cl_2,\text{hydrolysis}}} + k_{11}\alpha_1\beta_1 + k_{12}\alpha_1\beta_2 + k_{13}\alpha_1\beta_3$	(12.13)

The mathematic expressions of α and β values are discussed in the text.
For TMP: k_{H^+} (in M^{-2}s^{-1}) represents the rate constant for the acid-catalyzed reaction between HOCl and TMP, $k_{Cl2(aq)}$ (in M^{-1}s^{-1}) represents the apparent second-order rate constant for the bulk reaction of $Cl_{2(aq)}$ with all three TMP species, and $K_{Cl_2,\text{hydrolysis}}$ is the equilibrium constant for the hydrolysis of $Cl_{2(aq)}$ (Equation 12.2).

<5 could be attributable to generation of molecular chlorine, $Cl_{2(aq)}$, at increasing acidities,[57,64] via the equilibrium represented by Equation 12.2. In this case, a term

$(k_{Cl_{2(aq)}} \alpha_1 \dfrac{[Cl^-][H^+]}{K_{Cl_2,hydrolysis}}$) for the bulk reaction of $Cl_{2(aq)}$ with all three TMP species (i.e.,

$TMP + Cl_{2(aq)} \xrightarrow{k_{Cl2,hydrogen}}$ products) can be added to Equation 12.7 to yield Equation 12.13.

The kinetic models for each of the above reactions were fitted to experimental data using nonlinear Marquardt-Levenberg regression (SigmaPlot 2002, SPSS Software). As shown in Figure 12.2, these kinetic models describe the experimental data measured for each antibacterial very well. In the case of TMP, the two alternative models presented above (using an estimated $[Cl^-] = 1.7 \times 10^{-5}$ M for Equation 12.13) appeared to be quite similar and resulted in a nearly perfect overlap. However, based on evidence presented by Cherney et al. on the role of $Cl_{2(aq)}$ in acidic chlorination reactions,[64] the latter model may be more plausible.

All of the antibacterials reacted fairly quickly with aqueous chlorine at circumneutral pH values. Based on the k_{app} values obtained (Table 12.1 and Table 12.2), the half-lives (pH 7 to 7.2, 25°C) of CDX, CIP, ENR, SMX, and TMP at 2.5 mg/L as Cl_2 of $[FAC]_0$ are estimated to be 1.5 s, 0.03 s, 37.1 s, 13.1 s, and 6.8 min, respectively. These estimates indicate that substantial transformation of each antibacterial by aqueous chlorine can be expected during contact times typical of drinking water (1 to 17 h)[65] and wastewater (30 to 120 min)[66] treatment.

12.4.2 IDENTIFICATION OF REACTIVE FUNCTIONAL GROUPS

The kinetic models discussed above provide useful tools for evaluating the importance of each antibacterial and oxidant species in the kinetics observed for each reaction and facilitate determination of reactive sites within each antibacterial structure. The experimental data suggest that attack of CDX, CIP/ENR, SMX, and TMP by free chlorine is closely related to protonation and deprotonation of their hydrazone linkage, piperazinyl N4 atom, anilinyl amino group, and 2,4-diamino-methylpyrimidinyl moiety, respectively. A clear inverse relationship between the reaction rate and the degree of protonation of the aforementioned functional groups is demonstrated in Figure 12.2. Such trends are consistent with the expectation that protonation leads to a decrease in reactivity of various functional groups toward free chlorine.

To verify reactive functional groups, examination of model compounds that resemble the hypothesized reactive and nonreactive moieties of each antibacterial in its reactions with chlorine was conducted, in addition to structural analysis of each antibacterial's reaction intermediates and products. As shown in Table 12.1 and Table 12.2, a range of model compounds was carefully selected and examined for each antibacterial structural class. The results shown in Table 12.1 indicate that both CDX and DCDX react with chlorine at comparable rates, whereas ODX, QX, QDX, QXO, and QNO do not react at pH 7.1 to 7.4 for up to 1 to 2 hours under excess free chlorine conditions, confirming that the *hydrazone* side-chain is the reactive site and the quinoxaline *N,N′*-dioxide and quinoxaline moieties are relatively inert to chlorine. The results that both CIP and ENR are reactive to chlorine; that the reactivity of CIP is about three orders of magnitude higher than that of ENR; and that FLU, which lacks the piperazine ring, does not react with chlorine clearly demonstrate the critical role of the *piperazinyl N4* atom in the reaction with chlorine. The comparison among SMX, MMIB, APMS, and DMI verifies that the *anilinyl amino* group is required for the reaction with chlorine. Table 12.2 shows that both DAMP and TMT—the two model compounds selected to evaluate the TMP structure's reactivity—are reactive to chlorine, but exhibit significantly different pH-dependency in their reactivities toward free chlorine. Like TMP, DAMP is most reactive toward chlorine at pH 7, and the magnitudes of $k_{app,DAMP}$ at pH 7 and 9 compare favorably with those of $k_{app,TMP}$. However, unlike TMP, DAMP exhibits very low reactivity toward chlorine at pH 4. In contrast, TMT reacts rapidly with chlorine at pH 4 but exhibits much decreased reactivity with increasing pH. This comparison strongly supports that TMP's *3,4,5-trimethoxybenzyl* moiety is primarily responsible for the observed increase in reaction rate at acidic pH, whereas the *2,4-diamino-methylpyrimidinyl* moiety governs TMP's reactivity at intermediate and alkaline pH.

12.4.3 REACTION PATHWAYS AND PRODUCTS' BIOLOGICAL IMPLICATIONS

Extensive efforts using LC/MS, GC/MS, ¹H-NMR, and other techniques were made to identify the most probable structures of reaction intermediates and products. Evolution of reaction intermediates and products over time was also monitored. These results were combined with the results of reaction kinetic modeling and reactive site identification to deduce the pathways by which the antibacterials are transformed in their reactions with FAC. Details of product characterization and determination of reaction pathways are omitted here for the sake of brevity and can be found in the previous publications.[29–32] The proposed pathways for reactions of FAC with CDX, CIP, ENR, SMX, and TMP (Scheme 12.1 through Scheme 12.5) are discussed briefly below along with possible biological implications of the reaction products.

CDX: Deprotonation of the hydrazone N-H apparently results in a negative charge that substantially enhances reactivity of CDX toward electrophilic chlorine. The reaction begins with an initial HOCl attack at the hydrazone N_β of the negatively charged CDX⁻, leading to a reactive N-chloro intermediate (Scheme 12.1). At acidic to slightly basic conditions, the imine carbon of the N-chloro intermediate is susceptible to intramolecular nucleophilic attack (by the oxygen of the γ-carbonyl group) or bimolecular nucleophilic attack by constituents of the water matrix (e.g., OH⁻, CH_3COO^-, and Cl⁻), to yield various nucleophile-parent adducts. Hydroxylated products (i.e., m/z M+16) resulting from bimolecular nucleophilic substitution may gradually decay. As pH increases, the much stronger OH⁻ nucleophile plays an increasingly greater role in product formation. At alkaline pH (8.9 to 9.0) and upon further excess chlorine oxidation, either the attack of the N-chloro intermediate by two OH⁻ or attack of the hydroxylated product (i.e., m/z M + 16) by one OH⁻ can yield an azoxy product (i.e., m/z M + 30).

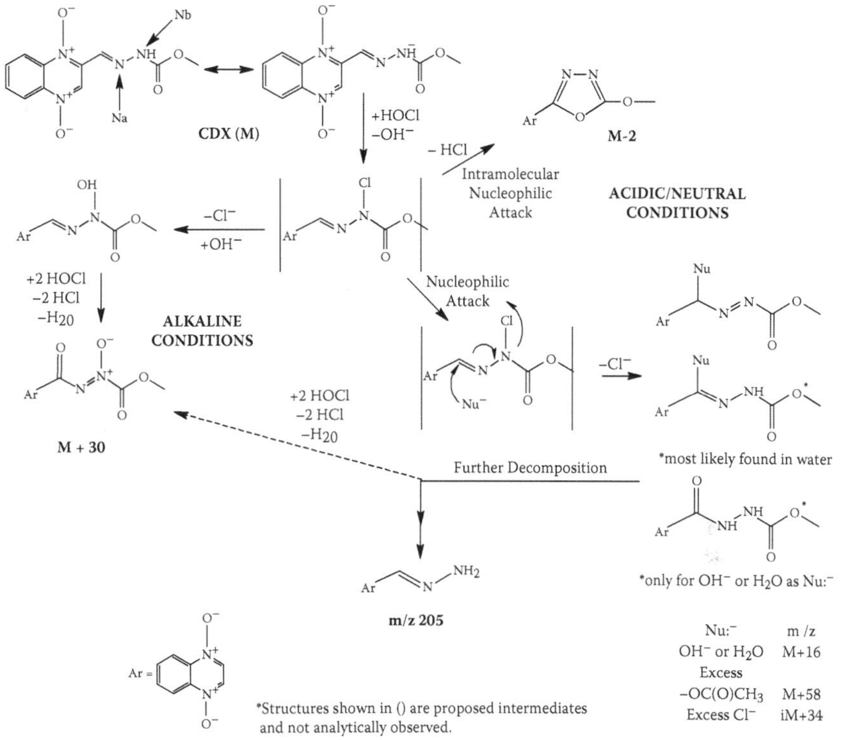

SCHEME 12.1 Proposed reaction pathways for CDX with free available chlorine. (Adapted from Reference 29.)

The antibacterial activity of CDX rests on the presence of the N-oxide groups which impair DNA synthesis by forming a radical upon N-oxide reduction.[67,68] However, the CDX N-oxide moiety was not transformed during reaction with FAC, and all products formed retain this functional group. These results imply little compromise in CDX's antibacterial activity by reaction with chlorine, irrespective of the rapid kinetics with which the CDX structure is transformed in this reaction.

CIP and ENR: Reaction of CIP with FAC proceeds primarily through an initial attack of HOCl on CIP's piperazinyl N4 atom, leading to the unstable intermediate, CIP-Ia1, which subsequently decays via piperazine ring fragmentation to the product CIP-Pa1 (Scheme 12.2). In the presence of excess chlorine, CIP-Pa1 reacts further via N-chlorination followed by decay in water to yield the fully dealkylated products (CIP-Pa2 and CIP-Pa3). Because of the N4 and secondary N1 amines' presumably high reactivity toward chlorine,[69,70] addition of Cl to the aromatic system of CIP is only likely to occur under conditions for which chlorination of the amines is unfavorable, namely, when they are protonated (e.g., in cationic CIP) or already chlorinated (e.g., in CIP-Ib1).

Although ENR differs from CIP only with respect to degree of N4 amine alkylation in the structures, it exhibits markedly different chlorine reaction pathways. As shown in Scheme 12.3, initially HOCl also attacks the tertiary N4 amine of ENR (but at a much slower rate than that with CIP), forming an apparently highly reactive

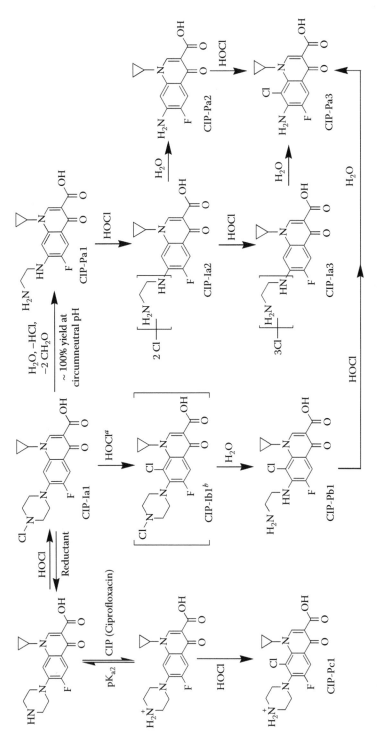

SCHEME 12.2 Proposed pathways for reactions of CIP with free available chlorine. (a) Pathway not likely to be important except at very high FAC concentrations. (b) CIP-Ib1 structure was proposed not explicitly observed. (Adapted from Reference 30.)

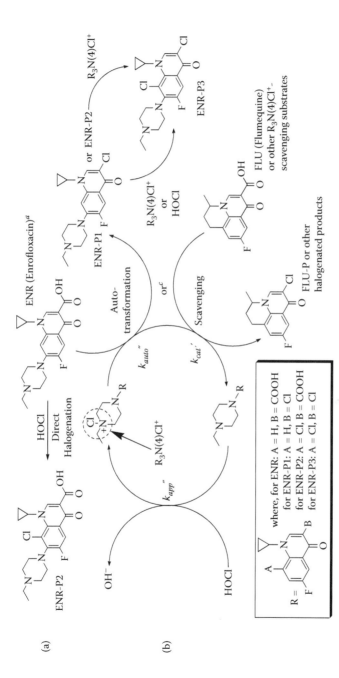

SCHEME 12.3 Proposed pathways for reactions of ENR with free available chlorine. (a) Direct halogenation and autotransformation pathways presumably apply to all four ENR acid-base species. (b) Catalytic autotransformation pathways should be suppressed by the presence of $R_3N(4)Cl^+$-scavenging substrates in real water reaction matrixes. (Adapted from Reference 30.)

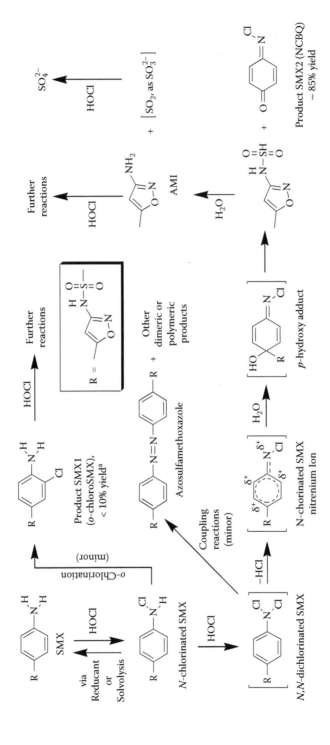

SCHEME 12.4 Proposed pathways for reactions of SMX with free available chlorine. Structures enclosed in brackets are given as probable intermediates. (a) Estimated yield obtained from data reported by Reference 73. (Adapted from Reference 31.)

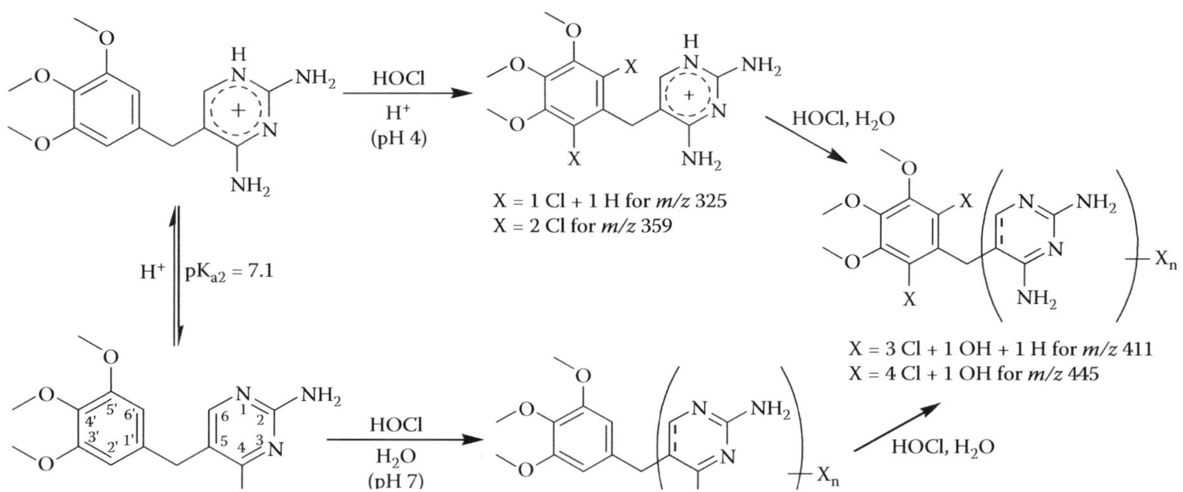

SCHEME 12.5 Proposed major products and transformation pathways for the reaction of TMP with free available chlorine. (Adapted from Reference 32.)

chlorammonium intermediate ($R_3N(4)Cl^+$) that can catalytically halogenate ENR or other substrates present in solution. FLU, which exhibits no apparent reactivity toward HOCl, was found to undergo facile halodecarboxylation in the presence of $R_3N(4)Cl^+$ species derived from ENR. Details for determination of this mechanism are described in Dodd et al.[30] These results with CIP and ENR demonstrate the two very different transformation pathways and kinetic trends that should be expected for secondary versus tertiary amine-containing fluoroquinolones in reaction with chlorine.

The ability of fluoroquinolones to inhibit bacterial DNA replication and repair is believed to be linked to their quinolone moieties,[71] suggesting that CIP's predominant transformation pathway (i.e., piperazine fragmentation) may not lead to complete elimination of antibacterial activity. Accordingly, even quinolone structures which lack the piperazine ring (e.g., nalidixic acid) can be active against various types of bacteria.[72] Direct halogenation ortho- to the piperazine ring is also unlikely to result in complete elimination of quinolone activity, as suggested by the considerable antibacterial potency of lomefloxacin,[72] which is substituted by a fluorine atom in the same position. Although modification of ENR's quinolone moiety via halodecarboxylation (Scheme 12.3) might interrupt the charge interactions and hydrogen-bonding required for quinolone-DNA binding, this reaction may be relatively unimportant in real waters (due to the presence of scavengers for the $R_3N(4)Cl^+$ intermediate of ENR in natural waters). On the basis of these observations, one can infer that a majority of the transformation products likely to result from passage of fluoroquinolones through water chlorination processes could remain biochemically active.

SMX: The reaction of SMX and FAC begins with an initial attack of HOCl on the anilinyl amino N of SMX (Scheme 12.4). Under conditions where $[FAC]_0:[SMX]_0$ <1, the N-chlorinated SMX slowly rearranges to yield *ortho*-chlorination of SMX's aniline[73] or reverts to the parent SMX structure. When $[FAC]_0:[SMX]_0$ >1 (i.e., excess chlorine), further chlorination of *N*-chlorinated SMX is believed to lead to formation of a *N,N*-dichlorinated SMX intermediate, which subsequently decays via facile cleavage of the S-C bond to yield the N-chlorimine NCBQ (*N*-chloro-*p*-benzoquinoneimine), as well as AMI (3-amino-5-methylisoxazole) and SO_2.

The antibacterial activity of SMX is derived from its antagonistic competition with *p*-aminobenzoic acid for dihydropteroate synthase enzyme, which is necessary for bacterial folic acid synthesis.[74] According to this mode of activity, sulfonamide activity is derived from the characteristic anilinyl moiety. In consideration of the reaction pathways identified for SMX, the critical role of the anilinyl moiety implies that SMX/FAC reaction will likely contribute to at least partial reduction or elimination of SMX's antibacterial activity, since these pathways lead to extensive disruption of the sulfonamide structure. However, the effects of aromatic ring-chlorination—in which the *p*-aminobenzenesulfonamide structure is essentially conserved—are less clear. Furthermore, other products formed in this reaction, particularly, *N*-chloro-*p*-benzoquinoneimine (NCBQ), might possess higher acute toxicity[75] to aquatic organisms than the parent substrates.

TMP: As described in the previous section, the two aromatic moieties of TMP each react with FAC, but with markedly different pH dependency. At acidic pH, HOCl attacks TMP's 3,4,5-trimethoxybenzyl moiety, leading to sequential mono-

and dichlorination of the trimethoxybenzyl ring (Scheme 12.5). In the presence of excess chlorine, slower halogenation of the dihalogenated product's protonated 2,4-diaminopyrimidinyl moiety can apparently occur to yield the tetrahalogenated reaction product (m/z 445). At intermediate to alkaline pH, HOCl reacts primarily with TMP's 2,4-diamino-methylpyrimidinyl moiety, by halogenation or oxidation, to yield a wide range of Cl- and OH-substituted products. Slow halogenation of these products' 3,4,5-trimethoxybenzyl moiety may occur in the presence of excess chlorine.

It is evident from these findings that the TMP structure is not substantially degraded upon reactions with FAC. Instead, a wide variety of (multi)chlorinated and hydroxylated products are formed. The effects of such ring substitution on the TMP's antibacterial activity are not currently clear, suggesting the need for additional biological studies designed to evaluate the chlorinated products' activities.

12.4.4 Reaction in Real Water Matrices

Additional experiments were conducted to validate the applicability of measured FAC reaction kinetics to modeling reactions in real municipal wastewater and surface water samples (Table 12.3). Reactions of ENR and SMX in real water samples were modeled by assuming two different reaction steps: one characterized by rapid decay of FAC during the first 10 s (due to rapid consumption by highly reactive substrates such as phenols, reduced sulfur compounds, amines, etc., in the water samples), and one by slower FAC decay during the remaining reaction time. FAC decay during the two reaction periods was treated as pseudo-first-order, and was modeled using FAC residual measurements taken at various time intervals during the experimental monitoring periods. ENR and SMX losses were in turn modeled by inserting modeled FAC concentrations into pseudo-first-order expressions for antibacterial losses. Observed oxidation kinetics for SMX loss corresponded quite closely to those predicted on the basis of reaction kinetics measured in clean water systems (Figure 12.3a). Rates of ENR transformation were moderately slower in environmental matrices than predicted on the basis of reaction kinetics measured in clean water systems (Figure 12.3b). As discussed in the previous section, ENR reacts with HOCl via formation of a highly reactive chloroammonium intermediate ($R_3N(4)Cl^+$). The slower rate of ENR reaction with FAC in the real water experiments matrices is likely due in part to suppression of ENR auto-transformation by $R_3N(4)Cl^+$-scavenging substrates such as NOM present in the real water matrices.

The initial reaction of CIP with FAC in real water matrices was too rapid to follow directly. However, the kinetics of its piperazine ring's fragmentation could be evaluated indirectly by monitoring loss of the N-chlorinated intermediate of CIP (i.e., CIP-Ia1 in Scheme 12.2). Model results confirmed the applicability of the kinetics measured in clean water systems to the real water samples (Figure 12.3c). Details of these measurements and the corresponding model evaluation can be referred to Dodd et al.[30]

For real water experiments involving TMP, the FAC loss to matrix constituents over time was also modeled using first-order decay constants from two separate kinetic regimes in the drinking water and three separate kinetic regimes in the wastewater. The equations describing FAC's pseudo-first-order decay were then

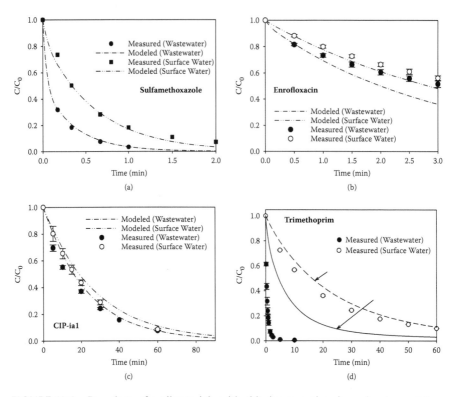

FIGURE 12.3 Reactions of antibacterials with chlorine over time in real water matrices. Real water samples: SMX (Atlanta WW; Atlanta SW 1); ENR (Zurich WW 1; Lake Zurich SW 1); CIP (Atlanta WW; Atlanta SW1); TMP (Zurich WW 2; Lake Zurich SW 2). (Adapted from Reference 30 through Reference 32.)

inserted into a pseudo-first-order expression for TMP loss and integrated (see the detailed descriptions in Dodd and Huang[32]) to permit modeling of TMP loss over time. As shown in Figure 12.3d, TMP depletion rates in the drinking water matrix could be accurately described. However, model predictions substantially underestimated the rate of TMP loss during chlorination of the wastewater matrix. A definite explanation for this discrepancy is not currently available, although it does not appear to be due to any change in reaction mechanism, as product formation in the real waters was similar to that observed in reagent water systems. Additional studies with a wider range of wastewater matrices would be necessary to more thoroughly evaluate the effect of wastewater composition (and perhaps individual wastewater constituents) on the magnitude of these deviations from predicted TMP loss rates. In any case, the rates at which TMP was oxidized in each real water sample indicate that substantial losses of the parent compound can be expected for residence times typical of drinking water and wastewater disinfection processes.

CDX was assessed for its reaction rate with FAC in a surface water sample (Atlanta SW2, Table 12.3) under conditions of excess free chlorine ($10 \times [CDX]_0$) and was found to undergo complete transformation in less than 30 s (i.e., the parent CDX was not detectable after 30 s). Although the reaction rate was too fast to monitor accurately by batch methods, these results confirmed the expectation that CDX will be rapidly transformed during chlorination of real water matrices.

In addition to kinetic evaluation, reaction product identities and evolution patterns for the target antibacterials in the environmental matrix experiments were found to be similar to those observed in the clean water systems. Overall, real water studies confirmed that the laboratory observations obtained from reagent water systems can be applied with reasonable accuracy for modeling antibacterial loss rates in real waters under realistic treatment conditions.

12.5 CONCLUSION

Carbadox (CDX), ciprofloxacin (CIP) and enrofloxacin (ENR), sulfamethoxazole (SMX), and trimethoprim (TMP) represent four structural classes (quinoxaline N,N'-dioxide, fluoroquinolone, sulfonamide, and pyrimidine) of widely used antibacterial agents, respectively. In assessing the susceptibility of the above antibacterials toward reactions with chlorine under typical water chlorination conditions, this study focused on obtaining a fundamental understanding of the reaction kinetics, products, and mechanisms between free available chlorine (FAC, i.e., $HOCl/OCl^-$) and antibacterials. Overall, the reaction kinetics are highly pH-dependent and can be successfully described using a second-order kinetic model that incorporates the acid-base speciation of reactants. The estimated half-lives of the four antibacterials ranged from 1.5 s to 6.7 min under conditions typical of drinking water treatment conditions (i.e., at 2.5 mg/L as Cl_2 of FAC, pH 7, and 25°C). Studies conducted in real wastewater and surface water matrices confirm that rapid transformation of these antibacterials can be expected during contact times typical of drinking water (1 to 17 h) and wastewater (30 to 120 min) chlorination. Notably, structural identification of many reaction products of antibacterials in this study provides critical information for assessing the potential biological activity of transformation products that are also likely present in finished drinking water and surface streams due to discharged wastewater effluent. Through a combination of effective research methodologies, the reactive sites and reaction mechanisms via which the antibacterials react with free chlorine were elucidated. The obtained fundamental knowledge can be applied to many other antibacterial compounds in the similar structural classes and also provides the basis for predicting the fate and transformation of other structurally related organic contaminants during water chlorination processes.

REFERENCES

1. Daughton, C.G. and Ternes, T.A. Pharmaceuticals and personal care products in the environment: agents of subtle change? *Environ. Health Perspectives,* 107, 907, 1999.
2. National Research Council. *The Use of Drugs in Food Animals*, National Academy Press, Washington, D.C., 1999.
3. NASS, National Agricultural Statistics Service. *Agricultural Chemical Usage, Fruit, and Nut Summary 1999*, USDA Economics and Statistics System, 2000.
4. Mellon, M., Benbrook, C., and Benbrook, K.L. *Hogging It. Estimates of Antimicrobial Abuse in Livestock*, Union of Concerned Scientists Publications, 2001.

5. Walsh, C. *Antibiotics: Actions, Origins, Resistance,* ASM Press, Washington, D.C., 2003.
6. Kolpin, D.W. et al. Pharmaceuticals, hormones, and other organic wastewater contaminants in U.S. Streams, 1999–2000: A national reconnaissance, *Environ. Sci. Technol.,* 36, 1202, 2002.
7. Renew, J.E. and Huang, C.-H., Simultaneous determination of fluoroquinolone, sulfonamide, and trimethoprim antibiotics in wastewater using tandem solid phase extraction and liquid chromatography—Electrospray mass spectrometry, *J. Chromatogr. A,* 1042, 113, 2004.
8. Glassmeyer, S.T. et al. Transport of chemical and microbial compounds from known wastewater discharges: Potential for use as indicators of human fecal contamination, *Environ. Sci. Technol.,* 39, 5157, 2005.
9. Ternes, T.A. Occurrence of drugs in German sewage treatment plants and rivers, *Water Res.,* 32, 3245, 1998.
10. Hartmann, A. et al. Identification of fluoroquinolone antibiotics as the main source of umuC genotoxicity in native hospital wastewater, *Environ. Toxicol. Chem.,* 17, 377, 1998.
11. Hartmann, A. et al. Primary DNA damage but not mutagenicity correlates with ciprofloxacin concentrations in German hospital wastewaters, *Arch. Environ. Contam. Toxicol.,* 36, 115, 1999.
12. Hirsch, R. et al. Occurrence of antibiotics in the environment, *Sci. Total Environ.,* 225, 109, 1999.
13. Hartig, C., Storm, T., and Jekel, M. Detection and identification of sulphonamide drugs in municipal waste water by liquid chromatography coupled with electrospray ionisation tandem mass spectrometry, *J. Chromatogr. A,* 854, 163, 1999.
14. Ternes, T.A. Analytical methods for the determination of pharmaceuticals in aqueous environmental samples, *Trends Anal. Chem.,* 20, 419, 2001.
15. Golet, E.M., Alder, A.C., and Giger, W. Environmental exposure and risk assessment of fluoroquinolone antibacterial agents in wastewater and river water of the Glatt Valley watershed, Switzerland, *Environ. Sci. Technol.,* 36, 3645, 2002.
16. Golet, E.M. et al. Environmental exposure assessment of fluoroquinolone antibacterial agents from sewage to soil, *Environ. Sci. Technol.,* 37, 3243, 2003.
17. Miao, X.-S. et al. Occurrence of antimicrobials in the final effluents of wastewater treatment plants in Canada, *Environ. Sci. Technol.,* 38, 3542, 2004.
18. Göbel, A. et al. Occurrence and sorption behavior of sulfonamides, macrolides, and trimethoprim in activated sludge treatment, *Environ. Sci. Technol.,* 39, 3981, 2005.
19. Lindberg, R.H. et al. Screening of human antibiotic substances and determination of weekly mass flows in five sewage treatment plants in Sweden, *Environ. Sci. Technol.,* 39, 3421, 2005.
20. Levy, S.B. et al. High-frequency of antimicrobial resistance in human fecal flora, *Antimicrob. Agents Chemother.,* 32, 1801, 1988.
21. Levy, S.B. and Marshall, B. Antibacterial resistance worldwide: causes, challenges and responses, *Nature Med.,* 10, S122, 2004.
22. Bouma, J.E. and Lenski, R.E. Evolution of a bacteria/plasmid association, *Nature,* 335, 1988.
23. Schrag, S.J., Perrot, V., and Levin, B.R. Adaptation to the fitness costs of antibiotic resistance in *Escherichia coli, Proc. Biol. Sci.,* 264, 1287, 1997.
24. Jones, M.E. et al. Multiple mutations conferring ciprofloxacin resistance in *Staphylococcus aureus* demonstrate long-term stability in an antibiotic-free environment, *J. Antimicrob. Chemother.,* 45, 353, 2000.
25. Smith-Adam, H.J. et al. Stability of fluoroquinolone resistance in Streptococcus pneumoniae clinical isolates and laboratory-derived mutants, *Antimicrob. Agents Chemother.,* 49, 846, 2005.

26. FAO/WHO. Joint expert committee on food additives. Evaluation of certain veterinary drug residues in food, *Tech. Ser.,* 799, 45, 1990.

27. WHO. Toxicological evaluation of certain veterinary drug residues in food, *WHO Food Additives Ser.,* 27, 1991.

28. JECFA 60th meeting of the joint FAO/WHO Expert Committee on Food Additives, Toxicological evaluation of certain veterinary drug residues in food, *WHO Food Additives Ser.,* 51, 50, 2003.

29. Shah, A.D., Kim, J.H., and Huang, C.-H. Reaction kinetics and transformation of Carbadox and structurally related compounds with aqueous chlorine, *Environ. Sci. Technol.,* 40, 7228, 2006.

30. Dodd, M.C. et al. Interactions of fluoroquinolone antibacterial agents with aqueous chlorine: reaction kinetics, mechanisms, and transformation pathways, *Environ. Sci. Technol.,* 39, 7065, 2005.

31. Dodd, M.C. and Huang, C.-H. Transformation of the antibacterial agent sulfamethoxazole in reactions with chlorine: kinetics, mechanisms, and pathways, *Environ. Sci. Technol.,* 38, 5607, 2004.

32. Dodd, M.C. and Huang, C.-H. Aqueous chlorination of the antibacterial agent trimethoprim: Reaction kinetics and pathways, *Water Res.,* 41, 647, 2007.

33. Commission Regulation No. 2788/98, *Off. J. Eur. Commun.,* 1998.

34. Initial decision on proposed withdrawal of Baytril poultry NADA, http://www.fda.gov/cvm/CVM-updates/baytrilup.htm, U.S. Food and Drug Administration, Center for Veterinary Medicine: Rockville, MD. (Accessed August 16, 2007.)

35. Eliopoulos, G.M. and Moellering, R.C.J. Antimicrobial combinations, in *Antibiotics in Laboratory Medicine,* Lorian, V., Ed., Waverly and Wilkins, Baltimore, MD, 1996, 331.

36. Strock, T.J., Sassman, S.A., and Lee, L.S. Sorption and related properties of the swine antibiotic carbadox and associated *N*-oxide reduced metabolites, *Environ. Sci. Technol.,* 39, 3134, 2005.

37. Vázquez, J.L. et al. Determination by fluorimetric titration of the ionization constants of ciprofloxacin in solution and in the presence of liposomes, *Photochem. Photobiol.,* 73, 14, 2001.

38. Barbosa, J. et al. Comparison between capillary electrophoresis, liquid chromatography, potentiometric, and spectrophotometric techniques for evaluation of pK_a values of zwitterionic drugs in acetonitrile-water mixtures, *Anal. Chim. Acta,* 437, 309, 2001.

39. Lucida, H., Parkin, J.E., and Sunderland, V.B. Kinetic study of the reaction of sulfamethoxazole and glucose under acidic conditions I. Effect of pH and temperature, *Int. J. Pharm.,* 202, 47, 2000.

40. Qiang, Z.M. and Adams, C. Potentiometric determination of acid dissociation constants (pK(a)) for human and veterinary antibiotics, *Water Res.,* 38, 2874, 2004.

41. Roth, B. and Strelitz, J.Z. The protonation of 2,4-diaminopyrimidines. I. Dissociation constants and substituent effects, *J. Org. Chem.,* 34, 821, 1969.

42. Singer, P.C. and Reckhow, D.A. Chemical oxidation, in *Water Quality and Treatment: A Handbook of Community Water Supplies,* Letterman, R.D., Ed., McGraw-Hill, New York, 1999, chap. 12.

43. Tchobanoglous, G. and Burton, F.L., *Wastewater Engineering: Treatment, Disposal, and Reuse,* 3rd ed., McGraw-Hill, New York, 1991, 353.

44. *CRC Handbook of Chemistry and Physics,* 87th ed., Lide, D.R., Ed., CRC Press, Boca Raton, FL, 2006.

45. Morris, J.C. Acid ionization constant of HOCl from 5 to 35°C, *J. Phys. Chem.,* 70, 3798, 1966.

46. Wang, T.X. and Margerum, D.W. Kinetics of reversible chlorine hydrolysis: temperature dependence and general-acid/base-assisted mechanisms, *Inorg. Chem.,* 33, 1050, 1994.

47. Larson, R.A. and Weber, E.J. Reactions with disinfectants, in *Reaction Mechanisms in Environmental Organic Chemistry*, CRC Press, Boca Raton, FL, 1994, chap. 5.

48. Alum, A. et al. Oxidation of bisphenol A, 17β-estradiol, and 17α-ethynyl estradiol and byproduct estrogenicity, *Environ. Toxicol.*, 19, 257, 2004.

49. Deborde, M. et al. Aqueous chlorination kinetics of some endocrine disruptors, *Environ. Sci. Technol.*, 38, 5577, 2004.

50. Pinkston, K.E. and Sedlak, D.L. Transformation of aromatic ether-and amine-containing pharmaceuticals during chlorine disinfection, *Environ. Sci. Technol.*, 38, 4019, 2004.

51. Adams, C. et al. Removal of antibiotics from surface and distilled water in conventional water treatment processes, *J. Environ. Eng.*, 128, 253, 2002.

52. Chamberlain, E. and Adams, C. Oxidation of sulfonamides, macrolides, and carbadox with free chlorine and monochloramine, *Water Res.*, 40, 2517, 2006.

53. Hu, J. et al. Products of aqueous chlorination of 17β-estradiol and their estrogenic activities, *Environ. Sci. Technol.*, 37, 5665, 2003.

54. Bedner, M. and MacCrehan, W.A. Transformation of acetaminophen by chlorination produces the toxicants 1,4-benzoquinone and *N*-Acetyl-*p*-benzoquinone imine, *Environ. Sci. Technol.*, 40, 516, 2006.

55. Boyd, G.R., Zhang, S.Z., and Grimm, D.A. Naproxen removal from water by chlorination and biofilm processes, *Water Res.*, 39, 668, 2005.

56. Gould, J.P. and Richards, J.T. The kinetics and products of the chlorination of caffeine in aqueous solution, *Water Res.*, 18, 1001, 1984.

57. Rule, K.L., Ebbett, V.R., and Vikesland, P.J. Formation of chloroform and chlorinated organics by free-chlorine-mediated oxidation of triclosan, *Environ. Sci. Technol.*, 39, 3176, 2005.

58. Westerhoff, P. et al. Fate of endocrine-disruptor pharmaceutical, and personal care product chemicals during simulated drinking water treatment processes, *Environ. Sci. Technol.*, 39, 6649, 2005.

59. Gallard, H. and von Gunten, U. Chlorination of phenols: kinetics and formation of chloroform, *Environ. Sci. Technol.*, 36, 884, 2002.

60. Zhang, H. and Huang, C.-H. Reactivity and transformation of antibacterial N-oxides in the presence of manganese oxide, *Environ. Sci. Technol.,* 39, 593, 2005.

61. APHA, *Standard Methods for the Examination of Water and Wastewater*, APHA, AWWA, WPCF, Washington, D.C., 1998.

62. Rebenne, L.M., Gonzalez, A.C., and Olson, T.M. Aqueous chlorination kinetics and mechanism of substituted dihydroxybenzenes, *Environ. Sci. Technol.*, 30, 2235, 1996.

63. Gerritsen, C.M. and Margerum, D.W. Non-metal redox kinetics: hypochlorite and hypochlorous acid reactions with cyanide, *Inorg. Chem.*, 29, 2757, 1990.

64. Cherney, D.P. et al. Monitoring the speciation of aqueous free chlorine from pH 1 to 12 with Raman spectroscopy to determine the identity of the potent low-pH oxidant, *Appl. Spectrosc.*, 60, 764, 2006.

65. Haas, C.N. et al. Survey of water utilities disinfection practices, *J. AWWA*, 84, 121, 1992.

66. Tchobanoglous, G., Burton, F.L., and Stensel, H.D. *Wastewater Engineering-Treatment and Reuse*, 4th ed., McGraw-Hill, Boston, 2003.

67. Kim, H.K. et al. Nitrones. 7. Alpha-quinoxalinyl-*N*-substituted nitrone 1,4-dioxides, *J. Med. Chem.,* 20, 557, 1977.

68. Suter, W., Rosselet, A., and Knusel, F. Mode of action of quindoxin and substituted quinoxaline-di-N-oxides on escherichia-coli, *Antimicrob. Agents Ch.,* 13, 770, 1978.

69. Abia, L. et al. Oxidation of aliphatic amines by aqueous chlorine, *Tetrahedron,* 54, 521, 1998.

70. Morris, J.C. Kinetics of reactions between aqueous chlorine and nitrogen compounds, in *Principles and Applications of Water Chemistry*, Faust, S.D. and Hunter, J.V., Eds., John Wiley & Sons, Inc., New York, 1967, 23.

71. Wiedemann, B. and Grimm, H. Susceptibility to antibiotics: Species incidence and trends, in *Antibiotics in Laboratory Medicine,* Lorian, V., Ed., Williams and Wilkins, Baltimore, MD, 1996, 900.
72. Shen, L.L. et al. Mechanism of inhibition of DNA gyrase by quinolone antibacterials: a cooperative drug-DNA binding model, *Biochemistry,* 28, 3886, 1989.
73. Uetrecht, J.P., Shear, N.H., and Zahid, N. *N*-Chlorination of sulfamethoxazole and dapsone by the myeloperoxidase system, *Drug Metab. Dispos.*, 21, 830, 1993.
74. Stratton, C.W. Mechanisms of action for antimicrobial agents: general principles and mechanisms for selected classes of antibiotics, in *Antibiotics in Laboratory Medicine,* Lorian, V., Ed., Williams and Wilkins, Baltimore, MD, 1996, 579.
75. McCloskey, P. et al. Resistance of three immortalized human hepatocyte cell lines to acetaminophen and *N*-acetyl-*p*-benzoquinoneimine toxicity, *J. Hepatology,* 31, 841, 1999.

13 Hormones in Waste from Concentrated Animal Feeding Operations

Z. Zhao, K.F. Knowlton, and N.G. Love

Contents

13.1   INTRODUCTION

Concentrated animal feeding operations (CAFOs) are the largest of the 238,000 live-stock farms or animal feeding operations (AFOs) in the United States.[29] The U.S. Environmental Protection Agency (EPA) defined AFOs as livestock operations where animals are maintained or confined for more than 45 days in 1-year period,[19] and AFOs that meet certain animals threshold numbers and on-farm situations are catego-rized as CAFOs. An estimated 15,500 AFOs (7% of the total) constitute CAFOs.[29]

Nitrogen (N) and phosphorus (P) pollution, pathogens, and odor of manure discharging from CAFOs have been a great concern for several decades.[57,72,113] In recent years, hormones (estrogens, androgens, progesterone, and various synthetic hormones) contained in the manure from CAFOs have generated wide interest because of their endocrine disrupting effects.[48,79,90] When manure is land applied, part of these hormones may enter water systems through runoff or leaching [24,37,63,102] and may cause developmental and reproductive impairment in aquatic animals. The effective levels of these hormones may be as low as nanograms per liter of water.[17] A high incidence of intersexuality (feminization) was observed in a wide population of male roaches in the United Kingdom.[64] After exposure to cattle feedlot effluent, both testosterone synthesis and testis size decreased in male fathead minnows.[105]

This chapter focuses on the occurrence, persistence, treatment, and transforma-tion of natural and exogenous hormones in waste from CAFOs, including dairy, beef, poultry, swine, and horse farms. The biochemistry, physiological functions, excretion, degradation, and environmental effects of these hormones are discussed.

13.2   BACKGROUND

13.2.1   CONCENTRATED ANIMAL FEEDING OPERATIONS (CAFOS) IN THE UNITED STATES

AFOs in the United States have been identified as one of the leading sources of impairment for all kinds of water bodies, because more than 500 million tons of animal manure need to be disposed from AFOs annually.[28] To ensure that the owners and managers of CAFOs take appropriate actions to manage manure effectively, CAFOs are subject to the National Pollution Discharge Elimination System permitting requirements and the Effluent and Limitations Guidelines and Standards under the Clean Water Act.[29]

CAFOs are divided into large and medium categories (Table 13.1). Large CAFOs are defined only by animal numbers, while medium CAFOs are defined by both animal numbers and on-farm conditions.[29] In addition to animal number requirement, medium CAFOs are also "discharging pollutants directly or indirectly via either a man-made ditch, flushing system, or other similar man-made devices into waters that originate outside of and pass over, across, or through the facility or the animals confined in the operations have direct contact with water."[29] Swine CAFOs are further divided by animal size, and poultry operations with both wet and dry manure handling systems are included.

The estimated numbers of CAFOs that need permits under the revised CAFO regulations announced by EPA in 2003 are listed in Table 13.2.[29] Sixty percent of the CAFOs are hog and dairy farms. CAFOs only account for a small percentage of all livestock farms, but they contribute about 50% animal manure production (Table 13.3).[68]

13.2.2   HORMONES AND CAFO

Hormones are synthesized from specialized glands in the endocrine system and are excreted at very low quantities in urine and feces.[86,95] The hormones in animal

TABLE 13.1

Size Thresholds of CAFOs for Different Species of Livestock

Animal Type	Medium CAFOs	Large CAFOs
Mature dairy cows	200–699 (milked or dry)	>700 (milked or dry)
Veal calves	300–999	>1000
Other cattle[1]	300–999	>1000
Swine (>55lbs)	750–2499	>2500
Swine (<55 lbs)	3000–9999	>10,000
Horses	150–499	>500
Sheep or lambs	3000–9999	>10,000
Turkeys	16,500–54,999	>55,000
Laying hens or broilers[2]	9000–29,999	>30,000
Chickens (other than laying hens)[3]	37,500–124,999	>125,000
Laying hens[4]	25,000–81,999	>82,000
Ducks[5]	10,000–29,999	>30,000
Ducks[6]	1500–4999	>5000

[1]Cattle includes heifers, steers, bulls, and cow/calf pairs
[2,6] If the AFO uses a liquid manure handling system
[3,4,5] If the AFO uses other than a liquid manure handling system
Source: Adapted from U.S. EPA [Reference 29].

TABLE 13.2
Numbers of CAFOs in the United States for Different Types of Operations

Animals	Medium CAFOs	Large CAFOs
Dairy	1949	1450
Veal	230	12
Fed cattle	174	1766
Hogs	1485	3924
Turkeys	37	388
Broilers	520	1632
Laying hens (dry manure)	26	729
Laying hens (wet manure)	24	383
Total	4452	10526

Source: Adapted from U.S. EPA (Reference 29).

manure that have important environmental effects include estrogens (estrone, estradiol, and estriol), androgens (testosterone), and progestagens (progesterone). If these hormones enter the water system through runoff and leaching following manure land application, they may alter or disrupt the functions of the endocrine system and cause adverse effects to organisms.[27] The adverse effects may include mimicking or blocking receptor binding, or altering the rate of hormone synthesis or metabolism through interactions with the endocrine system.[95]

It was predicted that about 1500 kg estrone and estradiol are excreted each year by farm animals in the United Kingdom (Table 13.4),[66] about four times more than the total estrogens from humans. Forty-nine tons of estrogens, 4.4 tons of androgens, and 279 tons of gestagens were excreted by farm animals in the United States in 2002,[79] and cattle production contributes about 90% of estrogens and gestagens and 40% of androgens. These estimates are questionable as data available may not

TABLE 13.3
Manure Available for Land Application

Animals	Total Manure (Billion lbs)	By CAFOs (Billion lbs)	Manure from CAFO (% of total)
Dairy	45.5	10.5	23
Cattle	32.9	27.3	83
Swine	16.3	9.0	55
Poultry	33.5	16.4	49
Total	128.2	63.2	49.4

Source: Adapted from Kellogg et al. (Reference 68).

TABLE 13.4

Predicted Total Excretion of Estrogens from the Human and Farm Animal Populations in the United Kingdom (2004)

Type	Population (million)	Estrone (kg/year)	Estradiol (kg/year)	Discharge Percentage (%)[1]
Human	59	219	146	17
Dairy cattle	2.2	693	365	49
Pigs	5	367	19	18
Broiler chickens	112	15	34	2
Laying hens[2]	29.2	NC[3]	NC	12
Breeding ewes	7.6	19	6	1.9
Nonbreeding sheep	1.5	1.6	0.4	0.1
Total farm animals	157.5	1096	424	83
Total	216.5	1315	570	100

[1] Based on the total amount of estrone and estradiol

[2] The combined amount of estrone and estradiol is 260 kg/year

[3] NC—not calculated for insufficient data

Source: Adapted from Johnson et al. (Reference 66).

be sufficient for accurate calculation of the total mass of estrogens excreted.[48] It is clear, however, that large amounts of hormones from CAFOs are released into the environment each year.

Synthetic hormones considered to be endocrine disruptors are commonly used in CAFOs for different purposes. Although banned in the European Union, hormonal growth promoters (HGPs) are widely used by the largest cattle-producing countries in the world, including the United States, Australia, Argentina, and Canada.[139] Three synthetic hormones, zearalanol, trenbolone acetate (TBA), and melengestrol acetate (MGA), have been licensed for animal production. Also used are the natural hormones testosterone, 17β-estradiol, and progesterone.[81]

These HGPs are primarily used in the beef cattle industry, as exogenous androgens and estrogens have little efficacy in pigs.[14] In the dairy industry, progesterone-releasing implant devices are approved in lactating cows for estrous synchronization.[114] The use of these hormones in CAFOs has raised concerns with the increased exposure of endocrine disruptors to the environment.[126,160]

13.3 HORMONES IN CAFOS

13.3.1 Natural Estrogens

Estrogens are hormones that are mainly responsible for the development of female sex organs and the secondary sex characteristics (Figure 13.1). The naturally occurring estrogens in livestock include estrone or E1 (3β-Hydroxyestra-1,3,5(10)-trien-

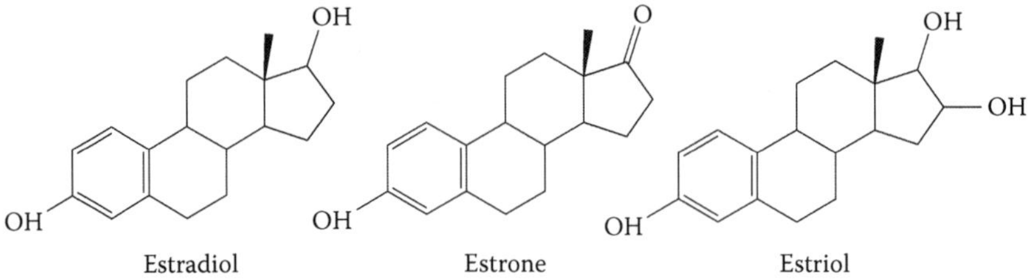

FIGURE 13.1 Molecular structures of natural estrogens.

17-one), 17β-estradiol or E2 (estra-1,3,5(10)-triene-3,17β-diol), and estriol or E3 (estra-1,3,5(10)-triene-3,16β,17β-triol). 17α-estradiol is the optical isomer of 17β-estradiol with the hydroxyl group at C-17 pointing downward from the molecule, and 17β-estradiol has the hydroxyl group pointing upward.[48] Natural estrogens are slightly soluble in water, moderately hydrophobic, and are weak acids with low volatility (Table 13.5). 17β-estradiol is the most potent natural estrogen, and the relative estrogenic potencies of estrone and estriol relative to 17β-estradiol (1.0) are 0.38 and 2.4×10^{-3}, respectively, based on the yeast estrogen screen assay.[122]

13.3.2 BIOSYNTHESIS OF ESTROGENS

The major source of estrogens in nonpregnant females is the ovary (granulosa cells), and the placenta produces the majority of estrogens during pregnancy.[54] Other sources may include the adrenal cortex, adipose tissue, muscle, kidney, liver, and hypothalamus. The main pathways of biosynthesis of estrogens are shown in Figure 13.2.[38] The synthesis starts with cholesterol, which is converted to pregnenolone, and subsequently 4-androstenedione and testosterone. The two androgens are then hydroxylated at C-19, and the resulted 19-hydroxyl groups are oxidized and further removed. The final products will be estrone and 17β-estradiol from androstenedione and testosterone, respectively. 17β-estradiol can then be converted to estriol

TABLE 13.5
Physicochemical Properties of Natural Estrogens

Property	Estrone	Estradiol	Estriol
Formula	$C_{18}H_{22}O_2$	$C_{18}H_{24}O_2$	$C_{18}H_{24}O_3$
Molecular weight (g/mol)	270.4	272.4	288.4
Sw^1 (mg/L)	0.8–12.4	3.9–13.3	3.2–13.3
Vapor pressure (Pa)	3×10^{-8}	3×10^{-8}	9×10^{-13}
Log K_{ow}^2	3.1–3.4	3.1–4.0	2.6–2.8
pKa^3	10.3–10.8	10.5–10.7	10.4

^1S_w—solubility in water
$^2K_{ow}$—octanol-water partition coefficient
^3PK_a—acid ionization constant
Source: Adapted from Hanselman et al. (Reference 48).

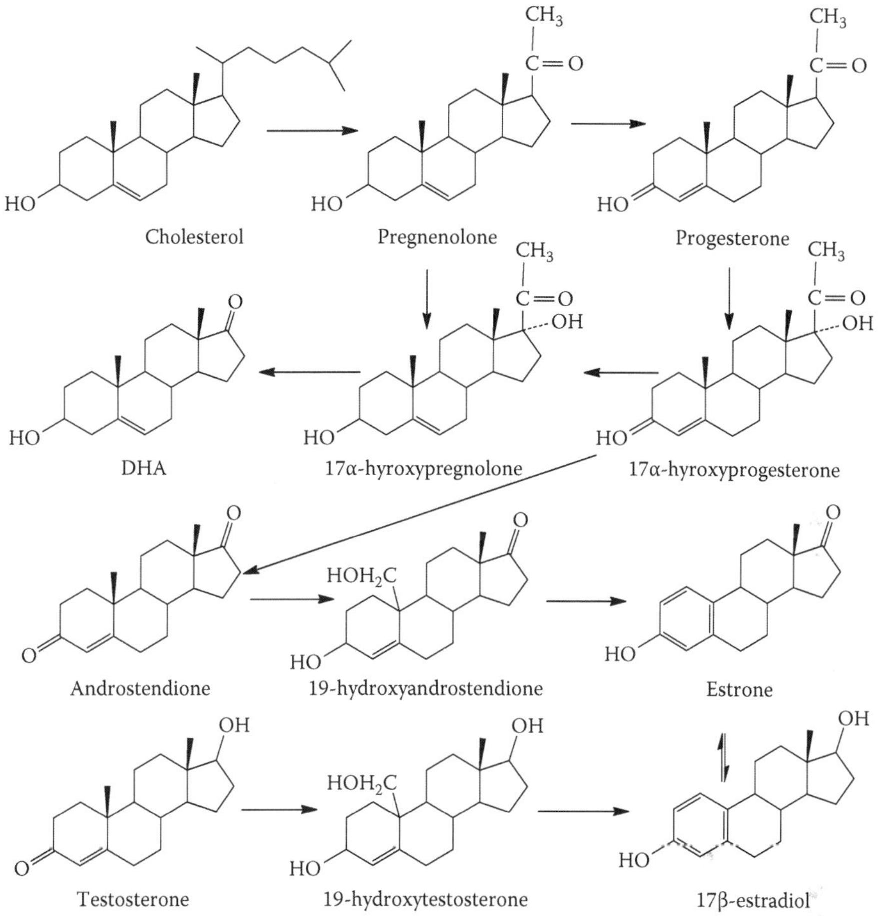

FIGURE 13.2   Biosynthesis pathways of natural estrogens. (Adapted from Fotherby.[38])

by 16β-hydroxylase. The biosynthesis pathways of 17α-estradiol are not entirely known.[148] Generally it is synthesized from aromatization of epitestosterone by the cytochrome P450 aromatase.[36]

13.3.3   Metabolism of Estrogens

After synthesis and secretion, estrogens go through a series of metabolic pathways in the liver, kidney, gastrointestinal tract, and target tissues. Estrone and 17β-estradiol are interconvertible by 17β-hydroxysteroid dehydrogenase.[120] All estrogens may be converted to glucuronides or sulfates by UDP-glucuronosyltransferase and sulfo-transferase, and glucuronidase and sulfatase will hydrolyze estrogen conjugates back to free forms, as shown in Figure 13.3.[115]

These conjugated estrogens are not involved in estrogen receptor-mediated activity.[166] Estrogen sulfates have a much longer half-life and higher concentrations in human circulation than the free forms,[108] so sulfation and desulfation may be

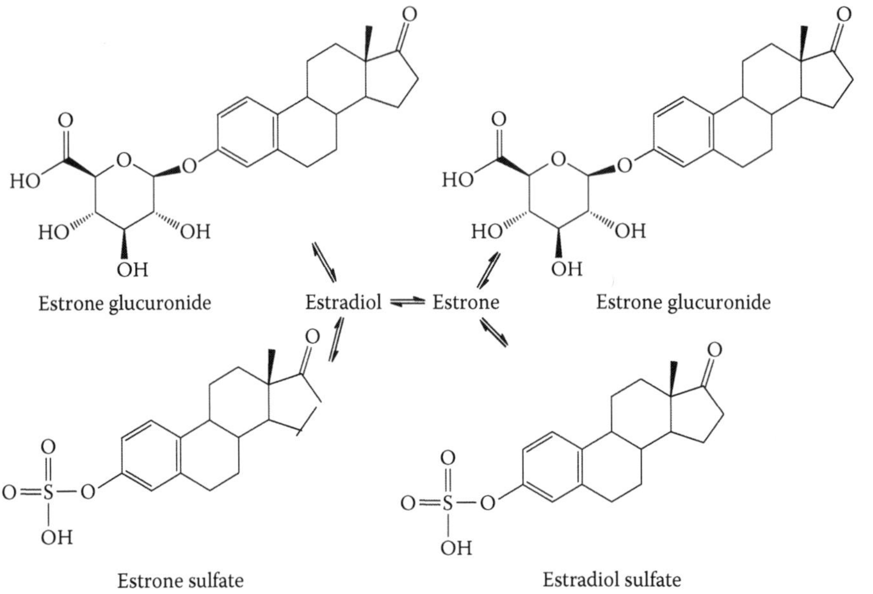

FIGURE 13.3 Conjugation and deconjugation pathways of estrogens. (Adapted from Rafto-gianis et al.[115])

important in the regulation of biologically active or free estrogens in the body.[52] Compared with estrogen sulfates, much less attention has been paid to estrogen glucuronides.[166]

Hepatic hydroxylation of estrogens at C-2 and C-4 forms catechol estrogens (CEs) with less hydroxylation at C-4 than at C-2 in humans and most mammals.[103] The hydroxylated estrogens can form sulfates and glucuronides, or with the presence of catechol-O-methyltransferase, the CEs are methylated to form 2- or 4-O-methylethers and then excreted as shown in Figure 13.4.[115] The CEs can also be oxidized further into quinones or semiquinones with subsequent glutathione conjugation.[9,124]

In humans about half of the estrogen conjugates that enter or are formed in the liver will be excreted into the intestine through bile for enteroheptatic circulation (intestinal reabsorption and reentering the liver for metabolism) or excretion in feces.[41] Generally estrogen conjugates are hydrolyzed considerably in the intestine, and estradiol is converted to estrone to a large extent.[1] Reconjugation can occur in the intestinal mucosal cells, and some of the conjugated and free estrogens are then reabsorbed to the bloodstream and either reenter bile or are transported to the kidney for urinary excretion.[2]

13.3.4 EXCRETION OF ESTROGENS

Estrogens are mainly excreted through urine and feces. In feces, estrogens mainly exist in free forms, while urinary estrogens are mostly conjugated.[138] 17β-estra-diol, 17ß-estradiol, and estrone (free and conjugated) account for more than 90% of the excreted estrogens in cattle,[48] but 17β-estradiol is rarely excreted by swine and

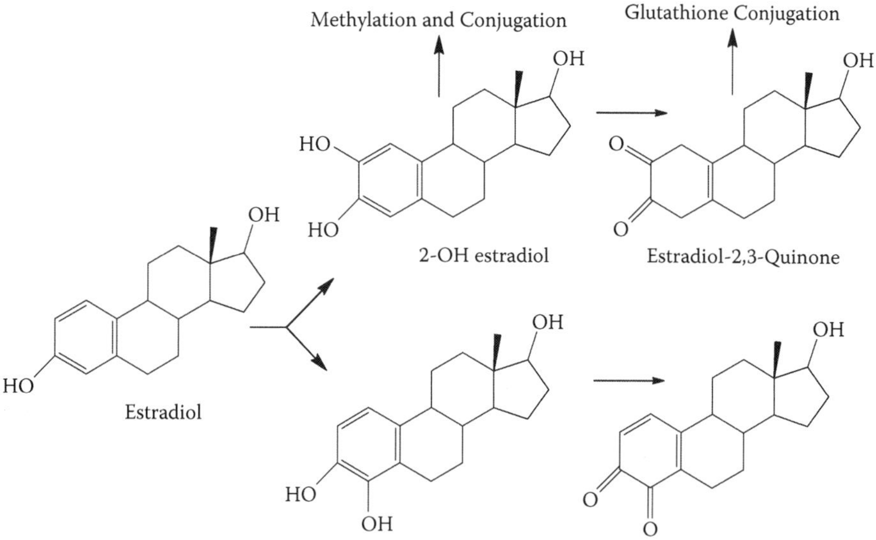

FIGURE 13.4 Conjugation of catechol estrogens and estrogen quinines. (Adapted from Raftogianis et al.[115])

poultry. In cattle 58% of the total estrogen excretion is via the feces,[59] while swine and poultry excrete 96% and 69% of estrogens in urine, respectively.[3,106]

Total estrogen excretion by cattle is clearly quantitatively significant, but relatively few data are available. 174 μg per day of total estrogens was excreted through cow urine in days 6 to 25 of lactation.[30] In cattle, 11.6 ng/g estrone, 60.0 ng/g 17β-estradiol, and 33.6 ng/g 17ß-estradiol, respectively, were found in the feces 5 days before parturition (giving birth), but no daily excretion rate data for estrogens were available.[53] Total estrogen content (estradiol plus estrone) was reported to be 28 ng/g of dry broiler litter. ("Litter" is the term used for the combination of manure and wood shaving bedding material.[137]) Similarly, 30 ng/g of 17β-estradiol was found in broiler chicken manure.[10] Table 13.6 shows the calculated total daily excretion

TABLE 13.6
Estimation of Total Daily Estrogen Excretion by Farm Animals

Species	Category	Fecal Excretion μg/day	Urinary Excretion	Total Excretion
Cattle	Cycling cows	200	99	299
	Bulls	360	180	540
Pigs	Cycling sows	14	100	114
Sheep	Cycling ewes	20	3	23

Source: Adapted from Lange et al. (Reference 79).

of estrogens for cattle, pigs, and sheep.[79] Several factors, such as age, diet, season, health status, and diurnal variation may contribute to variation in excretion rates.[130]

13.3.5 DEGRADATION OF ESTROGENS

Degradation of estrogens is a complicated process and may include deconjugation, dissipation, and mineralization.[61,83,145] A rapid biodegradation of 17β-estradiol and its related metabolites were reported by sewage bacteria under both aerobic and anaerobic conditions.[82] The same kind of result occurred during aerobic batch incubation with activated sludge from a sewage treatment plant.[145]

The hydrolysis or cleavage of sulfate or glucuronide of conjugated estrogens is called deconjugation. It was suggested that some natural fecal bacterial and enzymes may degrade estrogen metabolites several hours after sampling if no preservative is added or samples are not put into cold storage.[70] The fecal microorganism *Escherichia coli* (ubiquitous in the digestive tract) is capable of producing large quantities of β-glucuronidase.[20,60] and has been considered to be responsible for estrogen glucuronides deconjugation to some extent. *Escherichia coli* has a weak arysulfatase activity,[131] so it is possible that portions of estrogen sulfates are left intact. This may explain why several studies reported estrogen sulfates in sewers,[145] sewage treatment plants,[20] and river water.[121] No estrogen glucuronides were detected in sow feces, and after incubation of estrone conjugates at 20°C for 30 min in fecal suspension, 90% of estrone glucuronide was deconjugated, but estrone sulfate was not hydrolyzed.[156]

The dissipation of estrogens refers to the decrease in extractable concentrations, and the possible dissipation pathways include conversion of estradiol to estrone and subsequent formation of nonextractable residues.[18] Mineralization is the final degradation of estrogens to CO_2, water, and other compounds through cleavage of the phenolic ring. The degradation pathways for dissipation and mineralization have not been clearly understood, but both biotic and abiotic pathways are possible for estrogen degradation as shown in Figure 13.5.[71] Under aerobic conditions, the introduction of hydroxyl groups by mono- and dioxygenase ring cleavage, and final decarboxylation, are the key steps for degradation of phenolic compounds.[128] Anaerobes may degrade phenolic compounds through hydroxylation and carboxylation followed by β-oxidation to CO_2.[11] With the assistance of TiO_2, estrogens are degraded chemically at the phenol ring first and then DEO (10ε-17β-dihydroxy-1,4-estradien-3-one) is produced as an intermediate for further degradation.[104] Because of the similarity in their basic structures, the degradation of androgens and progesterone is expected to go through the same pathways in terms of the final mineralization.

Photodegradation has also been suggested as another mechanism of estrogen dissipation and mineralization.[35,97] Most reported work focuses on estrogens in manure-amended soil and biosolids from wastewater treatment systems; only a few studies have explored degradation of estrogens in stored manure. Estrogen concentrations decreased significantly in broiler litter at pH 5 and 7 after 1 week incubation.[137] The total estrogens were reduced by 80% in cattle feces following 12 weeks of incubation at 20 to 23°C.[129] Substantial losses of 17β-estradiol (90%) and total estrogens (40%) occurred without acidification and cold storage in press-cake samples of dairy manure.[116]

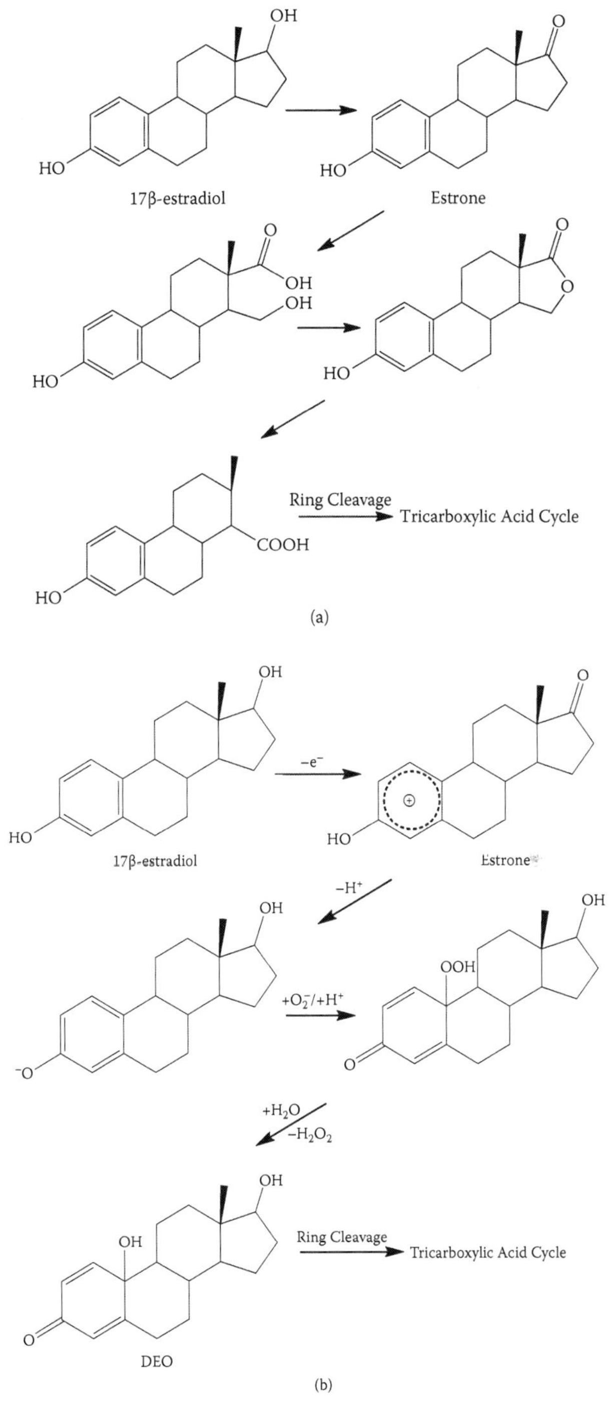

FIGURE 13.5 Estrogen degradation by sewage bacteria (a) and TiO_2 photocatalyst (b). (Adapted from Ohko et al.[104])

13.4 NATURAL ANDROGENS

Natural androgens are C-19 steroids that possess androgenic activities (stimulating and maintaining masculine characteristics), including testosterone, 5α-dihydrotestosterone, 5α-androstane-3β,17β-diol, and three weakly androgenic steroids—4-androstenedione, dehyroepiandrosterone (DHA), and androsterone—as shown in Figure 13.6.[43] Testosterone and androsterone are less soluble in water and more hydrophobic compared to estrogens as shown in Table 13.7.[86] The 17β-hydroxyl group accounts for most of the androgenic activities, and oxidation of the 17β-hydroxyl group to a 17-oxo group will cause an 80% loss of the androgenic potency.[42] Epitestosterone (17β-testosterone) is a natural optical isomer of testosterone (17β-testosterone), but its biological activity has not been fully clarified.[142] Androgens are important for the development of male sex organs and the maintenance of the secondary sex characteristics.[8]

13.4.1 BIOSYNTHESIS OF ANDROGENS

Androgens are largely synthesized in the testes, secondly in the adrenal cortex, and to a limited extent in the ovaries and placenta. As shown in Figure 13.2, pregnenolone is also the precursor for androgen biosynthesis, and it goes through hydroxylation at C-17 and subsequent removal of the side-chain.[42] Then the resulted DHA is metabolized into testosterone through either 5α-androstane-3β,17β-diol, or 4-androstenedione. Another pathway starts with progesterone converted from pregnenolone, and then hydroxylation at C-17 and removal of the side-chain gives 4-androstenedione. Finally, testosterone is formed from 4-androstenedione by 17β-hydroxysteroid dehydrogenase.

13.4.2 METABOLISM OF ANDROGENS

The liver is mainly responsible for androgen catabolism through oxidation, hydroxylation, and conjugation. In the liver and other target organs and tissues (skin, prostate),

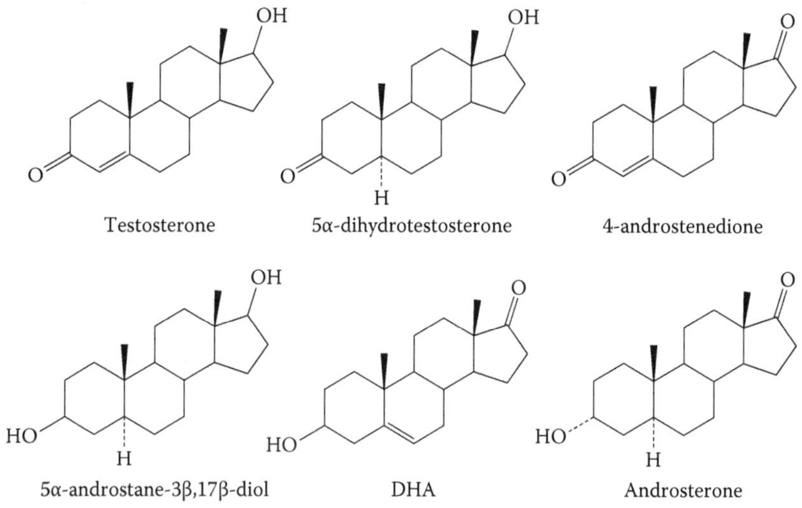

| Testosterone | 5α-dihydrotestosterone | 4-androstenedione |

| 5α-androstane-3β,17β-diol | DHA | Androsterone |

FIGURE 13.6 Chemical structures of natural androgens.

TABLE 13.7

Physicochemical Properties of Natural Androgens

Property	Testosterone	Androsterone
Formula	$C_{19}H_{28}O_2$	$C_{19}H_{30}O_2$
Molecular weight (g/mol)	288.4	290.4
Sw (mg/L)	5.57	8.75
Log K_{ow}	3.32	3.69

Source: Adapted from Lintelmann et al. (Reference 86).

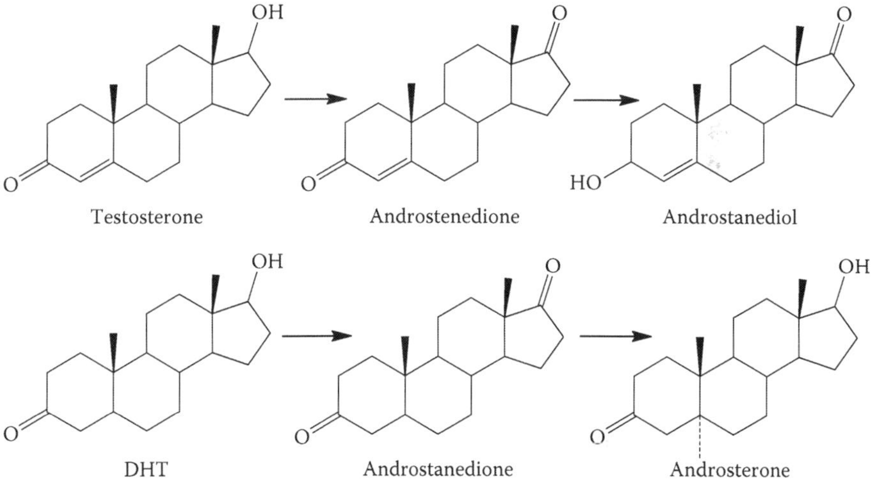

FIGURE 13.7   Degradation pathways of testosterone and DHT.

testosterone is converted to 5α- and/or 5β-dihydrotestosterone (DHT) as shown in Figure 13.7.[143] 5a-DHT binds to human androgen receptor with twofold higher affinity than testosterone, while testosterone dissociates from the receptor fivefold faster than 5α-DHT.[45] The primary pathway to inactivate testosterone and DHT is the oxidation of the 17-hydroxy group resulting in androstanedione and androstenedione.[42] Furthermore, the reduction of the keto group at C-3 produces four metabolites (epi)androstanediol and (epi)androsterone. As with estrogens, testosterone can also be conjugated to glucuronic acid or sulfate directly or following hydroxylation.[159,161]

13.4.3   EXCRETION OF ANDROGENS

Similar to estrogen metabolites, the metabolites of androgens formed in the liver have three destinies: returning back to the circulation, excretion in the urine, or excretion through bile for enterohepatic circulation and fecal excretion. In cattle urine, the main metabolites of testosterone are three isomers of androstane-3,17-

diol, 5β-androstan-3α-ol-17-one, and epietiocholanolone.[123] In the intestine, androgen metabolites are subjected to hydrolase, dehydroxylase, reductase, and epimerase activities of the bacteria.[123]

Little data are available about androgens in livestock urine and feces; 133 μg and 250 μg testosterone were reported in per kg of broiler litter (both sexes) and breeder litter,[137] but only 30 μg/kg and 20 to 30 μg/kg of testosterone equivalents were detected in broiler litter (both sexes) and breeder layers, respectively.[90] These different observations may reflect the differences associated with chickens (breed, age) and manure treatment. The androgenic activity in manure from pregnant dairy cows was determined to be 1737 ng testosterone equivalents/g dry weight.[90] The estimated yearly total excretion of androgens is 120 mg, 390 mg, 670 mg, and 3.4 mg for male calves, bulls, boars, and laying hens, respectively.[79] Based on these excretion data, about 4.4 Mg of androgens were excreted to the environment from farm animals in the United States in 2000.[79]

13.4.4 DEGRADATION OF ANDROGENS

As with estrogens, testosterone may be degraded into extractable or nonextractable products, or completely mineralized into carbon dioxide.[40] The degradation process depends on the organic matter, moisture, temperature, and oxygen availability of the matrix. After 23 weeks incubation in poultry litter with water potential –24 MPa, an average of less than 2, 11, and 27% of the radiolabelled testosterone was mineralized to $^{14}CO_2$ at 45, 35, and 25°C, respectively.[51] Aerobic composting for 139 days decreased the average concentration of testosterone in poultry manure from 115 ng/g to 11 ng/g (dry basis), which means a 90% reduction in potent hormones.[47]

Microbial activity plays an important role in testosterone degradation. Jacobsen et al.[61] observed rapid conversion of testosterone to 4-androstene-3,17-dione within the mix of soil and nonsterilized swine manure, and this effect was totally absent in soil amended with sterilized manure. Mineralization of ^{14}C-testosterone decreased in manured soil, and did not occur in sterilized soil.[61] *Comamonas testosterone*, a gram-negative bacterium, can metabolize testosterone as its sole carbon and energy source.[56] The possible mechanisms include dehydrogenation of the 17β-hydroxyl group, desaturation of the A-ring, hydroxylation, and final metacleavage of the bone structure.

13.5 NATURAL PROGESTAGENS

Progestagens (also called progestogens or gestagens) are hormones that produce effects similar to progesterone, the only natural progestagen. All other progestagens are synthetic and are often referred to as progestins.

13.5.1 BIOSYNTHESIS OF PROGESTERONE

The two main sources of progesterone in female livestock are the corpus luteum (or "yellow body," a gland that forms on the surface of the ovary following ovulation) and the placenta. During the estrous cycle, the ovarian corpus luteum produces more progesterone as it matures. If pregnancy does not occur, prostaglandin $F_{2\alpha}$ secreted by the uterus will cause corpus luteum to regress, triggering a decline

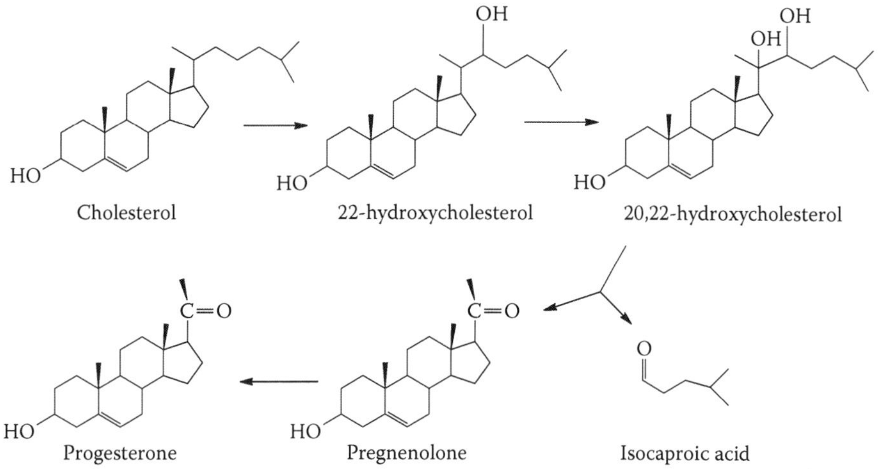

FIGURE 13.8   Biosynthesis pathways of progesterone from cholesterol.

in progesterone production. During gestation, the placenta is the dominant source of progesterone to maintain normal pregnancy. As shown in Figure 13.2 and Figure 13.8,[42] progesterone is also synthesized from cholesterol via pregnenolone. First cholesterol is hydroxylated at C-20 and C-22 to form the subsequent intermediate 20,22-dihydroxycholesterol, and then pregnenolone and isocaproic aldehyde are formed through side-chain cleavage. Finally, pregnenolone is converted to progesterone by hydroxysteroid dehydrogenase. The physiochemical properties of pregnenolone and progesterone are shown in Table 13.8.[26,87]

13.5.2   Metabolism of Progesterone

About 95% of all progesterone is metabolized in the liver through hydroxylation and conjugation.[107] In *in vitro* study with bovine liver, 46% of the progesterone metabolite was pregnanediol, and other metabolites were mainly monohydroxylated and dihydroxylated products.[16] In that study, glucuronides of progesterone metabolites were

TABLE 13.8
Physiochemical Properties of Pregnenolone and Progesterone

Property	Pregnenolone	Progesterone
Formula	$C_{21}H_{32}O_2$	$C_{21}H_{30}O_2$
Molecular weight (g/mol)	316.4	314.4
Sw (mg/L)	33	1–7
Log K_{ow}	3.89	3.67

Source: Adapted from Elkins and Mullis (Reference 26) and Loftsson and Hreinsdóttir (Reference 87).

also determined, and more than 55% of the conjugated metabolites was 5β-pregnane-3α,20α-diol glucuronide. 3β-hydroxy-5α-pregnan-20-one 3 sulfate was reported to be the major metabolite in pregnant sheep plasma after injection of ^{14}C-progesterone.[150]

Progesterone also serves as a precursor for estrogen and testosterone synthesis, as shown in Figure 13.2. Hydroxylation on progesterone may happen in the adrenal gland, leading to cortisol and aldosterone production.[76]

13.5.3 EXCRETION OF PROGESTERONE

Progesterone in the urine and feces has been used to monitor the estrous and pregnancy status of livestock.[58,80,99] Two pregnenolones and two pregnanediols were detected in pregnant sow urine.[25] Significant positive correlations were found between fecal and plasma progesterone concentrations during the ovarian cycle in beef cattle ($r = 0.7$), and the fecal progesterone ranged between 9 and 139 ng/g, depending on the cycle phases.[58] An even higher correlation ($r = 0.98$) was found between plasma progesterone and fecal gestagens (progesterone and its metabolites) in sows, and the average concentration of fecal gestagens was 71 to 497 ng/g during the ovarian cycle.[99] Fifty percent of the radiolabeled progesterone was recovered in bile/feces in 1 to 2 days following intravenous injection, and only 3% of the dose was found in urine.[31]

13.5.4 DEGRADATION OF PROGESTERONE

Data regarding progesterone degradation in the environment are scarce. Plourde et al.[110] reported the biotransformation of progesterone by spores and vegetative cells of microorganisms found in soil. Some fungi contained in *Rhizopus nigricans* transformed progesterone into a mixture of 11α-hydroxy-4-androstene-3,17-dione and 11α-hydroxy-1,4-androstadiene-3,17-dione.[111] The side-chain progesterone was degraded by some species of *Aspergillus flavus;* 4-androstene-3,17-dione, testosterone, and testololactone were the main degraded metabolites.[100] Progesterone was also detected up to 6 ng/g in river sediments in some Spanish rivers.[88]

13.6 HORMONE GROWTH PROMOTERS

In both humans and animals, sex steroids such as estrogens, testosterone, and progesterone regulate growth and development. This has led to the use of hormonal growth promoters (HGPs), natural sex steroids or their synthetic counterparts, in meat animals to increase feed efficiency and weight gain. The first approved synthetic estrogen was diethylstilbestrol (DES) in 1954. Because of its carcinogenic potential, DES was banned for all the use in cattle production by the U.S. Food and Drug Administration (FDA) in 1979.[49]

Currently, six different hormones are approved for such use in the United States, including three natural hormones (17β-estradiol, testosterone, and progesterone), and three synthetic compounds that mimic the functions of these hormones (trenbolone acetate, zeranol, and melengestrol acetate). The chronology of the use of these hormones in cattle in the United States is listed in Table 13.9.[118]

TABLE 13.9

Chronology of the Use of Anabolic Agents in the U.S. Cattle Industry

Year	Issue
1956	Estradiol benzoate/progesterone implants approved for steers
1958	Estradiol benzoate/testosterone propionate implants approved for beef heifers
1968	Oral melengestrol acetate approved for beef heifers
1969	Zearnol implants (36 mg) approved for cattle
1982	Silastic estradiol implant approved for cattle
1984	Estradiol benzoate/progesterone implants approved for beef calves
1987	Trenbolone acetate implants approved for cattle
1991	Estradiol/trenbolone acetate implants approved for steers
1993	Bovine somatotropin approved for use in lactating cows
1994	Estradiol/trenbolone acetate implants approved for use in heifers
1995	72-mg zeranol implants approved for beef cattle
1996	Estradiol/trenbolone acetate implants approved for stocker cattle

Source: From Raun and Preston (Reference 118).

13.6.1 NATURALLY OCCURRING HORMONAL GROWTH PROMOTERS (HGPS)

The direct effects of exogenous 17β-estradiol administration on farm animals (calves, heifers, steers, lambs) include enhanced protein deposition in skeletal muscle and reduced nitrogen excretion[94]; growth performance is increased by 5 to 15%. Testosterone or other androgens are less active as compared to estrogens in cattle or lambs, probably because there are fewer androgen receptors than estrogen receptors.[125]

The anabolic mode of action of steroidal hormones has been well established in the past two decades (Figure 13.9). The anabolic effects are mediated directly by several organs and tissues (liver, bone, skin, and other tissues) and indirectly via the somatotropic axis involving growth hormone and insulin-like growth factor I.

13.6.2 SYNTHETIC HGPS

The structures of TBA (17β-acetoxyestra-4,9,11-triene-3-one), MGA (17β-acetoxy-6-methyl-16-methylene-pregna-4,6-diene-3,20-dione), and zeranol are shown in Figure 13.10. The physiochemical properties of the three synthetic hormones are shown in Table 13.10.[32–34] Compared to estrogens, TBA is more soluble in water, and MGA and zeranol have a much higher vapor pressure.

TBA mimics the activity of testosterone and is administered as a subcutaneous implant either alone or coupled with 17β-estradiol. The anabolic effect of TBA is based on its androgenic and antiglucocorticoid activity.[93] In heifers the major pathway of TBA metabolism is formation of 17α- and 17β-trenbolone (TBOH) through hydroxylation on C-17, and subsequent oxidation and reduction on the formed hydroxyl.[112] TBA has 8 to 10 times greater anabolic effects than testosterone,[126] but

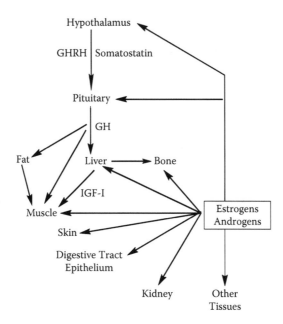

FIGURE 13.9 Anabolic modes of action of steroid hormones. (Adapted from Meyer.[94]) GHRH = Growth hormone releasing hormone; GH = Growth hormone; IGF-I = Insulin-like growth factor-I.

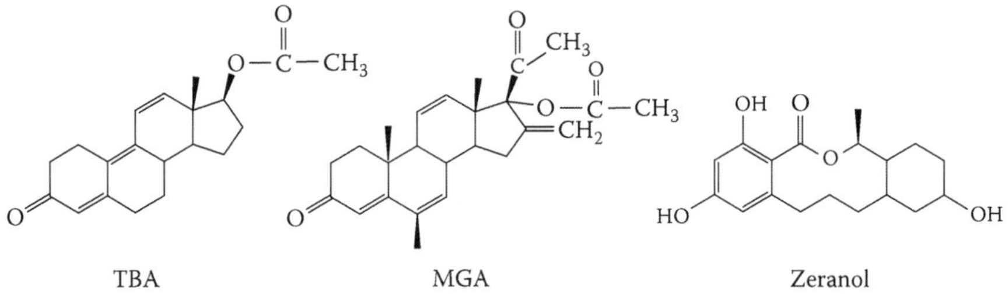

FIGURE 13.10 Chemical structures of TBA, MGA, and Zeranol.

TABLE 13.10
Physiochemical Properties of Synthetic Hormones

Property	TBA	MGA	Zeranol
Formula	$C_{20}H_{24}O_3$	$C_{25}H_{32}O_4$	$C_{18}H_{26}O_5$
Molecular weight (g/mol)	312.4	396.5	322.4
Sw (mg/L)	17–21	1.06	4.13–5.14
Vapor pressure (Pa)	1×10^{-9}	$<1 \times 10^{-7}$	3.9×10^{-9}
Log K_{ow}	NA	3.98	3.13–3.47
pKa	NA	NA	8.44

Source: Adapted from the FDA (References 32,33,34).

17α-TBOH has only 2% of the androgenic potency, 5% of the anabolic effect,[112] and 4% the affinity to the recombinant human androgen receptor compared to that of TBA.[7]

As with the natural steroidal hormones, little data are available on the excretion or fate of synthetic HGPs in manure storage. 17α-TBOH and 17β-TBOH ranged from 10 to 120 ng/L and 10 to 20 ng/L in the liquid discharge from a beef feedlot.[23] The half-lives of the two metabolites were about 260 days in liquid cattle manure.[126]

Zeranol, also called β-zearalanol, is a resorcylic acid lactone. The related compounds of zeranol are shown in Figure 13.11, including zearalanone, taleranol (β-zearalanol), β-zearalenol, zearalenon (mycotoxin), and β-zearalenol.[140] These metabolites can all be metabolized or converted into all the other compounds with varying efficiencies.[96,147] Zeranol mimics the action of 17 β-estradiol and is often implanted alone. It is about equally potent to 17 β-estradiol in inducing expression of endogenous estrogen-regulated genes in human breast cancer MCF-7 cells.[84]

Zeranol and its metabolites do appear in manure, with concentrations highest immediately after implantation. β-zearalanol, zeranol, and zearalanone were all detected in the urine of zeranol-implanted male veal calves, and β-zearalanol was the major metabolite after 3 days following implantation, while zeranol was a minor component during the whole excretion period (14 days).[62] This is in contrast to the observation that zearalanone was the major metabolite in adult cattle.[6] In steers treated with 36 mg zeranol, the concentrations of zeranol in urine and feces were 13 ng/ml (peaked at day 8) for urine and from 10 to 15 ng/g (peaked at day 20) in feces.[22] Zeranol concentrations in manure declined steadily during the

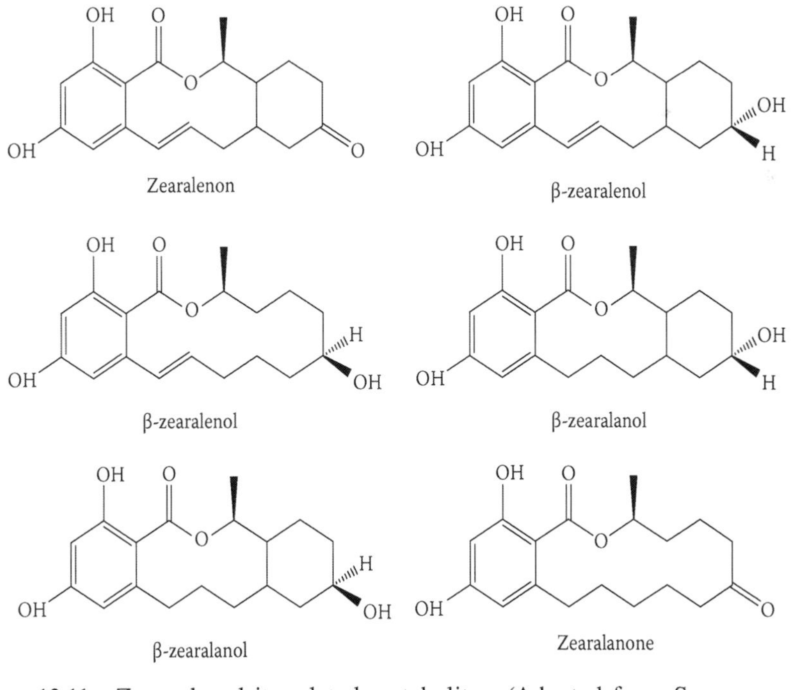

Figure 13.11   Zeranol and its related metabolites. (Adapted from Songsermsakul et al.[140])

period 20 to 70 days, but were still detectable 120 days after implantation. About 10 and 45% of the implanted zeranol in steers were excreted through urine and feces, respectively.[133]

MGA is administered as a feed additive, while all the other HGPs are administered as implants. MGA can be used for estrus synchronization or lactating induction in cattle as an active gestagen.[39] It is also fed to feedlot heifers to improve feed efficiency and weight gain.[77] The progestinal activity of MGA measured by inhibition of estrous in cattle is about 125 times greater than that of progesterone.[80] MGA binds the progesterone receptor at 11.1-fold higher affinity than progesterone.[109] Administration of 0.5 mg/day results in 30 pg/mL circulating MGA, sufficient for suppression of the positive estrogen feedback and ovulation.[21] This concentration is too low, however, to exert any significant effect on estrogen receptors,[46,81] so its anabolic effect is assumed to be due to stimulation on the ovarian synthesis of endogenous estrogens, rather than direct effects on estrogen receptors.

Excretion of MGA is primarily through the feces, but little data are available on persistence of excretion following cessation of administration. The radioactivity of ^{3}H-MGA was eliminated via the feces and urine at a 6:1 ratio in MGA fed heifers.[78] MGA excretion varies with dose. Concentration of MGA in feedlot cattle feces was 2.5, 6.5, and 18.5 ng/g 24 hours after feeding with doses as 0.5, 1.5, and 5 mg/day, respectively.[126]

13.7   ROUTES OF HORMONE LOSS FROM CAFOs

With the application of manure in farmland, hormones from CAFOs may enter surface and groundwater through runoff or leaching. The half-lives of the hormones vary with environmental matrix, such as in stored manure, soil, rivers, and sediments (Table 13.11), and half-lives of hormones in manure and soil are longer than in the aquatic environment.

TABLE 13.11
The Half-Lives of Hormones in the Environment

Hormone	Matrix	Half-Life	Reference
17β-estradiol	Chicken manure compost	69 d	Hakk et al. (2005)
	Anaerobic soil	24 d	Ying and Kookana (2005)
	River	0.2–9 d	Jurgens et al. (2002)
Testosterone	Clay-amended compost	43 d	Hakk et al. (2005)
Progesterone	Soil	28 d	FDA (2001)
Trenbolone	Liquid manure	267 d	Schiffer et al. (2001)
MGA	Water	3.84–25.3 h	FDA (1996)
Zeranol	Manure	56 d	FDA (1994)
	Soil	49–91 d	

13.7.1 Soil and Runoff

All the hormones discussed above will experience further degradation following manure land application. The oxidation of 17β-estadiol to estrone in soil was not dependent on living microorganisms, but estrone was stable in the absence of microorganisms, indicating its dependence on microorganisms for degradation.[18] Estrogens were degraded to be below the detection limit from the initial 20 to 25 mg/L after 14 days incubation with activated sludge or night soil-composting microorganisms.[136] Their degradation products were unknown but did not pose estrogenic activities. Temperature, moisture, oxygen, and pH also exert influence on estrogen degradation.[164]

Estrogens may be strongly sorbed to soil, with a sorption equilibrium attained in 1 to 2 days, depending on the soil matrix and estrogen concentrations.[165] Sorption reduces the potential of leaching and runoff. Because manure from CAFOs is land applied rather than discharged directly into waterways, the likely risk of hormones from CAFOs should be lower compared to hormones discharged from wastewater treatment plants (WWTP). Estrogens still have mobility to runoff from soil, however, and much more research is needed on sorption of estrogens and other steroid hormones to agricultural soils.

Published research has indicated contamination of water resources with steroidal hormones from CAFOs. 0.5 to 5 ng/L of estrogens (total estradiol and estrone) and 1 to 28 ng/L testosterone were reported in small streams draining from farm fields following application of poultry litter.[137] The effect of broiler litter application on 17β-estradiol concentrations was evaluated in surface runoff.[102] The 17β-estradiol concentrations increased with increasing application rate, reaching a maximum of 1280 ng/L at an application rate of 7.05 tons litter ha^{-1}. In another study,[37] runoff concentrations of 17β-estradiol ranged between 20 and 2330 ng/L, depending on broiler litter application rates and time between application and runoff, and soil concentrations of 17β-estradiol reached 675 ng/kg. 3300 ± 700 ng/L of 17ß-estradiol in runoff was reported from plots amended with dairy manure.[24] The concentration in the runoff reached 41 ng/L and 29 ng/L when manure was applied at a nitrogen or phosphorus-based rate, respectively, but only 2.2 ng/L was found in the runoff from plots without manure application.

4-androstene-3,17-dione, 5α-androstane- 3,17-dione, and 1,4-androstadiene-3,17-dione were detected as the major metabolites of C-14 labeled testosterone in the soil.[89] Heat speeded testosterone degradation, as the extractable C-14 decreased quickly at 30°C with the moisture ranging from 7 to 39%. Runoff concentrations of testosterone following poultry litter application to grassland grazed by cattle ranged between 10 and 1830 ng/L, and the soil testosterone concentration reached 165 ng/kg.[37]

Less data are available on the environmental fate of the synthetic hormones. Zeranol may be moderately sorbed to soil following manure application. Only 50% of zeranol was left after 56 days of manure storage.[32] This degradation in manure continued after field application, and 50% of the applied zeranol was mineralized to CO_2 in 90 days after field application.[32]

The mobility of 17β-TBOH and MGA in agricultural soil was determined by means of column experiments.[127] Both hormones exhibited high affinities to the organic matter, and only a small proportion of TBA and MGA passed the columns quickly. However, the application of these results is limited due to the difference from the real soil matrix. Aerobic degradation of 17β-trenbolone to trendione was reported when spiked into soil microorganisms, and the half-life of 17β-trenbolone increased from a few hours to 1 day, with the spiked concentration increasing from 1 mg/kg to 12 mg/kg.[69] TBOH and its metabolites were still detected in the soil even 30 days after fresh dairy manure application, and MGA was traceable in soil samples for more than 6 months after stored manure fertilization.[126]

13.7.2   STREAMS AND RIVERS

Endocrine disruptors in streams and rivers are of great interest and concern because of the direct exposure of wild lives and humans. Concentrations of the steroid hormones discussed above were monitored in a national survey in 2002, and are listed in Table 13.12.[73] Generally, hormones are detected in 3 to 20% of the river samples. Although the maximum concentrations are well below 1 μg/L for all hormones, this does not eliminate of the possible interactions of these hormones and other contaminants in the environment.[73] In a river adjacent to a feedlot in eastern Nebraska, the concentrations of estrone, 17α-estradiol, and 17β-estradiol were 900 pg/L, 35 pg/L, and 84 pg/L, respectively, 80 km downstream from the feedlot.[141]

13.7.3   GROUNDWATER

Sources of groundwater contamination with reproductive hormones may include field leaching and runoff, leaking from sewage systems, and percolation of domestic water.[61] Pesticides, herbicides, and pharmaceuticals and their metabolites have been identified in groundwater.[50,74,149] Sex hormones in groundwater are rarely reported, but in sampling locations near a residential septic system, Swartz et al.[144] detected 0.2 to 45 ng/L of 17β-estradiol and 0.4 to 120 ng/L of estrone in the groundwater,

---

TABLE 13.12
Steroid Hormones Reported in U.S. Rivers

Hormone	N[1]	RL[2] (μg/L)	Freq[3] (%)	Max[4] (μg/L)	Med[4] (μg/L)
17β-estradiol	85	0.5	10.6	0.2	0.16
17α-estradiol	70	0.005	5.7	0.074	0.03
Estrone	70	0.005	7.1	0.112	0.027
Estriol	70	0.005	21.4	0.051	0.019
Testosterone	70	0.005	2.8	0.214	0.116
Progesterone	70	0.005	4.3	0.199	0.11
Cis-androsterone	70	0.005	14.4	0.214	0.017

[1]N: number of samples
[2]RL: reporting level
[3]Freq: frequency of detection
[4]Max/Med: maximum/median detectable concentration
Source: Adapted from Kolpin et al. (Reference 73).

depending on the distance and depth of sampling. The U.S. Geological Survey[153] reported a 60% frequency of detecting steroid hormones in groundwater from 47 sampling locations susceptible to contamination from either animal wastes or human wastewaters. The concentrations of 17β-estradiol ranged from 13 to 80 ng/L in eight springs draining a karstic aquifer.[157]

13.8   FATE OF HORMONES DURING MANURE STORAGE, TREATMENT, AND LAND APPLICATION

In order to decrease the discharge of nutrients from CAFOs, various manure treatment methods have been utilized in animal farming practices for some years, and more innovative practices are being evaluated.

13.8.1   CONVENTIONAL MANURE STORAGE AND TREATMENT SYSTEMS

The ultimate fate of manure constituents (i.e., nutrients, hormones) is strongly influenced by how that manure is removed from the animal facility and how (and whether) it is stored. With increasingly stringent federal and state CAFO regulations, significant investment has been made in manure storage systems on livestock farms. At their simplest, manure storage systems allow the farmer to time manure application to match crop needs, and to prevent application to frozen or saturated ground. In many northern states, manure application is simply banned during winter months. More commonly, a minimum amount of manure storage (120 or 180 days) is required as part of the nutrient management plan on larger farms. Typical systems vary by species, and also by climate.

13.8.1.1   Manure Handling on Dairy Farms

Systems on dairy farms in colder climates usually involve scrape removal of manure from the barn once or twice a day, with manure stored in steel or concrete tanks or lined in-ground pits until land application. Manure slurry stored in this manner quickly becomes anaerobic, but these systems are not designed to treat manure. The impact on steroidal hormones is unknown, but the long retention time under anaerobic conditions should allow significant degradation.

Figure 13.12 shows a typical liquid manure system on a southern dairy farm, in which flushed dairy manure from a freestall barn is stored in a waste treatment lagoon and a waste storage pond either for irrigation or recycled as flush water.[151]

13.8.1.2   Poultry Farms

Most chickens and turkeys raised for meat are raised in large pens inside barns, and manure storage is typically "in-house," meaning that manure and bedding material (typically wood shavings) are accumulated during the growth cycle of two or three sequential flocks, then removed, stored temporarily in sheds, and then land applied. The fate of steroidal hormones in these systems likely depends primarily on the

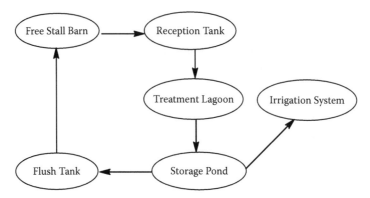

FIGURE 13.12 Freestall barn with flushing alleyway and irrigation system. (Adapted from USDA.[152])

duration of storage, but no data are available. The manure of laying hens is typically managed as a liquid, flushed to storage from beneath the pens.

13.8.1.3 Swine Farms

Swine farms have manure handling and treatment systems similar to that in Figure 13.12, with manure falling through slatted floors to a below-barn tank before being flushed to treatment lagoons. Lagoon facilities for manure treatment may be aerobic or anaerobic,[75] so their impact on estrogen degradation will be variable. A standard manure treatment system on swine farms showing promise for removal of estrogens in agricultural wastewater is the anaerobic treatment lagoon. Recent work shows that these may be highly effective in removing estrogens from swine wastewater. The estrogenic activity in primary anaerobic lagoons was 1% of the activity observed in the reception pit leaving the barn.[132] A survey of swine facilities in Tennessee indicated the highest concentrations of 17β-estradiol (20 ng/mL) in fresh slurry from farrowing barns (the barn where pregnant sows give birth) and much lower concentrations (4 ng/mL) in anaerobic treatment lagoons.[158]

13.8.1.4 Beef Cattle Operations

Manure excreted by beef cattle on feedlots typically dries quickly and accumulates in the feedlot, then is scraped, stacked, and land applied. Photodegradation may be an important mechanism for hormone breakdown in these systems.

Not all manure generated on livestock farms is stored. On smaller beef and dairy farms, manure may be collected and land applied daily, without storage. The environmental implications of year-round manure application in colder climates are obvious, but the cost of implementing manure storage is prohibitive for many small farms. Also, beef and dairy cattle are commonly grazed during certain stages of their life cycle, with manure directly deposited onto pastures. No data are available comparing hormone runoff from grazed fields compared to fields following application of stored manure, but it is a question of obvious importance.

13.8.2   INNOVATIVE MANURE TREATMENT SYSTEMS

More advanced treatment systems focused on nutrient removal, such as composting, anaerobic or aerobic digestion of manure, nitrification and denitrification reactors, and chemical phosphorus removal or biological phosphorus removal, are slowly being adopted on livestock farms or are being evaluated. These treatments may be coupled or combined together more intensively for their possible applicability to animal agriculture. The effects of these manure methods that are applied to manure management on estrogen degradation or reduction are discussed.

13.8.3   COMPOSTING

Composting decomposes organic matter in manure with proper temperature, moisture, and aerobic conditions.[98] The resulting compost is less odorous than fresh manure, its N is more stable, and its pathogen and weed content are dramatically reduced or eliminated. Consequently, composted manure is considered more marketable and transportable than fresh manure. These attributes make composting increasingly attractive to livestock farms in densely populated areas.

There are three basic composting methods: windrow, static pile, and in-vessel composting.[152] Method selection is based on cost, management and labor capability, site features, compost utilization, climate, and other factors. Composting should result in reduced hormone content due to aerobic degradation. Windrow composting for 139 days with original moisture of 60% decreased 84% of 17β-estradiol (13 ng/g vs. 83 ng/g) and 90% of testosterone (11 ng/g vs.115 ng/g) in poultry manure.[47] No data are available on the effect of composting on hormone content of other manures.

13.8.4   ANAEROBIC DIGESTION

In anaerobic digestion, facultative and strict anaerobes hydrolyze complex organics to form volatile organic acids that are ultimately metabolized to methane and CO_2.[92] Recent studies[55] showed that the 17β-estradiol equivalent (EEQ) determined with a yeast estrogen screen (YES) bioassay increased in response to anaerobic digestion, and subsequently decreased when further treated aerobically. The relative impact of sorption vs. biodegradation in the loss of estrogens in this study was not clarified. The authors of this chapter quantified the effects of an anaerobic digester receiving dairy manure on 17β-estradiol removal using an enzyme-linked immunosorbent assay (ELISA) protocol and found that the effluent had 40% less 17β-estradiol compared to the influent (9.9 ng/L vs. 15.9 ng/L, unpublished data). This ELISA method was performed on total manure and total effluent samples after a rigorous base-chloroform-toluene extraction procedure, and the results suggest that detected losses were due to biodegradation as opposed to adsorption.

13.8.5   NITRIFICATION AND DENITRIFICATION

Nitrification and denitrification are used for N removal from wastewater.[44] Nitrification is the biological process that converts ammonia (NH_3) to nitrate (NO_3^-) via nitrite (NO_2^-),[91] and it can be achieved in any aerobic biological process with low organic loadings and suitable environmental conditions. Denitrification is the biological process that converts nitrate (NO_3^-) to nitrogen (N_2) and other gaseous end products.[44]

Researchers have suggested that some endocrine-disrupting chemicals are preferentially transformed in WWTP bioreactors that have long hydraulic retention times, typical of nitrifying activated sludge cultures.[4,15,85,135] It is reasonable to assume that nitrifying ammonia oxidizing bacteria (AOBs) are capable of transforming these chemicals, given that they contain a broad specificity monooxygenase enzyme that is capable of oxidizing substituted[12,67,155] and polycyclic aromatic rings as well as aliphatic chlorinated solvents.[117,162] However, proof that AOBs are primarily or even significantly responsible for this transformation in full-scale WWTPs has not been shown definitively. Furthermore, many transformations generate byproducts that are frequently not monitored but may be biodegraded further if exposed to the correct environmental conditions.

The correlation between nitrogen removal and estrogen degradation in domestic wastewater treatment plants has been shown repeatedly. For example, 99% of the estrone was removed by activated sludge treatment (anoxic zone and nitrifying) in two European sewage plants.[65] In an Australian study the EEQ decreased to 5.1 ng/L after activated sludge treatment and nitrification/denitrification compared with 80 ng/L in the influent.[85] Nitrifying activated sludge systems are not currently used in CAFOs for manure treatment, but their obvious benefits in terms of nutrient removal have focused research attention on these systems. Based on results in WWTPs, hormone removal can reasonably be expected with application of this treatment in CAFOs, but more work needs to be done about its effects on hormone degradation and sorption.

13.8.6   CHEMICAL AND BIOLOGICAL PHOSPHORUS REMOVAL

Reducing P discharge from agricultural, municipal, and industrial wastewater is critical in preventing eutrophication of water bodies.[13] Both chemical and biological phosphorus removal processes have been applied in wastewater treatment.[44] The chemical process is to precipitate the inorganic forms of phosphate by adding coagulants, such as lime, aluminum sulfate, ferric chloride, or ferric sulfate.[163] Biological P removal systems result in the incorporation of P into cell biomass. The cell biomass is subsequently removed as sludge.[5,146] No data are available on the effects of P removal systems on the estrogen content of manure. Because of the removal of biosolids and the anaerobic/aeration process in biological P removal, estrogen degradation and removal would be expected.

13.8.7   AERATION OF DAIRY MANURE

While hog farms commonly implement advanced manure treatment systems such as anaerobic and aerobic lagoons, true manure treatment is less common on dairy farms. The authors quantified 17β-estradiol and estriol in each of six stages of an innovative manure handling system at the dairy center milking 140 cows at Virginia Tech. This system employs a mechanical separator to separate manure liquids from solids, a short retention time anaerobic settling basin to remove further solids, and three aerated tanks in sequence (Figure 13.13). The effluent from the third tank is

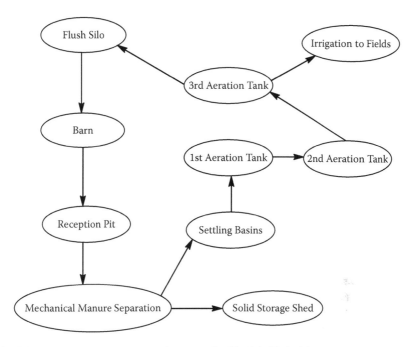

FIGURE 13.13   The manure system layout at the Virginia Tech dairy center.

reused to flush the barn or is land applied to crops via irrigation. Total retention time in the system is approximately 180 days.

17β-estradiol and estriol in the system were assayed using an ELISA method. Based on results from monthly samples taken between August 2005 and July 2006 (unpublished data), mechanical separation did not affect the total mass of 17β-estradiol and estriol (45.7 mg/day and 28.1 mg/day, respectively), but the effluent from the third aeration tank had a significantly reduced total flux of 17β-estradiol (16.3 mg/day out versus 53.3 mg/day in) and estriol (5.46 mg/day out versus 38.4 mg/day in) compared to the original slurry (Figure 13.14).

13.9   OTHER BEST MANAGEMENT PRACTICES TO REDUCE HORMONE LOSS FROM CAFOs

Ultimately manure is applied to cropland or pastures as a fertilizer or soil amendment, with the goal of recycling manure nutrients through crops. A variety of best management practices (BMPs) are used by farmers to reduce nutrient losses from their farms. In many cases the effects of these BMPs on the fate of endocrine-disrupting chemicals (EDCs) are not known, but inferences can be drawn. Most states provide significant cost-share funds or tax credit programs to support the implementation of BMPs. Their implementation is seen as a key component of strategies to reduce nonpoint source nutrient pollution. In Virginia, for instance, 48 BMPs qualify for tax credit and cost-share programs.

Common BMPs include buffer strips, constructed wetlands, and stream-bank fencing. The effects of these practices on the fate of excreted hormones (reported or

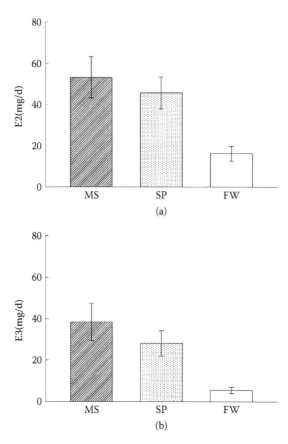

FIGURE 13.14 Mass balance of 17b-estradiol (a) and estriol (b) at the Virginia Tech dairy center. MS: main slurry from the barn; SP; separator effluent; FW: effluent from the third tank. Error bars stand for standard error (n = 12).

inferred) are summarized below. Because these BMPs are so widely implemented and represent a significant investment of public funds, where effects on hormone fate are truly unknown, more research is merited.

13.9.1 Constructed Wetlands

Constructed wetlands are marshes built to treat sewage; storm water; leachate; and increasingly, agricultural wastewater. Shallow depths and dense vegetation make these systems effective in the removal of biological oxygen demand, nutrients, metals, and toxic organic compounds.[119] Nitrogen is removed from influent through plant and microbial uptake, nitrification, ammonia volatilization, and denitrification, while phosphorus may be taken up by plants and microbes or sorbed to soil. The constructed wetlands reduced TP loads by 41%, with retention increasing with increased surface-area/watershed-area ratio.[154]

The same mechanisms that make constructed wetlands effective in nutrient and sediment removal should logically also affect the fate of steroidal hormones from agricultural wastewater, but little data are available. Recently it was reported that estrogenic activity in swine wastewater (measured with the *in vitro* E-screen assay) was reduced by 83 to 93% with constructed wetlands, and wetland outflow concentrations were below 3 ng/L.[132] Wetlands require a consistent and adequate supply of water, making them well suited to dairy and swine farms where water use is high. Constructed wetlands are still relatively rare on livestock farms, but are receiving more interest, typically as a component of conventional manure management systems, such as lagoons.

13.9.2   BUFFER STRIPS

Buffer strips are narrow strips of a permanent vegetative cover planted across a slope along a waterway. They are used to reduce runoff of nutrients, organic matter, and pathogens from manure-amended fields to water bodies. Field runoff is decreased through infiltration, deposition, and sorption by grasses in the buffer area.[101] Few data are available about hormone removal by buffer strips, but a positive impact has been reported.[101] Simulated rainfall techniques were used to evaluate buffer strips around poultry-litter amended pastures. Compared to no-buffer controls, estrogen mass loss was reduced by 79, 90, and 98% with buffer strips of 6, 12, and 18 m in width (effect of buffer width was not significant). Buffer strips of varying width are implemented on farms; 10-m buffers are fairly standard.

13.9.3   CONTROLLED STREAM ACCESS

The BMP probably receiving most emphasis on smaller farms is preventing or reducing access of cattle to streams and other waterways. Obviously, streams and ponds provide an easy source of drinking water and cooling for cattle, but equally obviously, allowing cattle access to these waterways results in direct deposition of nutrients, pathogens, and sex hormones in manure. Streambank fencing is the most common BMP to prevent exclusion, but the industry often resists this approach because of economic impacts. Pastures are often far from water supply lines, and seasonal variation in stream width forces a choice; installing fencing at the widest typical flow path results in loss of pasture that is useful much of the year.

Interestingly, cattle prefer to drink from off-stream water sources when offered a choice. Researchers have found that simply providing an alternative water source will reduce the time cattle spend in streams by up to 90%, even without fencing.[134] Streambank erosion was reduced by 77%, and concentrations of nitrogen and phosphorus in the stream were reduced by 54 and 81%, respectively. Significant reduction in manure deposition to streams can also be achieved by establishing controlled access points, wide enough to allow one animal to drink from the stream at a time but narrow enough to discourage loitering. These access points are typically reinforced with geotextile material or aggregate.

While no data exist on the effect of preventing or limiting access to waterways on steroidal hormone deposition in streams, it is reasonable to expect a significant benefit. The farther the cattle are from the waterbody, the more time for soil sorption and degradation of sex hormones by soil microbes.

13.10   FUTURE RESEARCH NEEDS REGARDING MANURE TREATMENT

Little research has been conducted on the effect of standard manure storage and treatment systems on the hormonal content and activity of manure. Systems designed to achieve nutrient management goals will almost certainly influence content and activity of hormones in manure. Conditions of storage or treatment (moisture content, oxygen availability, duration) vary widely. Livestock wastes differ in many ways from municipal waste, so applying assumptions from that body of literature to predict the fate of steroidal hormones in livestock manures may not be appropriate.

Also, as with most research focused on WWTPs, the limited research that has been conducted with livestock wastes focuses exclusively on overall percent removal, without consideration of the mechanism (degradation, sorption) of removal and ignoring the nature of transformation products that remain in the treated wastewater. This focus on percent removal often leads to misinterpretations regarding the causal mechanisms at work that define the fate of hormones. It is important to recognize that nonbiodegraded hormones that are sorbed to biomass can become bioavailable when localized conditions (sometimes influenced by the biomass) are changed such that desorption becomes favorable. Also, if the reduction in estrogenic activity in the effluent of some of these systems is due primarily to sorption of these compounds to the organic matter retained in the system, the problem of contamination of water resources has only been delayed. Eventually the sludge from the bottom of even the largest anaerobic lagoon must be removed. Clearly, additional research is needed on the fate of excreted EDCs during manure storage on CAFOs.

13.11   CONCLUSION

The environmental effects associated with hormones in the waste from CAFOs have induced great interest and concerns in scientists, governments, and the general public. Estrogens, androgens, and progestagens are hormones that may cause endocrine-disrupting effects, especially in aquatic species. There were 49, 4.4, and 279 tons of natural estrogens, androgens, and gestagens, respectively, excreted by farm animals in the United States in 2002.[79] Although hormone removal is possible before and after manure land application, part of these hormones will finally enter surface and groundwater and therefore pose endocrine-disrupting effects on living creatures. These three hormones have been reported to be detectable in some U.S. rivers.[73]

While it is unclear whether the hormones in rivers are of municipal, industrial, or agricultural origins, it is important to decrease hormones available in animal manure before land application. Within the several manure treatments discussed above, the removal of estrogens depends substantially on matrix conditions, such as oxygen, temperature, microorganism populations, and nutrients. Nutrient removal is generally the most important issue driving implementation of innovative manure treatment and other BMPs, but improved understanding of the effects of these systems on the environmental fate, potency, and biological effects of hormones originating from CAFOs is needed.

REFERENCES

1. Adlercreutz, H. and Martin, F. 1980. Biliary excretion and intestinal metabolism of progesterone and estrogens in man. *J Steroid Biochem* 13:231–244.
2. Adlercreutz, H., Martin, F., Jarvenpaa, P., and Fotsis, T. 1979. Steroid absorption and enterohepatic recycling. *Contraception* 20:201–223.
3. Ainsworth, L., Common, R.H., and Carter, R.L. 1962. A chromatographic study of some conversion products of estrone-16-C-14 in the urine and feces of the laying hen. *Can J Biochem Physiol* 40:123–135.
4. Andersen, H., Siegrist, H., Halling-Sorensen, B., and Ternes, T.A. 2003. Fate of estrogens in a municipal sewage treatment plant. *Environ Sci Technol* 37:4021–4026.
5. Arvin, E. 1985. Biological removal of phosphate from wastewater. CRC *Crit Rev Environ Control* 15:25–64.
6. Baldwin, R.S., Williams, R.D., and Terry, M.K. 1983. Zeranol: a review of the metabolism, toxicology, and analytical methods for detection of tissue residues. *Regul Toxicol Pharmacol* 3:9–25.
7. Bauer, E.R., Daxenberger, A., Petri, T., Sauerwein, H., and Meyer, H.H. 2000. Characterisation of the affinity of different anabolics and synthetic hormones to the human androgen receptor, human sex hormone binding globulin and to the bovine progestin receptor. *APMIS* 108:838–846.
8. Brooks, A.V. 1975. Androgens. *Clin Endocrinol Metab* 4:503–520.
9. Butterworth, M., Lau, S.S., and Monks, T.J. 1996. 17 beta-estradiol metabolism by hamster hepatic microsomes: comparison of catechol estrogen O-methylation with catechol estrogen oxidation and glutathione conjugation. *Chem Res Toxicol* 9:793–799.
10. Casey, F.X., Hakk, H., Simunek, J., and Larsen, G.L. 2004. Fate and transport of testosterone in agricultural soils. *Environ Sci Technol* 38:790–798.
11. Chakraborty, R. and Coates, J.D. 2005. Hydroxylation and carboxylation—two crucial steps of anaerobic benzene degradation by dechloromonas strain RCB. *Appl Environ Microbiol* 71:5427–5432.
12. Chang, S.W., Hyman, M.R., and Williamson, K.J. 2003. Cooxidation of napthalene and other polycyclic aromatic hydrocarbons of the nitrifying bacterium, Nitrosomonas europaea. *Biodegradation* 13:373–381.
13. Chardon, W.J. and Koopmans, G.F. 2005. Phosphorus Workshop. *J Environ Qual* 34:2091–2092.
14. Chaudhary, Z.I. and Price, M.A. 1987. Effects of castration and exogenous estradiol or testosterone on limb bone growth and some performance traits in young male pigs. *Can J Anim Sci* 67:681–688.
15. Clara, M., Kreuzinger, N., Strenn, B., Gans, O., and Kroiss, H. 2005. The solids retention time—A suitable design parameter to evaluate the capacity of wastewater treatment plants to remove micropollutants. *Water Res* 39:97–106.
16. Clemens, J.D. and Estergreen, V.L. 1982. Metabolism and conjugation of [4-14C] progesterone by bovine liver and adipose tissues, in vitro. *Steroids* 40:287–306.
17. Colborn, T., Vom Saal, F.S., and Soto, A.M. 1993. Developmental effects of endocrine-disrupting chemicals in wildlife and humans. *Environ Health Perspect* 101:378–384.
18. Colucci, M.S., Bork, H., and Topp, E. 2001. Persistence of estrogenic hormones in agricultural soils I. 17ß-estradiol and estrone. *J Environ Qual* 30:2070–2076.
19. Copeland, C. 2006. Animal waste and water quality: EPA regulation of concentrated animal feeding operations (CAFOs). Congressional Research Service & The Library of Congress.
20. D'Ascenzo, G., Di Corcia, A., Gentili, A., Mancini, R., Mastropasqua, R., Nazzari, M., and Samperi, R. 2003. Fate of natural estrogen conjugates in municipal sewage transport and treatment facilities. *Sci Total Environ* 302:199–209.

21. Daxenberger, A., Meyer, K., Hageleit, M., and Meyer, H.D. 1999. Detection of melengestrol acetate residues in plasma and edible tissues of heifers. *Vet Q* 21:154–158.
22. Dixon, S.N., Russell, K.L., Heitzman, R.J., and Mallinson, C.B. 1986. Radioimmunoassay of the anabolic agent zeranol. V. Residues of zeranol in the edible tissues, urine, faeces and bile of steers treated with Ralgro. *J Vet Pharmacol Ther* 9:353–358.
23. Durhan, E.J., Lambright, C.S., Makynen, E.A., Lazorchak, J., Hartig, P.C., Wilson, V.S., Gray, L.E., and Ankley, G.T. 2006. Identification of metabolites of trenbolone acetate in androgenic runoff from a beef feedlot. *Environ Health Perspect* 114 Suppl:65–68.
24. Dyer, A.R., Raman, D.R., Mullen, M.U., Burns, R.T., Moody, L.B., Layton, A.C., and Sayler, G.S. 2001. Determination of 17β-estradiol concentrations in runoff from plots receiving dairy manure. Presentation at 2001 ASAE annual international meeting sponsored by ASAE. Paper number: 01–2107.
25. Edgerton, L.A. and Erb, R.E. 1971. Metabolites of progesterone and estrogen in domestic sow urine. I. Effect of Pregnancy. *J Anim Sci* 32:515–524.
26. Elkins, C.A. and Mullis, L.B. 2006. Mammalian steroid hormones are substrates for the major RND- and MFS-type tripartite multidrug efflux pumps of Escherichia coli. *J Bacteriol* 188:1191–1195.
27. EPA, U. 1997. Special Report on Endocrine Disruption: An Effects Assessment and Analysis. EPA/630/R-96/012. Office of Research and Development, Washington, D.C.
28. EPA, U. 2002. Subject: EPA and agriculture working together to improve America's waters. http://www.epa.gov/npdes/caforule. Accessed 10/29/06.
29. EPA, U. 2003. National pollutant discharge elimination system permit regulation and effluent limitations guidelines and standards for concentrated animal feeding operations. *Federal Register* 68:7176–7274.
30. Erb, R.E., Chew, B.P., and Keller, H.F. 1977. Relative concentrations of estrogen and progesterone in milk and blood, and excretion of estrogen in urine. *J Anim Sci* 45:617–626.
31. European Medical Evaluation Agency. 1999. Subject: Committee for veterinary medicinal products: progesterone summary report. http://www.emea.eu.int/pdfs/vet/mrls/014696en.pdf. Accessed 12/22/06.
32. FDA. 1994. Environmental Assessment Report for Zeranol.
33. FDA. 1995. Environmental Assessment Report for NADA 034–254.
34. FDA. 1995. Environmental Assessment Report for NADA 141–043.
35. Feng, X., Ding, M., Tu, J., Wu, F., and Deng, N. 2005. Degradation of estrone in aqueous solution by photo-Fento system. *Sci Total Environ* 345:229–237.
36. Finkelstein, M., Weidenfeld, J., Ne'eman, Y., Samuni, A., Mizrachi, Y., and Ben-Uzilio, R. 1981. Comparative studies of the aromatization of testosterone and epitestosterone by human placental aromatase. *Endocrinology* 108:943–947.
37. Finlay-Moore, O., Hartel, P.G., and Cabrera, M.L. 2000. 17 beta-estradiol and testosterone in soil and runoff from grasslands amended with broiler litter. *J Environ Qual* 29:1604–1611.
38. Fotherby, K. 1984. Biosynthesis of the estrogens. In Making, H.L. (ed.) Biochemistry of Steroid Hormones. Blackwell Scientific, Oxford.
39. Funston, R.N., Ansotegui, R.P., Lipsey, R.J., and Geary, T.W. 2002. Synchronization of estrus in beef heifers using either melengesterol acetate (MGA)/prostaglandin or MGA/Select Synch. *Theriogenology* 57:1485–1491.
40. Gevao, B., Semple, K.T., and Jones, K.C. 2000. Bound pesticide residues in soils: a review. *Environ Pollut* 108:3–14.
41. Gorbach, S.L. and Goldin, B.R. 1987. Diet and the excretion and enterohepatic cycling of estrogens. *Prev Med* 16:525–531.
42. Gower, D.B. 1979. Steroid Hormones. Croom Helm London.

43. Gower, D.B. 1984. Biosynthesis of the androgens and other C19 steroids. Makin, H.L.J. (ed). In *Biochemistry of Steroid Hormones.* 2nd ed. Blackwell Scientific Publication, Oxford.
44. Grady, C.P.L., Daigger, G.T., and Lim, H.C. 1999. Biological Wastewater Treatment. 2nd ed. Marcel Dekker Inc., New York.
45. Grino, P.B., Griffin, J.E., and Wilson, J.D. 1990. Testosterone at high concentrations interacts with the human androgen receptor similarly to dihydrotestosterone. *Endocrinology* 126:1165–1172.
46. Hageleit, M., Daxenberger, A., Kraetzl, W.D., Kettler, A., and Meyer, H.D. 2000. Dose-dependent effects of melengestrol acetate (MGA) on plasma levels of estradiol, progesterone, and luteinizing hormone in cycling heifers and influences on oestrogen residues in edible tissues. *Acta Pathol Microbiol Immunol Scand* 108:847–854.
47. Hakk, H., Millner, P., and Larsen, G. 2005. Decrease in water-soluble 17beta-Estradiol and testosterone in composted poultry manure with time. *J Environ Qual* 34:943–950.
48. Hanselman, T.A., Graetz, D.A., and Wilkie, A.C. 2003. Manure-borne estrogens as potential environmental contaminants: a review. *Environ Sci Technol* 37:5471–5478.
49. Hayes, A.W. 2005. The precautionary principle. *Arh Hig Rada Toksikol* 56:161–166.
50. Heberer, T. 2002. Occurrence, fate, and removal of pharmaceutical residues in the aquatic environment: a review of recent research data. *Toxicol Lett* 131:5–17.
51. Hemmingsa, S.N.J. and Hartelb, P.G. 2006. Mineralization of hormones in breeder and broiler litters at different water potentials and temperatures. *J Environ Qual* 35:701–706.
52. Hobkirk, R. 1985. Steroid sulfotransferases and steroid sulfate sulfatases: characteristics and biological roles. *Can J Biochem Cell Biol* 63:1127–1144.
53. Hoffmann, B., Goes de Pinho, T., and Schuler, G. 1997. Determination of free and conjugated oestrogens in peripheral blood plasma, feces and urine of cattle throughout pregnancy. *Exp Clin Endocrinol Diabetes* 105:296–303.
54. Hoffmann, B. and Karg, H. 1976. Metabolic fate of anabolic agents in treated animals and residue levels in their meat. *Environ Qual Saf* Suppl 5:181–191.
55. Holbrook, R.D., Novak, J.T., Grizzard, T.J., and Love, N.G. 2002. Estrogen receptor agonist fate during wastewater and biosolids treatment processes: a mass balance analysis. *Environ Sci Technol* 36:4533–4539.
56. Horinouchi, M., Yamamoto, T., Taguchi, K., Arai, H., and Kudo, T. 2001. Meta-cleavage enzyme gene test is necessary for testosterone degradation in Comamonas testosteroni TA441. *Microbiology* 147:3367–3375.
57. Hutchison, M.L., Walters, L.D., Avery, S.M., Munro, F., and Moore, A. 2005. Analyses of livestock production, waste storage, and pathogen levels and prevalences in farm manures. *Appl Environ Microbiol* 71:1231–1236.
58. Isobe, N., Akita, M., Nakao, T., Yamashiro, H., and Kubota, H. 2005. Pregnancy diagnosis based on the fecal progesterone concentration in beef and dairy heifers and beef cows. *Anim Reprod Sci* 90:211–218.
59. Ivie, G.W., Christopher, R.J., Munger, C.E., and Coppock, C.E. 1986. Fate and residues of [4-14C] estradiol-17 beta after intramuscular injection into Holstein steer calves. *J Anim Sci* 62:681–690.
60. Jackson, L., Langlois, B.E., and Dawson, K.A. 1992. Beta-glucuronidase activities of fecal isolates from healthy swine. *J Clin Microbiol* 30:2113–2117.
61. Jacobsen, A.M., Lorenzen, A., Chapman, R., and Topp, E. 2005. Persistence of testosterone and 17beta-estradiol in soils receiving swine manure or municipal biosolids. *J Environ Qual* 34:861–871.
62. Jansen, E.H., van den Berg, R.H., Zomer, G., Enkelaar-Willemsen, C., and Stephany, R.W. 1986. A chemiluminescent immunoassay for zeranol and its metabolites. *J Vet Pharmacol Ther* 9:101–108.

63. Jenkins, M.B., Endale, D.M., Schomberg, H.H., and Sharpe, R.R. 2006. Fecal bacteria and sex hormones in soil and runoff from cropped watersheds amended with poultry litter. *Sci Total Environ* 358:164–177.
64. Jobling, S., Nolan, M., Tyler, C.R., Brighty, G., and Sumpter, J.P. 1998. Widespread sexual disruption in wild fish. *Environ Sci Technol* 32:2498–2506.
65. Johnson, A.C., Aerni, H.R., Gerritsen, A., Gibert, M., Giger, W., Hylland, K., Jurgens, M., Nakari, T., Pickering, A., Suter, M.J., Svenson, A., and Wettstein, F.E. 2005. Comparing steroid estrogen and nonylphenol content across a range of European sewage plants with different treatment and management practices. *Water Res* 39:47–58.
66. Johnson, A.C., Williams, R.J., and Matthiessen, P. 2006. The potential steroid hormone contribution of farm animals to freshwaters, the United Kingdom as a case study. *Sci Total Environ* 362:166–178.
67. Keener, W.K. and Arp, D.J. 1994. Transformations of aromatic compounds by Nitrosomonas europaea. *Appl Environ Microbiol* 60:1914–1920.
68. Kellogg, R.L., Lander, C.H., Moffitt, D.C., and Gollehon, N. 2000. Subject: manure nutrients relative to the capacity of cropland and pastureland to assimilate nutrients: spatial and temporal trends for the United States. http://www.nrcs.usda.gov/technical/land/pubs/manntr.pdf. Accessed 11/09/06.
69. Khan, B., Sassman, S., and Lee, L. 2006. Subject: Degradation of 17beta-trenbolone and trendione in agricultural soil. http://a-c-s.confex.com/crops/2006am/techprogram/P25513.HTM. Accessed 10/26/06.
70. Khan, M.Z., Altmann, J., Isani, S.S., and Yu, J. 2002. A matter of time: evaluating the storage of fecal samples for steroid analysis. *Gen Comp Endocrinol* 128:57–64.
71. Khanal, S.K., Xie, B., Thompson, M.L., Sung, S., Ong, S.K., and Leeuwan, J.V. 2006. Fate, transport, and biodegradation of natural estrogens in the environment and engineered systems. *Environ Sci Technol* 40:6537–6546.
72. Koelsch, R. 2005. Evaluating livestock system environmental performance with whole-farm nutrient balance. *J Environ Qual* 34:149–155.
73. Kolpin, D.W., Furlong, E.T., Meyer, M.T., Thurman, E.M., Zaugg, S.D., Barber, L.B., and Buxton, H.T. 2002. Pharmaceuticals, hormones, and other organic wastewater contaminants in U.S. streams, 1999–2000: a national reconnaissance. *Environ Sci Technol* 36:1202–1211.
74. Kolpin, D.W., Thurman, E.M., and Linharta, S.M. 2000. Finding minimal herbicide concentrations in ground water? Try looking for their degradates. *Sci Total Environ* 248:115–122.
75. Kolz, A.C., Moorman, T.B., Ong, S.K., Scoggin, K.D., and Douglass, E.A. 2005. Degradation and metabolite production of tyrosin in anaerobic and aerobic swine-manure lagoons. *Water Environ Res* 77:49–56.
76. Kominami, S., Ochi, H., Kobayashi, Y., and Takemori, S. 1980. Studies on the steroid hydroxylation system in adrenal cortex microsomes. *J Biol Chem* 256:3386–3394.
77. Kreikemeier, W.M. and Mader, T.L. 2004. Effects of growth-promoting agents and season on yearling feedlot heifer performance. *J Anim Sci* 82:2481–2488.
78. Krzeminski, L.F., Cox, B.L., and Gosline, R.E. 1981. Fate of radioactive melengestrol acetate in the bovine. *J Agric Food Chem* 29:387–391.
79. Lange, I.G., Daxenberger, A., Schiffer, B., Witters, H., Ibarreta, D., and Meyer, H. 2002. Sex hormones originating from different livestock production systems: fate and potential disrupting activity in the environment. *Analytica Chimica Acta* 473:27–37.
80. Lauderdale, J.W. 1983. Use of MGA® (melengestrol acetate) in Animal Production. Office International des Epizooties, Paris, France.
81. Le Guevel, R. and Pakdel, F. 2001. Assessment of oestrogenic potency of chemicals used as growth promoter by in-vitro methods. *Hum Reprod* 16:1030–1036.

82. Lee, H.B. and Liu, D. 2002. Degradation of 17beta-estradiol and its metabolites by sewage bacteria. *Water Air Soil Pollut* 134:353–368.

83. Lee, L.S., Strock, T.J., Sarmah, A.K., and Rao, P.S. 2003. Sorption and dissipation of testosterone, estrogens, and their primary transformation products in soils and sediment. *Environ Sci Technol* 37:4098–4105.

84. Leffers, H., Naesby, M., Vendelbo, B., Skakkebaek, N.E., and Jorgensen, M. 2001. Oestrogenic potencies of Zeranol, oestradiol, diethylstilboestrol, Bisphenol-A and genistein: implications for exposure assessment of potential endocrine disrupters. *Hum Reprod* 16:1037–1045.

85. Leusch, F.D., Chapman, H.F., Korner, W., Gooneratne, S.R., and Tremblay, L.A. 2005. Efficacy of an advanced sewage treatment plant in southeast Queensland, Australia, to remove estrogenic chemicals. *Environ Sci Technol* 39:5781–5786.

86. Lintelmann, J., Katayama, A., Kurihara, N., Shore, L., and Wenzel, A. 2003. Endocrine disruptors in the environment. *Pure Appl Chem* 75:631–681.

87. Loftsson, T. and Hreinsdóttir, D. 2006. Subject: Determination of aqueous solubility by heating and equilibration: a technical note. http://www.aapspharmscitech.org. Accessed 01/10/07.

88. Lopez de Alda, M.J., Gil, A., Paz, E., and Barcelo, D. 2002. Occurrence and analysis of estrogens and progestogens in river sediments by liquid chromatography-electrospray-mass spectrometry. *Analyst* 127:1299–1304.

89. Lorenzen, A., Chapman, R., Hendel, J.G., and Topp, E. 2005. Persistence and pathways of testosterone dissipation in agricultural soil. *J Environ Qual* 34:854–860.

90. Lorenzen, A., Hendel, J.G., Conn, K.L., Bittman, S., Kwabiah, A.B., Lazarovitz, G., Masse, D., McAllister, T.A., and Topp, E. 2004. Survey of hormone activities in municipal biosolids and animal manures. *Environ Toxicol* 19:216–225.

91. Madigan, M.T. and Martinko, J.M. 2006. Brock: Biology of Microorganisms. 11th ed. Pearson Prentice Hall, Upper Saddle River, NJ.

92. Mergaert, K. and Verstraete, W. 1987. Microbial parameters and their control in anaerobic digestion. *Microbiol Sci* 4:348–351.

93. Meyer, H.H. and Rapp, M. 1985. Estrogen receptor in bovine skeletal muscle. *J Anim Sci* 60:294–300.

94. Meyer, H.H.D. 2001. Biochemistry and physiology of anabolic hormones used for improvement of meat production. *APMIS* 109:1–8.

95. Meyers, J.P., Krimsky, S., and Zoeller, R.T. 2001. Endocrine disruptors—A controversy in science and policy: Session III summary and research needs. *Neuro Toxicol* 22:557–558.

96. Migdalof, B.H., Dugger, H.A., Heider, J.G., Coombs, R.A., and Terry, M.K. 1983. Biotransformation of zeranol: disposition and metabolism in the female rat, rabbit, dog, monkey and man. *Xenobiotica* 13:209–221.

97. Mitamura, K., Narukawa, H., Mizuguchi, T., and Shimada, K. 2004. Degradation of estrogen conjugates using titanium dioxide as a photocatalyst. *Anal Sci* 20:3–4.

98. Miyatake, F. and Iwabuchi, K. 2006. Effect of compost temperature on oxygen uptake rate, specific growth rate and enzymatic activity of microorganisms in dairy cattle manure. *Bioresour Technol* 97:961–965.

99. Moriyoshi, M., Nozoki, K., Ohtaki, T., Nakada, K., Nakao, T., and Kawata, K. 1997. Measurement of gestagen concentration in feces using a bovine milk progesterone quantitative test EIA kit and its application to early pregnancy diagnosis in the sow. *J Vet Med Sci* 59:695–701.

100. Mostafa, M.E. and Zohri, A.A. 2000. Progesterone side-chain degradation by some species of Aspergillus flavus group. *Folia Microbiol* (Praha) 45:243–247.

101. Nichols, D.J., Daniel, T.C., Edwards, D.R., Moore, P.A., and Pote, D.H. 1998. Use of grass filter strips to reduce 17β-estradiol in runoff from fescue-applied poultry litter. *J Soil Water Cons* 53:74–77.

102. Nichols, D.J., Daniel, T.C., Moore, P.A., Edwards, D.R., and Pote, D.H. 1997. Runoff of estrogen hormone 17ß-estradiol from poultry litter applied to pasture. *J Environ Qual* 26:1002–1006.

103. Ohe, T., Hirobe, M., and Mashino, T. 2000. Novel metabolic pathway of estrone and 17β-estradiol catalyzed by cytochrome. *Drug Metab Dispos* 28:110–112.

104. Ohko, Y., Iuchi, K., Niwa, C., Tatsuma, T., Nakashima, T., Iguchi, T., Kubota, Y., and Fujishima, A. 2002. 17β-estradiol degradation by TiO2 photocatalysis as a means of reducing estrogenic activity. *Environ Sci Technol* 36:4175–4181.

105. Orlando, E.F., Kolok, A.S., Binzcik, G.A., Gates, J.L., Horton, M.K., Lambright, C.S., Gray Jr., L.E., Soto, A.M., and Guillette Jr., L.J. 2004. Endocrine-disrupting effects of cattle feedlot effluent on an aquatic sentinel species, the fathead minnow. *Environ Health Perspect* 112:353–358.

106. Palme, R., Fischer, P., Schildofer, H., and Ismail, M.N. 1996. Excretion of infused 14C-steroid hormones via faeces and urine in domestic livestock. *Anim Reprod Sci* 43:43–63.

107. Parr, R.A., Davis, I.F., Miles, M.A., and Squires, T.J. 1993. Liver blood flow and metabolic clearance rate of progesterone in sheep. *Res Vet Sci* 55:311–316.

108. Pasqualini, J.R., Gelly, C., Nguyen, B.L., and Vella, C. 1989. Importance of estrogen sulfates in breast cancer. *J Steroid Biochem* 34:155–163.

109. Perry, G.A., Welshons, W.V., Bott, R.C., and Smith, M.F. 2005. Basis of melengestrol acetate action as a progestin. *Domest Anim Endocrinol* 28:147–161.

110. Plourde, J.R., Hafez-Zedan, H., and Lemoine, J.P. 1974. Biotransformation of progesterone by spores and vegetative cells of micro-organisms found in the soil of Quebec. *Rev Can Biol* 33:111–116.

111. Pokorna, J. and Kasal, A. 1990. Progesterone side-chain degradation beside hydroxylation with Rhizopus nigricans depends on the presence of nutrients. *J Steroid Biochem* 35:155–156.

112. Pottier, J., Cousty, C., Heitzman, R.J., and Reynolds, J.P. 1981. Differences in the biotransformation of a 17P-hydroxylated steroid, trenbolone acetate, in rat and cow. *Xenobiotica* 11:489–500.

113. Powers, W.J. 1999. Odor control for livestock systems. *J Anim Sci* 77 Suppl 2:169–176.

114. Price, C.R. and Webb, R. 1988. Steroid control of gonadotropin secretion and ovarian function in heifers. *Endocrinology* 122:2222–2231.

115. Raftogianis, R., Creveling, C., Weinshilboum, R., and Weisz, J. 2000. Chapter 6: Estrogen metabolism by conjugation. *J Natl Cancer Inst Monographs* 27:113–124.

116. Raman, D.R., Layton, A.C., Moody, L.B., and Easter, J.P. 2001. Degradation of estrogens in dairy waste solids: effects of acidification and temperature. Trans. *ASAE* 44:1881–1888.

117. Rasche, M.E., Hyman, M.R., and Arp, D.J. 1991. Factors limiting aliphatic chlorocarbon degradation by Nitrosomonas europaea: cometabolic inactivation of ammonia monooxygenase and substrate specificity. *Appl Environ Microbiol* 57:2986–2994.

118. Raun, A.P. and Preston, R.L. 2002. Subject: history of diethylstilbestrol use in cattle. http://www.asas.org/Bios/Raunhist.pdf. Accessed 10/26/06.

119. Reddy, K.R. and D'Angelo, E.M. 1997. Biogeochemical indicators to evaluate pollutant removal efficiency in constructed wetlands. *Water Sci Technol.* 35:1–10.

120. Reed, M.J., Rea, D., Duncan, L.J., and Parker, M.G. 1994. Regulation of estradiol 17 beta-hydroxysteroid dehydrogenase expression and activity by retinoic acid in T47D breast cancer cells. *Endocrinology* 135:4–9.

121. Rodriguez-Mozaz, S., de Alda, M.J., and Barcelo, D. 2004. Monitoring of estrogens, pesticides and bisphenol A in natural waters and drinking water treatment plants by solid-phase extraction-liquid chromatography-mass spectrometry. *J Chromatogr A* 1045:85–92.

122. Rutishauser, B.V., Pesonen, M., Escher, B.I., Ackermann, G.E., Aerni, H.R., Suter, M.J., and Eggen, R.I. 2004. Comparative analysis of estrogenic activity in sewage treatment plant effluents involving three in vitro assays and chemical analysis of steroids. *Environ Toxicol Chem* 23:857–864.

123. Samuels, T.P., Nedderman, A., Seymour, M.A., and Houghton, E. 1998. Study of the metabolism of testosterone, nandrolone and estradiol in cattle. *Analyst* 123:2401–2404.

124. Sarabia, S.F., Zhu, B.T., Kurosawa, T., Tohma, M., and Liehr, J.G. 1997. Metabolism of cytochrome P450-catalyzed aromatic hydroxylation of estrogens. *Chem Res Toxicol* 10:767–771.

125. Sauerwein, H. and Meyer, H.H.D. 1989. Androgen and estrogen receptors in bovine skeletal muscle: Relation to steroid induced allometric muscle growth. *J Anim Sci* 67:206–212.

126. Schiffer, B., Daxenberger, A., Meyer, K., and Meyer, H.H. 2001. The fate of trenbolone acetate and melengestrol acetate after application as growth promoters in cattle: environmental studies. *Environ Health Perspect* 109:1145–1151.

127. Schiffer, B., Totsche, K.U., Jann, S., Kogel-Knabner, I., Meyer, K., and Meyer, H.H. 2004. Mobility of the growth promoters trenbolone and melengestrol acetate in agricultural soil: column studies. *Sci Total Environ* 326:225–237.

128. Schink, B., Philipp, B., and Muller, J. 2000. Anaerobic degradation of phenolic compounds. *Naturwissenschaften* 87:12–23.

129. Schlenker, G., Muller, W., and Glatzel, P. 1998. Continuing studies on the stability of sex steroids in the feces of cows over 12 weeks. *Berliner Münchener tierärztliche Wochenschrift* 111:248–252.

130. Schwarzenberger, F., Möstl, E., Palme, R., and Bamberg, E. 1996. Faecal steroid analysis for non-invasive monitoring of reproductive status in farm, wild and zoo animals. *Anim Reprod Sci* 42:515–526.

131. Shackleton, C.H., Irias, J., McDonald, C., and Imperato-McGinley, J. 1986. Late-onset 21-hydroxylase deficiency: reliable diagnosis by steroid analysis of random urine collections. *Steroids* 48:239–250.

132. Shappell, N.W., Billey, L.O., Forbes, D., Matheny, T.A., Poach, M.E., Reddy, G.B., and Hunt, P.G. 2007. Estrogenic activity and steroid hormones in swine wastewater through a lagoon constructed-wetland system. *Environ Sci Technol* 15:444–450.

133. Sharp, G.D. and Dyer, I.A. 1972. Zearalanol metabolism in steers. *J Anim Sci* 34:176–179.

134. Sheffield, R.E., Mostaghimi, S., Vaughan, D.H., Collins, E.R., and Allen, V.G. 1997. Off-stream water sources for grazing cattle as a stream bank stabilization and water quality BMP. Trans. *ASAE* 40:595–604.

135. Shi, J., Fujisawa, S., Nakai, S., and Hosomi, M. 2004. Biodegradation of natural and synthetic estrogens by nitrifying activated sludge and ammonia-oxidizing bacterium Nitrosomonas europaea. *Water Res* 38:2323–2330.

136. Shi, J.H., Suzuki, Y., Nakai, S., and Hosomi, M. 2004. Microbial degradation of estrogens using activated sludge and night soil-composting microorganisms. *Water Sci Technol* 50:153–159.

137. Shore, L.S., Correll, D.L., and Chakraborty, P.K. 1995. Relationship of fertilization with chicken manure and concentration of estrogens in small streams. In Steele, K. (ed.) *Animal Waste and the Land-Water Interface.* Lewis Publ., Boca Raton, FL, 155–162.

138. Shore, L.S. and Shemesh, M.. 2003. Naturally produced steroid hormones and their release into the environment. *Pure Appl Chem* 75:1859–1871.

139. Silence, M.N. 2004. Technologies for the control of fat and lean deposition in livestock. *Vet J* 167:242–257.

140. Songsermsakul, P., Sontag, G., Cichna-Markl, M., Zentek, J., and Razzazi-Fazeli, E. 2006. Determination of zearalenone and its metabolites in urine, plasma and faeces of horses by HPLC-APCI-MS. *J Chromatogr B Analyt Technol Biomed Life Sci* 843:252–261.

141. Soto, A.M., Calabro, J.M., Prechtl, N.V., Yau, A.Y., Orlando, E.F., Daxenberger, A., Kolok, A.S., Guillette, L.J., le Bizec, B., Lange, I.G., and Sonnenschein, C. 2004. Androgenic and estrogenic activity in water bodies receiving cattle feedlot effluent in Eastern Nebraska, USA. *Environ Health Perspect* 112:346–352.

142. Starka, L. 2003. Epitestosterone. *J Steroid Biochem Mol Biol* 87:27–34.

143. Sunahara, G.I. and Bellward, G.D. 1985. Testosterone metabolism in rat liver cytosolic androgen binding assays. *Drug Metab Dispos* 14:366–369.

144. Swartz, C.H., Reddy, S., Benotti, M.J., Yin, H., Barber, L.B., Brownawell, B.J., and Rudel, R.A. 2006. Steroid estrogens, nonylphenol ethoxylate metabolites, and other wastewater contaminants in groundwater affected by a residential septic system on Cape Cod, MA. *Environ Sci Technol* 40:4894–4902.

145. Ternes, T.A., Kreckel, P., and Mueller, J. 1999. Behavior and occurrence of estrogens in municipal sewage treatment plants—II. Aerobic batch experiments with activated sludge. *Sci Total Environ* 225:91–99.

146. Tetreault, M.J., Benedict, A.H., Kaempfer, C., and Barth, E.D. 1986. Biological phosphorus removal: a technology evaluation. 58:823–837.

147. Thouvenot, D., Morfin, R., Di Stefano, S., and Picart, D. 1981. Transformations of zearalenone and alpha-zearalanol by homogenates of human prostate glands. *Eur J Biochem* 121:139–145.

148. Toran-Allerand, C.D., Tinnikov, A.A., Singh, R.J., and Nethrapalli, I.S. 2005. 17α-estradiol: A brain-active estrogen? *Endocrinology* 146:3843–3850.

149. Trojan, M.D., Maloney, J.S., Stockinger, J.M., Eid, E.P., and Lahtinen, M.J. 2003. Effects of land use on ground water quality in the Anoka Sand Plain Aquifer of Minnesota. *Ground Water* 41:482–492.

150. Tsang, C.P. and Hackett, A.J. 1979. Metabolism of progesterone in the pregnant sheep near term: identification of 3 beta-hydroxy-5 alpha-pregnan-20-one 3-sulfate as a major metabolite. *Steroids* 33:577–588.

151. USDA. 1997. Agricultural Waste Management Field Handbook. Chapter 9: Agricultural Waste Management Systems.

152. USDA. 1997. Agricultural Waste Management Field Handbook. Chapter 10: Agricultural Waste Management System Component Design.

153. USGS. 2005. Subject: pharmaceuticals, hormones and other organic wastewater contaminants in ground water resources. http://toxics.usgs.gov/pubs/contaminant_studies_article.pdf. Accessed 12/20/06.

154. Uusi Kamppa, J., Braskerud, B., Jansson, H., Syversen, N., and Uusitalo, R. 2000. Buffer zones and constructed wetlands as filters for agricultural phosphorus. *J Environ Qual* 29:151–158.

155. Vannelli, T. and Hooper, A.B. 1995. NIH shift in the hydroxylation of aromatic compounds by the ammonia-oxidizing bacterium Nitrosomonas europaea. Evidence against an arene oxide intermediate. *Biochemistry* 34:11743–11749.

156. Vos, E.A. 1996. Direct elisa for estrone measurement in the feces of sows: prospects for rapid, sow-side pregnancy diagnosis. *Theriogenology* 46:211–231.

157. Wicks, C., Kelley, C., and Peterson, E. 2004. Estrogen in a karstic aquifer. *Ground Water* 42:384–389.

158. Williams, E.L., Raman, R., Burns, R.T., Layton, A.C., Daugherty, A.S., and Mullen, M.D. 2002. Estrogen concentrations in dairy and swine waste storage and treatment structures in and around Tennessee. In *ASAE Annual International Meeting*. Vol. Paper Number 024150. ASAE, Chicago, IL.

159. Wilson, J.D. and Foster, D.W. 1985. Textbook of Endocrinology. W.B. Saunders Company, Philadelphia, PA.

160. Wilson, T.W., Neuendorff, D.A., Lewis, A.W., and Randel, R.D. 2002. Effect of zeranol or melengestrol acetate (MGA) on testicular and antler development and aggression in farmed fallow bucks. *J Anim Sci* 80:1433–1441.

161. Wilson, V.S. and LeBlanc, G.A. 2000. The contribution of hepatic inactivation of testosterone to the lowering of serum testosterone levels by ketoconazole. *Toxicol Sci* 54:128–137.

162. Yang, L., Chang, Y.F., and Chou, M.S. 1999. Feasibility of bioremediation of trichloroethylene contaminated sites by nitrifying bacteria through cometabolism with ammonia. *J Hazard Mater* B69:111–126.

163. Yeoman, S., Stephenson, T., Lester, J.N., and Perry, R. 1988. The removal of phosphorus during wastewater treatment: a review. *Environ Pollut* 49:183–233.

164. Ying, G.G. and Kookana, R.S. 2005. Sorption and degradation of estrogen-like endocrine disrupting chemicals in soil. *Environ Toxicol Chem* 24:2640–2645.

165. Yu, Z., Xiao, B., Huang, W., and Peng, P. 2004. Sorption of steroid estrogens to soils and sediments. *Environ Toxicol Chem* 23:531–539.

166. Zhu, B.T. and Conney, A.H. 1998. Functional role of estrogen metabolism in target cells—Review and perspectives. *Carcinogenesis* 19:1–27.

14 Treatment of Antibiotics in Swine Wastewater

Craig D. Adams

Contents

14.1 INTRODUCTION

The formation and occurrence of antibiotic-resistant bacteria (especially pathogens) in the environment are of significant concern to society, and are the specific focus of the scientific and regulatory communities. In animal agriculture in the United States and elsewhere, antibiotics are provided to swine for therapeutic reasons, as well as for growth promotion. Many antibiotics that are fed to or injected into swine may pass through the swine unmetabolized and, therefore, end up in the swine manure that is passed into the treatment system. Accordingly, it is of considerable interest that an economical and effective means of treating these antibiotics prevent or minimize their introduction into the environment during their field application (Figure 14.1).

FIGURE 14.1   Sludge pump used to transfer lagoon slurry from lagoon to adjacent fields.

14.2   TYPICAL MANURE HANDLING SYSTEMS FOR SWINE

Swine manure is a mixture of urine and feces, which often contains significant concentrations of antibiotics and hormones. Swine manure may typically contain only 10 to 20% solids and, therefore, is generally in slurry form. Thus, swine manure generally cannot be handled using solids handling equipment. Lagoon slurry is usually discharged through direct land application to croplands. While the slurry has significant nutrient value, it may also contain antibiotics, hormones, antibiotic-resistant organisms, as well as excessive phosphorus and other problematic contaminants. Land application of liquid slurry from anaerobic lagoons or storage basins, and anaerobic digesters, is typically achieved by using irrigation-type equipment. These systems include stationary spray guns, sprinkler systems, and controlled flooding.

14.2.1   INTERIOR STORAGE

Swine manure from confined production facilities is often stored in either interior (underfloor) or exterior (lagoon) storage basins (Miner et al., 2000). The underfloor basin or pit is located directly beneath the slatted floor of the building housing the swine (Figure 14.2). The swine manure, along with excess food and other solids, falls or is periodically rinsed down into the underfloor pit. Maximum storage time in a typical underfloor pit may range from 5 to 12 months (Miner et al., 2000). Ventilation of the underfloor pits is critical to remove noxious gases such as hydrogen sulfide and ammonia, as well as carbon dioxide and methane, from the confinement building. Prior to removal from the pit, the manure must be agitated to homogenize it so that all of it can be completely removed from the pit, and to ensure that the

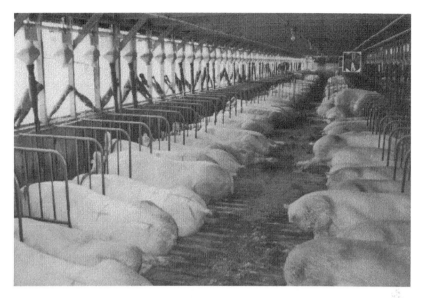

FIGURE 14.2   Inside a swine barn at a typical concentrated animal feed operation.

removed manure has uniform nutrient characteristics. Agitation is usually achieved using pumps placed in multiple locations along the pit wall (Miner et al., 2000). The discharge from underfloor storage is generally applied to surrounding fields as a fertilizer, although it is usually nonoptimal relative to nutrient p:n ratios.

14.2.2   EXTERIOR STORAGE AND TREATMENT

An alternative to an underfloor storage basin is the exterior storage basin, commonly referred to as a "lagoon." Occasionally, exterior storage of manure is instead accomplished by using a tank located outside the building; this is far less common, however, than the use of a lagoon. Generally, swine manure is collected using a slatted floor design, and then periodically (e.g., twice per day) flushed with water to move it from the building.

In the simplest design, the manure flows to a single- or two-stage lagoon for storage and treatment, followed by periodic land application. A variation of this system first provides for liquid-solid separation, after which the solids may be composted and the liquid passed to a lagoon for storage and treatment, prior to land application (Miner et al., 2000).

In a more sophisticated system, the waste from the barn is mixed and pumped into an anaerobic digester. Methane generated in the process provides for energy recovery. The effluent from the anaerobic digester is often pumped into an anaerobic lagoon for storage and treatment, followed by land application (Miner et al., 2000).

A variety of lagoon systems are used for swine wastewater treatment. A lagoon system may be a single cell, or may contain multiple cells in series. Generally, lagoons are not aerated and are, therefore, anaerobic. In some cases an aerated cell is used for enhancing ammonia removal.

In a standard anaerobic lagoon, swine manure is stored until the time of year that is amenable for field application, that is, when the nutrients are needed and the ground is not frozen. Additionally, some treatment of the manure is achieved in the lagoon and the quality of the slurry changes with time. In general, solids are decreased, which makes the slurry much more amenable to field application (Miner et al., 2000). Key differences between an anaerobic digester and an anaerobic lagoon for swine waste treatment are that lagoons have no temperature control and cannot capture methane for energy recovery (unless appropriately covered). Depending on the region, in the colder winter months, anaerobic activity in a lagoon may be very low relative to that in the warmer summer months. As temperatures increase during the transition from winter to summer, the excess stored organic matter may cause enhanced anaerobic activity until stored organic loads are reduced (Miner et al., 2000).

14.2.3 MULTICELL LAGOON SYSTEMS

Two- and three-celled lagoon systems, in series, are also used at some facilities. Typically, the first cell is operated as in a single-cell system. Most solids are retained in the first cell, which provides for additional solids decomposition. In the final cell, algae often may thrive, allowing better slurry treatment (Miner et al., 2000). Additionally, aeration is sometimes added to the final cell to improve effluent quality. Multicelled systems are more common when the lagoon water is to be used as the flushing water for the barns.

As an example of typical treatment parameters, characteristics of swine barn wastewater in two different lagoon systems studied by Qiang et al. (2006) are presented in Table 14.1. For Lagoon System A, comparison of the influent (A-INF) and effluent (A-EFF) from the first of two anaerobic cells showed a significant reduction in soluble chemical oxygen demand (SCOD) and dissolved organic carbon (DOC), while ammonia, alkalinity, pH, and UV adsorption all increased (Table 14.1). Comparison of the influent into the second (overflow) cell and its bulk concentration (A-OV) (which is periodically land applied), showed a further reduction of SCOD and COD, as well as ammonia (Table 14.1).

A second two-cell lagoon system was also studied (Lagoon B) that was similar to Lagoon System A, except that the first cell of the lagoon was aerated. Similar treatment was achieved for SCOD and DOC. However, a much lower ammonia concentration was achieved in the effluent of the first (aerated) cell (B-EFF), which helped achieve a very low final ammonia concentration (B-OV) in the second (nonaerated) cell. The slurry from this second cell is periodically land applied.

14.2.4 ANAEROBIC DIGESTION

Swine manure is also amenable to treatment in an anaerobic digester. An anaerobic digester is enclosed so as to capture product gases (e.g., hydrogen sulfide, ammonia, methane, and carbon dioxide) and to allow efficient treatment of the swine waste. However, because this level of efficient waste treatment is not required in the United States for CAFO wastes, the perceived cost associated with anaerobic digesters has limited their use for swine manure treatment in the United States. The use of

TABLE 14.1

Mean Physical-Chemical Properties of Wastewaters from Two Swine Lagoons

Properties	Lagoon A Raw Waste from Barn (AINF)	Lagoon A Bulk Slurry from 1st (Anaerobic) Cell (A-EFF)	Lagoon A Bulk Slurry from 2nd (Anaerobic) Cell (A-OV)	Lagoon B Raw Waste from Barn (BINF)	Lagoon B Bulk Slurry from 1st (Aerated) Cell (B-EFF)	Lagoon B Bulk Slurry from 2nd (Anaerobic) Cell (B-OV)
pH	7.5	8.1	8.3	7.4	8	8.5
Soluble COD (mg/L)	1102	302	145	104	248	126
Dissolved Organic Carbon (mg/L)	385	117	73	359	80	42.7 ± 5.2
Ammonia-N (mg/L)	141	390	165	120	20	14
Nitrate-N (mg/L)	2.9	3.8	3.7	3.2	3.97	3
Nitrite-N (mg/L)	ND*	ND*	0.6	ND*	ND*	0.1
Alkalinity (mg/L)	513	1405	620	539	723	2351
Total Dissolved Solids (mg/L)	1370	2510	1360	1430	1570	680
Conductivity (uS)	2060	3760	2035	2140	2350	1030
$UV_{254\,nm}$ (m^{-1})	1.8	2.2	1.1	1.7	1.5	0.7
Specific UV Absorbance (L/mg-m)	0.5	1.9	1.6	0.5	1.8	1.5

Source: Qiang et al., *Ozone Science and Engineering* 26, 1–13, 2006. (With permission.)

* ND = Not Detected

anaerobic digesters may increase, however, as more emphasis is placed on energy recovery, odor control, and more effective waste treatment than is provided by open storage basins or lagoons.

14.3 SORPTION OF ANTIBIOTICS IN SWINE LAGOONS

Sorption of antibiotics to biosolids is an important mechanism that affects whether an antibiotic would likely be in the aqueous phase vs. sorbed to biosolids. Furthermore, antibiotics that strongly sorb to biosolids may tend to exist in the settled solids, while weakly sorbed antibiotics may tend to exist predominantly in the slurry. If a lagoon is not mixed prior to land application, only the antibiotics in the slurry may be predominantly introduced to the environment. If the lagoon is mixed prior to land application, all antibiotics in the lagoon may, in that case, be released to the environment.

The linear sorption coefficient (K_D) (L/kg) between an antibiotic and biosolids in a treatment process represents the concentration of an antibiotic sorbed ($\mu g/kg$) relative to its concentration in the liquid phase ($\mu g/L$). The linear K_D model is very often used to model sorption of pharmaceuticals in sediment, solids, and soils due to the linearity of isotherms at low adsorbate concentrations. Because there are a variety of mechanisms for sorption of pharmaceuticals on the organic and inorganic solids in treatment processes, prediction of K_D is complex. Sorption onto solids can involve a variety of mechanisms, including absorption into organic carbon, adsorption to mineral surfaces, ion exchange, and chemical reactions (Schwarzenbach et al., 2003). Similarly, equilibrium solubility of the antibiotic in the aqueous phase can be affected by many factors, including temperature, dissolved solids, and pH.

Kurwadkar et al. (2007) investigated the effects of sorbate speciation on the sorption of selected sulfonamides in three loamy soils. Sulfonamides predominantly exist as anions at pH levels above their respective pK_2 values (5.3–7.5) (Qiang and Adams, 2004), as neutral species at pH between their respective pK_2 and pK_1 values (1.9–2.1) (Qiang and Adams, 2004), and as cations below their respective pK_1 values. An effective K_D can be estimated using a weighted K_D value approach (Schwarzenbach et al., 2003; Kurwadkar et al., 2007), that is:

$$K_{D,effective} = \alpha_{cationic} \cdot K_{D,cationic} + \alpha_{neutral} \cdot K_{D,neutral} + \alpha_{anionic} \cdot K_{D,anionic}$$

where $K_{D,cationic}$, $K_{D,neutral}$, and $K_{D,anionic}$ are the linear partition coefficients for the cationic, neutral, and anionic species, respectively, and $\alpha_{cationic}$, $\alpha_{neutral}$, and $\alpha_{anionic}$ are the fractions of each species present at a specific pH. While this study addressed sorption to soils rather than biosolids, the same general principles are likely to apply in biosolids. Extrapolating to biosolids, sulfonamides may be expected to sorb much less at a higher pH (above their pK_2) due to the predominance of the anionic form and much more at a lower pH (between their pK_1 and pK_2) where the neutral form predominates and sorption to organic carbon in biosolids may be more significant.

Typical values for log K_D are 4 tetracyclines, 3 for tylosin, and 1 for sulfonamides (Loftin et al., 2004). Therefore tetracyclines and tylosin would tend to sorb strongly to settled biosolids in a lagoon, whereas sulfonamides may appear in the aqueous phase to a much larger degree.

Related work by Vieno et al. (2007) examined removal of antibiotics, as well as other pharmaceuticals, in a municipal wastewater treatment plant in Finland. The study concluded that the ciprofloxacin is readily "eliminated" from wastewater by sorption to biosolids due to high K_D or K_{OW} values (e.g., >4).

14.4   HYDROLYSIS OF ANTIBIOTICS IN LAGOONS

Antibiotics have opportunities to hydrolyze in storage basins, anaerobic lagoons, and other treatment systems. Hydrolysis studies by Loftin et al. (2007) were conducted in deionized lab water and filtered lagoon slurry as a function of pH (2–11), temperature (7 to 35°C), and ionic strength. This study showed that lincomycin (LNC), trimethoprim (TRM), sulfadimethoxine (SDM), sulfathiazole (STZ), sulfachlorpyridazine (SCP), and tylosin A (TYL) were recalcitrant to hydrolysis in lagoon slurry for pH 5, 7, and 9. At a higher pH of 11, limited hydrolysis of TYL was observed.

On the other hand, the tetracyclines—oxytetracycline (OTC), chlorotetracycline (CTC), and tetracycline (TET)—were all readily hydrolyzed under anaerobic lagoon conditions at pH levels of 5, 7, 9, and 11 (Figure 14.3). Researchers, including Loftin et al. (2007), have noted that a wide range of hydrolysis byproducts of the tetracyclines occur under different conditions, including epi-, iso-, epi-iso-, anhydro-, and epi-anhydro-analogues. More study is warranted of the partitioning behavior of these compounds to estimate their mobility relative to the corresponding parent tetracyclines. For a temperature of 22°C or greater, half-lives of OTC, CTC, and TET were 16 hours or less (Loftin et al., 2007). At colder temperatures (e.g., 7°C), nearly an order-of-magnitude slower hydrolysis was observed. Due to the significant seasonal temperature fluctuations observed in many swine lagoons, a wide range of hydrolysis rates for tetracyclines would be expected, depending on both temperature and pH. However, due to long holding times, on the order of months, complete hydrolysis to below detection would often be expected.

Tylosin underwent no hydrolysis in the pH range 5 to 9, but was readily degraded or labile at alkaline pH (>11) at temperatures of 22°C or greater. Thus, tylosin would not be expected to undergo appreciable hydrolysis in swine lagoon pH levels. These results suggest that tylosin, lincomycin, and the sulfonamides would be expected to be recalcitrant to abiotic degradation (hydrolysis) in swine lagoons. In warmer seasons or locations, oxytetratracycline and related compounds might be expected to hydrolyze to some greater or lesser degree.

14.5   BIOLOGICAL TREATMENT OF ANTIBIOTICS IN CONVENTIONAL SWINE TREATMENT SYSTEMS

When antibiotics enter a treatment lagoon, there are many potential transformation and partitioning reactions that can potentially occur. Transformation reactions include anaerobic or aerobic biodegradation depending on redox conditions, hydrolysis, and photolysis. Partitioning reactions for antibiotics in a common treatment lagoon are primarily to suspended and settled solids. In the subsequent sections, these potential removal mechanisms are discussed in more detail.

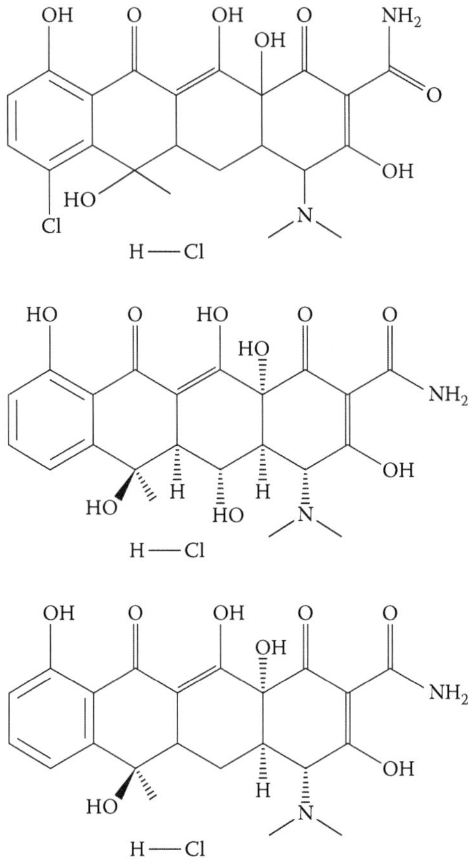

FIGURE 14.3 Structures of chlorotetracycline (top), oxytetracycline (middle), and tetracycline (bottom). (Courtesy of ChemFinder 2004, Cambridgesoft Corp.)

14.5.1 ANAEROBIC BIODEGRADATION

Very few studies have investigated the biodegradation of antibiotics in anaerobic swine lagoons. Anaerobic biodegradation in swine lagoon slurry was studied by Loftin et al. (2004) in laboratory microcosm experiments. In this work, soluble COD was readily removed. In these microcosms, sulfathiazole exhibited little degradation over a 2-month period (half-life = 222 days), suggesting that sulfathiazole would likely be biorecalcitrant under anaerobic conditions. This persistence also suggests overall concerns with sulfonamides in swine lagoons, that is, their presence in slurry applied to the environment. Similarly, lincomycin also was persistent with a half-life of 78 days in one slurry and no degradation in another.

Oxytetracycline, on the other hand, appears much more readily biodegradable under anaerobic conditions with half-lives of approximately 1 month in two different slurries. For example, with a 3-month treatment time, the concentrations of oxytetracycline decreased to only 12% of its initial value. Tylosine was observed to degrade even more readily under anaerobic conditions with a half-life of approximately 1 day in both swine lagoon slurries studies. Abiotic degradation rates were much slower (i.e., approximately 2 weeks) in the same autoclaved slurries.

Thus, the amount of anaerobic biodegradation treatment of antibiotics that would be expected is highly dependent on the nature of the antibiotics. While the sulfonamide and lincosamide were relatively recalcitrant, the tetracycline and macrolide were relatively easily degraded. More studies are needed in order to be able to realistically estimate and model the anaerobic biodegradation of antibiotics in swine lagoons.

14.5.2 Aerobic Biodegradation

Relatively little information is available on aerobic biodegradation of antibiotics in activated sludge, aerated lagoons, or other processes. Work by Ingerslev et al. (2001) suggests that antibiotics may generally degrade much more rapidly under aerobic conditions than under anaerobic conditions. They examined tylosin, oxytetracycline, metronidazole, and olaquindox in laboratory microcosms and demonstrated that for these antibiotics, aerobic biodegradation was significantly more rapid than anaerobic biodegradation.

The use of aerated lagoons, and aerated "caps" (an aerobic zone as the surface layer) on anaerobic lagoons, is a potentially viable option for more effectively treating antibiotics in swine wastewater treatment systems. An aerated "cap" is created by oxygenating the surface layer sufficiently to maintain dissolved oxygen at the surface on an otherwise anaerobic lagoon. More research in needed to better develop, optimize, and implement this technology.

14.5.3 Inhibition of Anaerobic Biodegradation by Antibiotics in Swine Lagoons

Swine manure consists of a complex mixture of fats, carbohydrates, and proteins. Anaerobic biodegradation of swine manure occurs by a series of metabolic steps, specifically: (1) conversion of complex organics to volatile fatty acids by fermentative organisms; (2) conversion of volatile fatty acids to acetate and hydrogen by fatty-acid oxidizing organisms; and (3) conversion of acetate and hydrogen to methane by methanogens (archea). Because antibiotics may often be present in swine manure and, hence, in the swine lagoon slurry, there is potential for the antibiotics to negatively impact the anaerobic biodegradation of other waste constituents in a lagoon.

Work by Loftin et al. (2005) investigated the inhibition of anaerobic biological activity in swine lagoon slurry in lab-scale microcosm experiments. This work monitored the impacts of varying concentrations of sulfonamides, tetracycline, lincomycin, and tylosin on the production of methane, hydrogen, and volatile fatty acids, including acetate.

These studies showed a significant (20 to 50%) inhibition of methane production for all of the antibiotics studied. Furthermore, antibiotic dosages of 1, 5, and 25 mg/L of a specific antibiotic all caused similar inhibitions, which in general plateaued at approximately 20 to 45%. Sanz et al. (1996) also saw a plateauing effect for ampicillin, novobiocin, penicillin, kanamycin, gentamicin, spectinomycin, streptomycin, tylosin, and tetracyclines over a wide range of inhibition (from 0 to 100%). This rapid plateauing in the inhibition of methane productions suggested that there exist certain bacterial subpopulations within the slurry that are greatly inhibited by antibiotics (even at low antibiotic concentrations [e.g., 1 mg/L]), while others are resistant to the effects of the antibiotics.

In the work by Loftin et al. (2004), no buildup of acetate or hydrogen was observed, suggesting that the methanogens were the microbial population most significantly inhibited as might most commonly be anticipated. Similarly, volatile fatty acids were not observed to build up in concentration, suggesting that the fatty-acid oxidizing organisms were not the most inhibited population. Thus this work suggested, but did not prove, that the fermentative organisms were the most significantly inhibited microbial population.

A key consequence of this observed inhibition of anaerobic metabolism in lagoons is that the presence of antibiotics may reduce the amount of manure degradation achieved in a swine waste treatment system. Furthermore, these findings suggest that if the amount of antibiotics entering a lagoon could be reduced then more effective treatment might be achieved. Finally this reduction in antibiotics could potentially be attained by pretreating the wastewater between a barn and the biological treatment system to remove antibiotics, or by reducing the application rate of antibiotics given to the swine. For example, by eliminating the use of antibiotics for growth promotion (where it is still practiced), the problem could possibly be minimized or reduced.

14.6   CHLORINE TREATMENT FOR ANTIBIOTICS IN SWINE WASTEWATER

14.6.1   ANTIBIOTIC REMOVAL

Chlorine treatment of wastewater from a barn, prior to discharge into a treatment process, is a potential means of removing antibiotics that could promote antibiotic-resistant bacterial growth within the lagoon, or may inhibit anaerobic activity within the lagoon. Chlorine treatment of treated wastewater (e.g., lagoon slurry) prior to field application is, similarly, a potential method for removing antibiotics, thereby preventing their introduction into the environment. Chlorine treatment prior to field application may also have potential for removing antibiotic-resistant bacteria, thereby preventing their release into the environment.

The pH has been shown to have a highly significant effect on the oxidation rates of selected antibiotics due to the speciation of the stronger hypochlorous acid (HOCl) to the weaker hypochlorite ion (OCl⁻) (Chamberlain and Adams, 2006). Because the acid dissociation constant for HOCl/OCl⁻ is approximately 7.6, at pH levels greater than 7.6, hypochlorite will be the most prevalent oxidant species of the two.

Oxidation of antibiotics and disinfection of antibiotic resistant bacteria by individual addition of free chlorine or monochloramine were studied by Qiang et al. (2006). The study looked at the oxidation of sulfonamides (sulfamethizole [SML], STZ, sulfamethazine [SMN], sulfamethoxazole [SMX], and SDM) in influents and effluents from two lagoon systems. Both lagoon systems had two cells in series, with one of the lagoons utilizing aeration in its first cell. In laboratory experiments, antibiotics were spiked and then treated with chlorine (or monochloramine) for three sample matrices: (1) the influent to the first cell from the barn (INF); (2) the effluent from the first cell and influent to the second cell (EFF); and (3) bulk slurry from the second cell of a swine lagoon system (OV), which is generally field applied. The characteristics of each stream for both systems are shown in Table 14.1.

Oxidation using free chlorine, or preformed monochloramine for comparison, was studied in detail to determine its effectiveness in removing of the sulfonamides. Addition of free chlorine ($HOCl/OCl^-$) to swine wastewater results in rapid conversion of the free chlorine to monochloramine, a much weaker oxidant and disinfectant than $HOCl$, even though it is similar to OCl^-. Thus, the ammonia concentration in swine waste plays a critical role in the oxidation of any antibiotics present due to its competition for oxidant, and its role in converting free chlorine to a weaker oxidant (monochloramine).

This work showed that the free chlorine concentration decreases very quickly after application due to reactions with ammonia and other wastewater constituents. The monochloramine formed thereafter decreases only slowly with time. These effects could be seen for the influent and effluent in the first lagoon of the nonaerated system, with the total chlorine concentration rapidly dropping from 500 mg/L to approximately 200 mg/L (as free chlorine), followed by a slow decline (as the monochloramine continued to react with wastewater constituents) (Figure 14.4B). As a result, rapid oxidation of the five sulfonamides was also observed during the initial

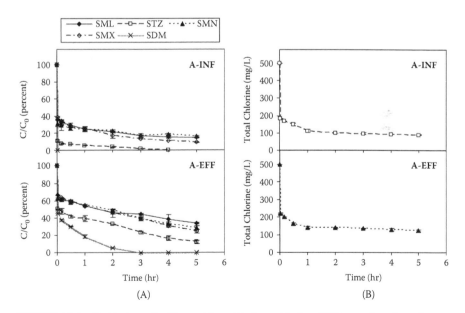

FIGURE 14.4 Oxidation of sulfonamides with free chlorine (FC) as a function of reaction time in the influent (A-INF) and effluent (A-EFF) of a swine lagoon system ("System A"): (A) decomposition of sulfonamides; (B) decay of total chlorine. Data are from duplicate runs (mean ± standard deviation). Experimental conditions: pH = 6.6; FC dose = 500 mg/L. (Reprinted with permission from Qiang et al., 2006. Copyright 2006 American Chemical Society.)

period in which free chlorine was dominant (Figure 14.4A). This was followed by slow further oxidation of the antibiotics by monochloramine of other oxidant species (Figure 14.4A).

The work by Qiang et al. (2006) suggested that application of chlorine dosages near the breakpoint (where all ammonia is converted to molecular nitrogen) is recommended for complete removal of both antibiotics and bacteria (including most antibiotic-resistant bacteria). The breakpoint generally occurs at a chlorine dosage near 7.6 mg/L of free chlorine per mg/L of ammonia present. This results in generally high chlorine dosages being required for oxidation of sulfonamides in swine wastewater. For example, chlorine dosages between 200 and 1000 mg/L of free chlorine were required to remove the five study sulfonamides (Figure 14.5). We hypothesize that significant concentrations of chlorinated oxidation byproducts would result from these oxidation reactions.

These results suggest that relatively high chlorine dosages could be used to remove antibiotics between a barn and the treatment lagoon and thereby reduce the antibiotics entering the lagoon. This could be accomplished in the pipe or conduit leading from the barn to the lagoon (or treatment system) with sufficient mixing and contact time. These results also suggest that lagoon slurry could be treated with chlorine in the pipe used to pump the slurry to adjacent fields for field application (in a plug flow mode) or in a tank used for lagoon slurry transport (in batch mode).

Free chlorine is rapidly consumed by ammonia and, therefore, limits the removal of antibiotics. The second two-cell swine lagoon (Lagoon B) studied by Qiang et al. (2006) had aeration in its first cell. The ammonia concentration in the effluent from the second cell was only 14 mg/L as NH_3-N (Table 14.1), while the ammonia concentrations were 120 and 200 mg/L as NH_3-N in the influent and effluent from the first cell, respectively. Therefore, much less chlorine was required to fully oxidize the ammonia in the slurry from both cells. A breakpoint dosage of approximately 100 mg/L of free chlorine was required to remove the 14 mg/L of ammonia (Figure 14.6A), although monochloramine was not fully removed until a larger dosage was administered. Furthermore, complete removal of all five sulfonamides studied was achieved with a dosage of 100 mg/L of chlorine, or less, in the second cell. Significantly larger chlorine dosages were required for the influent and effluent wastewater from the first lagoon due to the high ammonia concentrations (Figure 14.6B). For comparison, typical chlorine dosages for disinfecting septic tank effluent, and for municipally-treated activated sludge effluent, range from 20 to 60 mg/L and 2 to 30 mg/L as Cl_2, respectively (Metcalf and Eddy, 2003).

14.6.2 SIMULTANEOUS DISINFECTION

If achievable, it would be beneficial to reduce the introduction of antibiotic-resistant bacteria formed in swine, or in swine lagoons, prior to introduction into the environment. In one study by Qiang et al. (2006), free chlorine was observed to decrease the bacterial count (based on most probable number [MPN] methods) by approximately four orders of magnitude, with relatively low chlorine dosages (e.g., less than 50 mg/L) in both influent and effluent streams from a nonaerated lagoon (Figure 14.7).

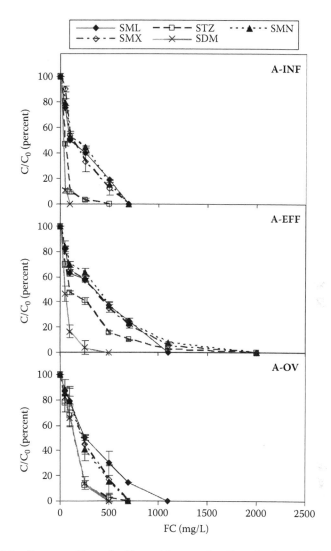

FIGURE 14.5 Decomposition of sulfonamides as a function of a free chlorine (FC) dose in the influent (A-INF) and effluent (A-EFF) from the first cell, and effluent (A-OV) from the second cell of a swine lagoon system ("System A"). Data are from duplicate runs (mean ± standard deviation). Experimental conditions: reaction time = 2.5 h, pH = 6.6. (Reprinted with permission from Qiang et al., 2006. Copyright 2006 American Chemical Society.)

However, much higher dosages of chlorine (up to 2000 mg/L) were not able to completely inactivate all bacteria present, even in filtered slurry.

This was suggestive that either (1) bacteria were shielded from chlorine disinfection by particles or (2) chlorine-resistant bacteria might be present in both the swine manure and swine lagoon slurry. While the shielding mechanism is well understood, further study by Macauley et al. (2006a) did find that chlorine-resistant bacteria were indeed present and appeared closely related to *Bacillus subtilis* and *Bacillus licheniformis*.

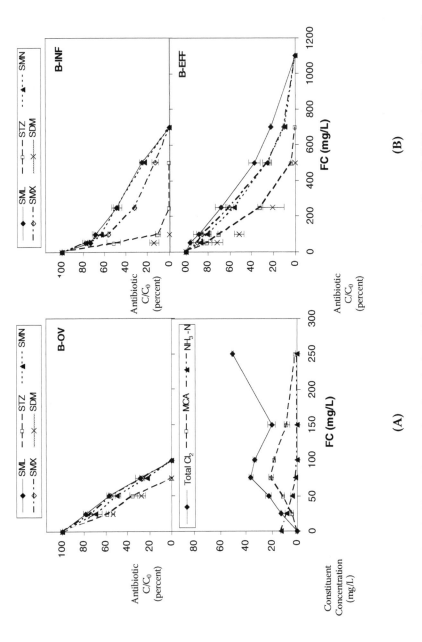

FIGURE 14.6 Decomposition of sulfonamides as a function of a free chlorine (FC) dose in the influent (B-INF) and effluent (B-EFF) from the first (aerated) cell, and effluent (B-OV) from the second cell. (Qiang et al., 2006. With permission)

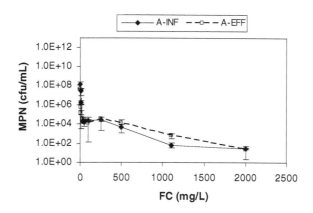

FIGURE 14.7   Inactivation of bacteria in the influent and effluent of a swine lagoon system (A) with free chlorine (FC). (Qiang et al., 2006. With permission.)

14.6.3   COMPARISON OF SELECTED CLASSES OF ANTIBIOTICS

In general, sulfonamides and lincomycin are expected to be persistent in lagoon slurry—that is, resistant to biological and abiotic degradation. Furthermore, sulfonamides are expected to sorb only weakly to lagoon solids. Thus, sulfonamides may pose an especially significant risk (with respect to exposure) due to their presence in lagoon slurry that is frequently discharged into the environment by field application. On the other hand, tetracyclines and tylosin appear to sorb more strongly to lagoon solids and to degrade more rapidly than the sulfonamides. Thus, tetracyclines and tylosin may not pose as significant a threat to the environment as these other compounds due to reduced concentrations in lagoon slurry (at least with respect to exposure).

While this treatment study focused on the oxidation of sulfonamides, clean water studies of the oxidation of other antibiotics with chlorine and chloramine provide some insight as to the relative efficiency for oxidation of other antibiotics. In work by Chamberlain and Adams (2006), carbodox and the macrolides (roxithromycin, erythromycin, and tylosin) were much more rapidly and fully oxidized by chlorine than were the sulfonamides. This rapid oxidation was observed over a wide pH range. Thus, while not tested, removal of these other classes of antibiotics in this manner may, in fact, be more efficient than when used for the sulfonamides.

14.7   OTHER TREATMENT APPROACHES

Physicochemical treatments other than chlorine oxidation may be potentially effective for treating antibiotics in swine waste, though few have been studied in detail. Chemical treatments that merit more study might include ozone, chlorine dioxide, and ultraviolet (UV) oxidation.

Dodd et al. (2006) studied ozonation of antibiotic compounds in secondary effluent from a municipal wastewater treatment plant. They found that a dosage of 3 mg/

L or greater of ozone was able to remove at least 99% of the "fast-reacting antibiotics" (i.e., with second-order rate constants of at least $5(10^4)$ L·mole^{-1}·sec^{-1}). A strong pH dependency was observed for the removal of many compounds due to speciation of the antibiotic or the hydroxide-ion-catalyzed decomposition of ozone to the less selective oxidant, hydroxyl radical. At pH 7.7 (typical for common anaerobic lagoons), Dodd et al. (2006) found that the "fast-reacting" antibiotics included macrolides (i.e., roxithromycin, azithromycin, tylosin), sulfonamides (i.e., sulfamethoxazole), fluoroquinolones (i.e., ciprofloxacin, enrofloxacin), trimethoprim, lincomycin, β-lactams (i.e., cephalexim), tetracycline, and vancomycin. Other antibiotics, including N(4)-acetylsulfamethoxazole, penicillin G, and amikacin, were not as readily removed (Dodd et al., 2006).

Macauley et al. (2006b) examined the disinfection of swine wastewater using ozone. Their results showed that bacterial disinfection based on MPN varied greatly between different anaerobic lagoons at swine CAFOs. For example, an ozone dosage of 20 mg/L resulted in almost no disinfection in one lagoon but nearly 2-log reduction in another lagoon. In both lagoon slurries studied, a relatively high ozone dosage of 100 mg/L resulted in 3.3- to 3.9-log reduction in MPN.

Huber et al. (2005) studied the oxidation of various pharmaceuticals using chlorine dioxide. Their work showed that, for a wide range of antibiotics (and other pharmaceuticals), the kinetic rate constants with chlorine dioxide were approximately two orders of magnitude lower than those for ozone and generally higher than those for chlorine. For the compounds examined in the Huber et al. (2005) study, the reactivity with chlorine dioxide was similar to that with chlorine. These antibiotics included both sulfamethoxazole and roxithromycin. More study is needed to establish its viability of using chlorine dioxide as a disinfectant in swine wastewater.

Macauley et al. (2006b) showed that UV was effective for disinfecting bacteria from swine lagoon wastewater. Macauley et al. (2006b) did not, however, report the concurrent removal of antibiotics during UV disinfection. Earlier work by Adams et al. (2002), however, demonstrated that UV dosages of 100 times greater than what is typically used for disinfection resulted in only a 50 to 80% reduction in the concentrations of seven common antibiotics (i.e., carbadox, sulfachlorpyridazine, sulfadimethoxine, sulfamerazine, sulfamethazine, sulfathiazole, and trimethoprim) in both laboratory water and filtered surface water. A key factor in this limited removal was the competitive absorbance of the UV radiation between the background constituents and the antibiotics themselves. In a swine lagoon, even less removal would generally be expected due to the relatively high UV absorbance of lagoon slurry or wastewater, as compared to drinking waters.

Also, a variety of membrane systems could be developed that might remove antibiotics present in swine wastewater. However, due to the costs associated with membrane systems, this approach can only be imagined to have limited application in large-scale animal agriculture in the current economic and regulatory climate.

14.8 CONCLUDING REMARKS

The most promising approaches to limiting the discharge of antibiotics into the environment during land application appear to include: (1) reducing of the use of antibiotics in animal agriculture, especially as it relates to growth promotion; (2) switching, where possible, from antibiotics that are more difficult to treat (e.g., sulfonamides) to more easily treatable options; and (3) using chlorination (especially after aerobic biological ammonia removal). Much more research is needed to fully address the feasibility of each of these (and other) options.

REFERENCES

Adams, C., Wang. Y., Loftin, K., and Meyer, M. (2002) Removal of antibiotics from surface and distilled water in conventional water treatment processes. *J. Environ. Engin.* 128:3, 253–260.

Chamberlain, E. and Adams, C. (2006) Oxidation of sulfonamides and macrolides with free chlorine and monochloramine. *Water Res.* 40, 2463–2592.

Dodd, M., Buffle, M.-O., and von Gunten, U. (2006) Oxidation of antibacterial molecules by aqueous ozone: moiety-specific reaction kinetics and applications to ozone-based wastewater treatment. *Environ. Sci. Technol.* 40, 1969–1977.

Huber, M., Korhonen, S., Ternes, T., and von Gunten, U. (2005) Oxidation of pharmaceuticals during water treatment with chlorine dioxide. *Water Research* 39, 3607–3617.

Ingerslev, F., Toräng, L., Loke, M-L, Halling-Sørensoen, B., and Nyholm, N. (2001) Primary biodegradation of veterinary antibiotics in aerobic and anaerobic surface water simulation systems. *Chemosphere* 44, 865–872.

Kurwadkar, S., Adams, C., Meyer, M., and Kolpin, D. (2007) Effects of sorbate speciation on sorption of selected sulfonamides in three loamy soils. *J. Agric. Food Chem.* 55, 1370–1376.

Loftin, K., Adams, C., and Surampalli, R. (2004) The fate and effects of selected veterinary antibiotics on anaerobic swine lagoons. *Water Environment Federation's 77th Annual Technical Exhibition and Conference (WEFTEC)*, Washington, D.C., (October, 2004).

Loftin, K., Adams, C., Meyer, M., and Surampalli, R. (2007) Effects of ionic strength, temperature, and pH on degradation of selected antibiotics (in review in *J. Environmental Quality*).

Loftin, K., Henny, C., Adams, C., Surampali, R., and Mormile, M. (2005) Inhibition of microbial metabolism in anaerobic lagoons by selected sulfonamides, tetracyclines, lincomycin, and tylosin tartrate. *Environ. Toxicol. Chem.* 24:4, 782–788.

Macauley, J., Adams, C., and Mormile, M. (2006a) Presence and absence of known tet-resistant genes in tetracycline resistant isolates from an anaerobic swine lagoon (in review at *Canadian J. Microbiol.*).

Macauley, J., Qiang, Z., Adams, C., Surampalli, R., and Mormile, M. (2006b) Disinfection of swine wastes using chlorine, ultraviolet light, and ozone. *Water Research* 40, 2017–2026.

Metcalf and Eddy (2003) *Wastewater Engineering: Treatment and Reuse,* 4th Edition, (Ed. Tchobanoglous, Burton, and Stensel), McGraw-Hill, New York.

Miner, J.R., Humenik, F., and Overcash, M. (2000) *Managing Livestock Wastes to Preserve Environmental Quality*, Iowa State University Press, Ames, IA.

Qiang, Z. and Adams, C. (2004) Potentiometric determination of acid dissociation constants (pKa) for human and veterinary antibiotics. *Water Res.* 38, 2874–2890.

Qiang, Z., Adams, C., and Surampalli, R. (2004) Determination of ozonation rate constants for lincomycin and spectinomycin. *Ozone Sci. Eng.* 26, 1–13.

Qiang, Z., MacCauley, J., Mormile, M., Surampalli, R., and Adams, C. (2006) Treatment of antibiotics and antibiotic resistant bacteria in swine wastewater with free chlorine. *J. Agric. Food Chem.* 54, 8144–8154.

Sanz, J., Rodriguez, N., and Amils, R. (1996) The action of antibiotics on the anaerobic digestion process. *App. Microbiol. Biotechnol.* 46, 587–592.

Schwarzenbach, R., Gschwend, P., and Imboden, D. (2003) *Environmental Organic Chemistry*, 2nd Edition, Wiley-Interscience, Hoboken, NJ.

Vieno, N., Tuhkanen, T., and Kronberg, L. (2007) Elimination of pharmaceuticals in sewage treatment plants in Finland. *Water Res.* 41, 1001–1012.

15 Removal of Pharmaceuticals in Biological Wastewater Treatment Plants

Sungpyo Kim, A. Scott Weber,
Angela Batt, and Diana S. Aga

Contents

15.1 INTRODUCTION

Providing sufficiently clean water to the public has become a challenging issue worldwide as the quality of water sources increasingly deteriorates.[1] As a consequence, treated wastewater has attracted attention as an alternative water resource, provided appropriate treatment can be applied.[2] Therefore, the removal of microcontaminants, such as pharmaceuticals and personal-care products, in wastewater is critical because many of these compounds survive conventional treatment.[3] In general, these compounds are present at parts per billion (ppb) levels or less in wastewater.[4,5] Although these concentrations are much lower than the levels of traditionally known organic pollutants (such as the persistent organic pollutants DDT, PCBs, and the like) the potential long-term effects of these compounds to humans and wildlife cannot be neglected. For example, several studies have shown that even parts per trillion (ppt) levels of ethinyl estradiol (the active ingredients of birth control pills) and natural estrogens can disrupt the hormone system of aquatic species.[6,7] In addition, low levels of antibiotics from the effluents of wastewater treatment plants (WWTPs) can promote antibiotic resistance in microorganisms that are exposed constantly to these compounds.[8–10]

349

Residues of human and veterinary pharmaceuticals are introduced into the environment via a number of pathways but primarily from discharges of wastewater treatment plants or land application of sewage sludge and animal manure. Most active ingredients in pharmaceuticals are transformed only partially in the body and thus are excreted as a mixture of metabolites and bioactive forms into sewage systems. Although WWTPs remove some pharmaceuticals during treatment,[4,11,12] removal efficiencies vary from plant to plant. Despite recent investigations documenting the occurrence of pharmaceuticals in the environment, important information on their fate and long-term effects is still lacking.[13]

15.2 FATE OF PHARMACEUTICALS IN BIOLOGICAL WASTEWATER TREATMENT PROCESS

Some antibiotics and other pharmaceutical compounds in wastewater can be reduced or eliminated in biological wastewater treatment systems using the activated sludge process, which is the most commonly used wastewater treatment process in the world. Wastewater treatment process generally consists of a primary, secondary, and sometimes an advanced treatment stage, with different biological, physical, and chemical processes available for each stage of treatment. A schematic diagram of a typical WWTP that employs the activated sludges for secondary treatment is shown in Figure 15.1. The primary treatment stage generally utilizes physical treatment, such as screens and a gravity settling process, typically referred to as sedimentation, to remove the solid contents in wastewater. Secondary treatment, which typically relies on microorganisms to biodegrade organic matter and/or other nutrients, can differ substantially. In some wastewater treatment facilities, the effluent also is disinfected before it is released into the environment, typically by chlorination or ultraviolet (UV) radiation. In addition, advanced waste treatment processes can be applied to remove nitrogen, phosphorus, and other pollutants or particles.[14]

Recent reports demonstrate that conventional WWTPs are not capable of removing pharmaceutical contaminants under typical operating conditions, which results in a discharge of these compounds into surface waters.[5,15–26] Accordingly, WWTPs are important point sources for antibiotic contamination of surface waters. [4,27–30]

15.3 PHARMACEUTICAL REMOVAL MECHANISMS

Several laboratory studies have been conducted to assess the efficiencies of various treatment technologies in removing antibiotics and other pharmaceuticals from

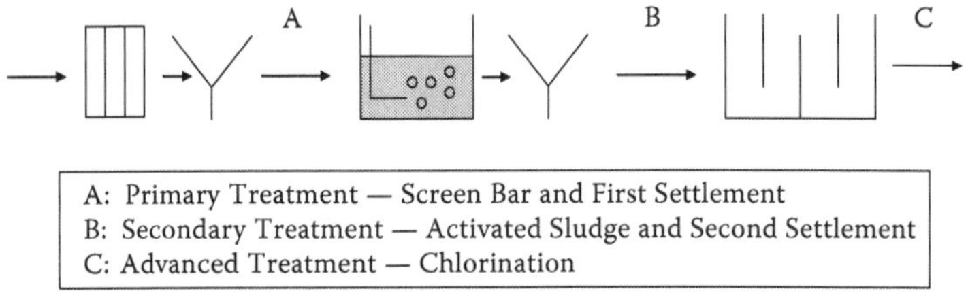

A: Primary Treatment — Screen Bar and First Settlement
B: Secondary Treatment — Activated Sludge and Second Settlement
C: Advanced Treatment — Chlorination

FIGURE 15.1 A schematic diagram of biological wastewater treatment process.

wastewater.[31-33] The primary pollutant removal mechanisms in conventional biological wastewater treatment processes are biodegradation and sorption by microorganisms. Therefore, it is reasonable to assume that the key component in biological WWTPs responsible for the removal of pharmaceutical pollutants is the aeration basin containing the microorganisms (activated sludge). Biodegradation and sorption also could take place in other unit processes, such as primary settling, but removal efficiencies at this stage are difficult to control.[14] Other removal mechanisms such as volatilization (due to aeration) or photodegradation (due to sunlight) are either negligible or nonexistent.[34] Disinfection processes, such as chlorination or UV treatment, which are intended to remove pathogens, not only reduce drug-resistant bacteria but may also contribute in the elimination of some pharmaceuticals in wastewater. However, not all WWTPs include a disinfection step, many facilities only disinfect treated effluents seasonally, and several studies reported that disinfection does not effectively remove a wide range of antibiotics.[18,35] Accordingly, in this chapter we will limit our discussion on the pharmaceutical removal by biodegradation and sorption. However, Chapters 10, 11, and 12 in this book examine the efficiencies of various disinfection processes in the removal of pharmaceuticals in drinking water.

15.3.1 Biodegradation and Biotransformations

During biological degradation in WWTPs, pharmaceutical contaminants could undergo (1) mineralization; (2) transformation to more hydrophobic compounds, which partition onto the solid portion of the activated sludge; and (3) transformation to more hydrophilic compounds, which remain in the liquid phase and are eventually discharged into surface waters.[13,36]

Despite the wide consortium of microorganisms present in the activated sludge, it is unlikely that pharmaceuticals present as microcontaminants in wastewater can be effectively removed by biodegradation alone. First, the relatively low concentration of pharmaceuticals relative to other pollutants in wastewater may be insufficient to induce enzymes that are capable of degrading pharmaceuticals.[3] Second, many of these compounds are bioactive, which can inhibit growth or metabolism of microorganisms. Thus, it is unlikely that pharmaceuticals will be favorable energy or carbon sources for microorganisms. Third, the degree of biodegradation will depend on the nature of each compound and on the operating conditions employed in WWTPs.

Joss et al.[34] provided a comprehensive and intensive study investigating the biodegradation of pharmaceuticals, hormones, and personal-care products in municipal wastewater treatment. Target compounds included antibiotics, antidepressants, antiepileptics, antiphlogistics, contrast agents, estrogens, lipid regulators, nootropics, and fragrances. Among them, only 4 (ibuprofen, paracetamol, 17β-estradiol, and estrone) of the 35 compounds studied were degraded by more than 90%, while 17 compounds (including macrolides and sulfonamides) were removed by less than 50% during biological wastewater treatment. The biodegradation of sulfonamides[31,33] and trimethoprim[33] has been evaluated in batch reactors, and they were found to be non-readily biodegradable and have the potential to persist in the aquatic environments.

Many biodegradation studies only report the disappearance of the parent compounds but do not elucidate the formation of metabolites, which also may be persistent

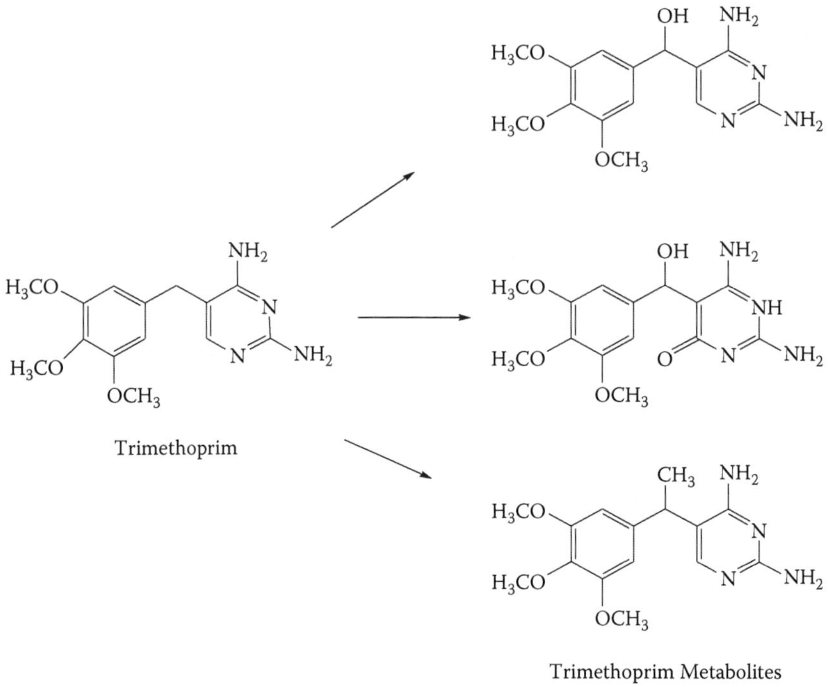

FIGURE 15.2   Trimethoprim and its biodegradation metabolites.

and may have similar ecotoxicological effects. Recent studies that attempted to identify the byproducts of biodegradation in wastewater indicated that metabolites of many pharmaceuticals are not very different from their parent compounds. For example, Ingerslev and Halling-Sørensen[31] reported biodegradation of several sulfonamide antibiotics in activated sludge, based on their disappearance over time, but quantities and identities of transformation product were not reported. In another study, the reported metabolites[37] of the antibiotic trimethoprim in activated sludge have structures that are only slightly modified compared to the parent compound (Figure 15.2). Whether these metabolites still exhibit the antibacterial activity of the parent compound or not remains to be tested. The lack of sensitive analytical tools able to detect low concentrations of unknown compounds in complex matrices has unfortunately limited the identification of pharmaceutical metabolites formed during biodegradation in WWTPs. This is a critical research need because risk assessments based only on the presence of parent compounds in wastewater could lead to underestimation of their risks to the aquatic environment.

15.3.2   Sorption

It is important to note that the main removal mechanism of some recalcitrant pharmaceuticals in biological WWTPs is sorption on activated sludge, rather than biodegradation. Sorption in WWTPs is more likely an adsorption process, which is the physical adherence onto activated sludge or bonding of ions and molecules onto the surface of microorganisms or microbial flocs. For example, ciprofloxacin and tetracycline are removed mainly by sorption to sludge.[38,39] A study was conducted by Kim et al.[39] to examine the relative importance of biodegradation and sorption in the removal of tetracycline in activated sludge. The similarity in the concentration profiles shown in Figure 15.3 obtained from two types of bioreactors, one of which was amended with 0.1% sodium azide to suppress microbial activity, reveals that tetracycline concentration decreases over time even in the "activity-inhibited" control conditions. This suggests that the decrease in concentration was not due to biodegradation. In fact, chemical analysis of the aqueous phases from these bioreactors showed no biodegradation products being formed. From this biodegradability test, the strong similarity between inhibited (spiked with tetracycline + 0.1% NaN$_3$) and noninhibited biomass (spiked with tetracycline only), and the lack of tetracycline metabolites, suggests that sorption is the primary mechanism for tetracycline removal in activated sludge.

The sorption isotherm of tetracycline on activated sludge was determined and is presented in Figure 15.4.[39] The calculated K_{ads} was 8400 ± 500 mL/g (standard error of slope). This adsorption coefficient in activated sludge is approximately three times that reported for the more polar oxytetracycline (3020 mL/g) and much higher than that of ciprofloxacin (K_d = 416.9 mL/g),[40] which was found to be 95% associated with the sludge or biosolids.[38] Therefore, it is reasonable to assume that tetracycline is mostly adsorbed in the activated sludge.

A study was conducted to compare the sorption kinetics of four selected antibiotics in activated sludge. To inhibit microbial activity, sodium azide was added into the mixed liquor. Also, caffeine was spiked into the test system to serve as indicator of residual biological activity. (Caffeine is known to be readily biodegradable and has no measurable sorption to sludge.) The experiment was conducted using 3600 mg/L of mixed liquor suspended solid (MLSS) obtained from a local municipal

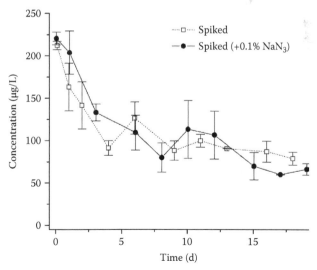

FIGURE 15.3   Removal of tetracycline under batch-activated sludge conditions with active and "activity-inhibited" biomass. (Reactor spiked with 200 µg/L.)

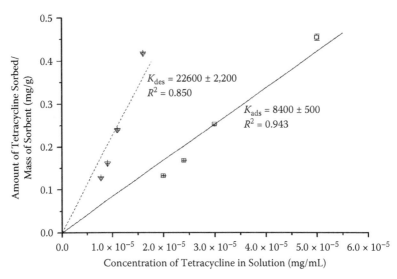

FIGURE 15.4 Adsorption and desorption isotherms for tetracycline on activated sludge. K_{ads}: sorption coefficient, K_{des}: desorption coefficient. (Error bars correspond to one standard deviation.)

WWTP. While more than 80% of ciprofloxacin and tetracycline were removed in the dissolved phase after a 5-h equilibration, only less than 20% of the more hydrophilic sulfamethoxazole and trimethoprim were removed (Figure 15.5). Removal of caffeine was not observed, indicating that the biological activity of the sludge was inhibited by the addition of azide. It can be inferred from these results that sorption is an important removal mechanism for both ciprofloxacin and tetracycline but not for sulfamethoxazole and trimethoprim. Sorption of antibiotics on activated sludge that is eventually land applied poses a special concern because these antibiotics may remain biologically active and thus have the potential to influence selection of antibiotic-resistant bacteria in the terrestrial environment.

15.4 INFLUENCE OF WASTEWATER TREATMENT PLANT (WWTP) OPERATING CONDITIONS

The performance efficiency of a biological WWTP depends highly on the operating conditions and design and may be affected by disturbances, such as high concentrations of pharmaceuticals or potentially toxic chemicals in influent wastewater. Usually the flow rate and pollutant concentrations of wastewater are time dependent and hard to control. For example, the antibiotic concentrations in a composite sample and in grab samples were compared in two different WWTPs (Amherst, New York, and Holland, New York). The populations served by Amherst and Holland WWTPs are 115,000 and 1,750, respectively. Grab and composite samples were obtained twice during the day (8 A.M. and 4 P.M.). Both grab and composite samples were analyzed for four selected antibiotics (ciprofloxacin, sulfamethoxazole, tetracycline, trimethoprim). Figure 15.6 shows the difference in concentration (% variation) for each antibiotic in the grab samples relative to the concentrations obtained in composite

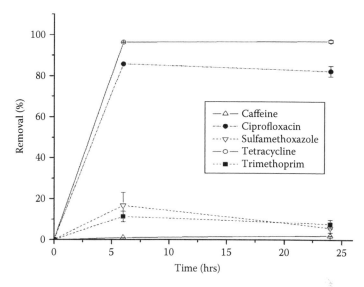

FIGURE 15.5   Antibiotic removal by adsorption in activated sludge. (Caffeine was used as marker for residual biological activity.)

samples. There is high variability in the antibiotic concentrations during the two sampling times in the smaller WWTP (Holland), showing up to 70% difference in concentrations between the grab samples and the composite samples. Joss et al.[41] also reported higher pharmaceutical loads in daytime composite samples (8:00 to 16:00) as compared with other sampling times. This trend is similar to the characteristics of conventional pollutant indicators such as suspended solid (SS) or biological oxygen demand (BOD).[14] This observation suggests that these conventional parameters may be good indicators for predicting the load of antibiotics in WWTPs.

Some degree of biological WWTP removal efficiency can be controlled by operating parameters such as solid retention time (SRT) and hydraulic retention time (HRT). Several studies reported that biological wastewater treatment processes with higher solids retention time (SRT) (>10 days) tend to have better removal efficiencies for pharmaceutical compounds compared to lower SRT processes.[35,42,43] This observation implies that there is an enhanced biodegradation ability or different sorption capacity for microcontaminants in sludge with a higher SRT.

Kim et al.[39] reported the influence of HRT and SRT on the removal of tetracycline in the activated sludge processes, using a sequencing batch reactor (SBR) spiked with 250 µg/L of tetracycline. Three different operating conditions were applied during the study (Phase 1—HRT: 24 h, SRT: 10 d; Phase 2—HRT: 7.4 h, SRT: 10 d; Phase 3—HRT: 7.4 h, SRT: 3 d). The removal efficiency of tetracycline in Phase 3 (78.4 ± 7.1%) was significantly lower than that observed in Phase 1 (86.4 ± 8.7%) and Phase 2 (85.1 ± 5.4%) at the 95% confidence level. The reduction of SRT in Phase 3 while maintaining a constant HRT decreased tetracycline sorption, resulting in decreased removal. To date, there is little evidence in the literature to suggest biodegradation as a likely removal mechanism for tetracycline. Because of the high sorption of tetracycline in sludge, the influence of SRT on the sorption behavior of

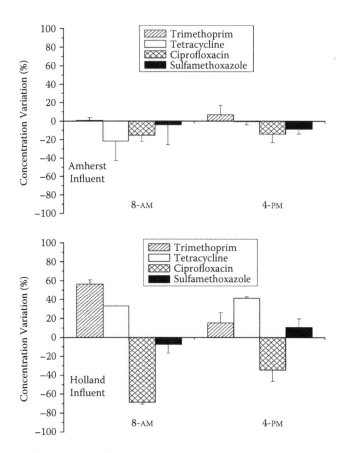

FIGURE 15.6 Variation of antibiotic concentrations during the day. (Error bars correspond to one standard deviation.)

pharmaceuticals is of interest, as sorption characteristics of the biomass may change with SRT. Several researchers have observed increased biomass hydrophobicity at higher SRTs.[44,45] In fact, recent work by Harper and Yi[46] has shown that bioreactor configuration can have a significant influence on biomass hydrophobicity and particle size, which can affect the bioavailability and fate of pharmaceuticals in WWTPs because of their impact on particle floc characteristics. Even though tetracycline has a low n-octanol/water partition coefficient, at certain pH values, hydrophobic interactions still play a role for the sorption of tetracycline on soil or clay.[47]

Batt et al.[48] explored the occurrence of ciprofloxacin, sulfamethoxazole, tetracycline, and trimethoprim antibiotics in four full-scale WWTPs. The WWTPs chosen utilized a variety of secondary removal processes, such as a two-stage activated sludge process with nitrification, extended aeration, rotating biological contactors, and pure oxygen activated sludge. In all four WWTPs, the highest reduction in antibiotic concentrations was observed after the secondary treatment processes, which is where the majority of the organic matter is eliminated and therefore is the most important processes for antibiotic removal. The extended aeration combined with ferrous chloride precipitation utilized at the East Aurora, New York, plant proved to be the most effective of the WWTP designs examined in terms of the overall removal of the four antibiotics. Extended aeration operates with the longest HRT among all the processes investigated (28 to 31 hours as opposed to 1 to 4 hours). Higher overall removal was observed at the Amherst, New York, plant than the remaining two (Holland, New York, and Lackawana, New York), with the second-stage activated sludge process at the Amherst operating with the longest SRT of the investigated WWTPs.

The enhanced nitrification activity under long SRT has been suggested to play an important role in the increased removal of micropollutants, such as pharmaceuticals, in WWTPs.[33,37,43] It was noted that ammonia oxidizing bacteria (AOB) can cometabolize various polyhalogenated ethanes[49] and monocyclic aromatic compounds.[50] It also has been reported that trimethoprim antibiotic can be removed more effectively in nitrifying activated sludge (high SRT) compared with that in conventional activated sludge (short SRT).[33,37,51] Batt et al.[51] investigated the fate of iopromide and trimethoprim under lab-scale nitrifying activated sludge. A significantly higher biodegradation of both iopromide and trimethoprim was observed in the bioreactor where the activity of nitrifying bacteria was not inhibited (Batch-1), relative to the bioreactor where nitrification was inhibited by addition of allylthiourea (Batch-2) (Figure 15.7). These results provide strong evidence that nitrifying bacteria play a key role in enhancing the biodegradation of pharmaceuticals in WWTPs.[33,37] It appears that prolonging SRT to achieve stable nitrification in the activated sludge has an added benefit of increasing the removal efficiencies of microcontaminants. A similar observation relating SRT and percent removal was reported recently for

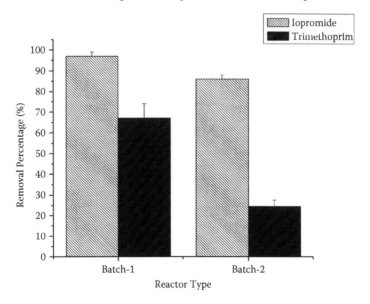

FIGURE 15.7   Iopromide and trimethoprim removal in nitrifying activated sludge: without amonia oxidizing material (AOB) inhibition (Batch-1), and with AOB inhibition (Batch-2). (Error bars correspond to one standard deviation.)

other pharmaceuticals and personal-care products in full-scale WWTPs with varying SRTs.[43]

15.5 FINAL REMARKS

It is clear from the existing knowledge that attention is needed on optimizing WWTP operation to achieve maximum removal efficiencies of pharmaceuticals in wastewater. It is known that current WTTP designs do not eliminate many micropollutants completely. While various treatment processes in drinking water production (such as activated carbon, ozonation, and membrane technologies) are effective in reducing concentration of micropollutant,[32] these technologies are not easily affordable at many municipal WWTP facilities. Therefore, prolonging SRT in WWTPs may be a simple solution to reduce the concentrations of pharmaceuticals in treated wastewater.

Knowledge of the identities of metabolites, particularly those that are similar in structures to the parent pharmaceuticals and are persistent in the environment, is critical. For a complete risk assessment of pharmaceuticals in the environment, it will be necessary to consider persistent transformation products in the equation because these compounds may pose their own ecological risks. To date, ecotoxicity data on metabolites and mixtures of pharmaceuticals are scarce.

In a study that aimed at removing organic contaminants from an industrial wastewater treatment, it was found that despite the complete removal of the only known toxic contaminant (diethanolamine) in the wastewater, the toxicity of the biologically treated effluents was higher than what was calculated based on the removal efficiency of the total organic carbon.[52] This implies that the majority of the observed effects after biological treatment must be due to the formation of metabolites, which were not identified in the study. In another study example, the photodegradation product of the diuretic drug furosemide was found to be more mutagenic than its parent compound,[53] further demonstrating the importance of considering byproducts in toxicity testing and risk assessment.

Based on a more realistic ecological and human health risk assessment, current water quality standards can be updated to set acceptable levels of micropollutants that determine how "clean" water should be before it can be discharged into the environment. This is particularly critical for recycled wastewater that will be used as a potable water resource. Reduction of pharmaceutical contaminants at the source (effluent of WWTP) is obviously needed if recycled water is to become a significant part of our domestic water supply.

REFERENCES

1. Saeijs, H.L. and Van Berkel, M.J., Global water crisis: the major issue of the 21st century, a growing and explosive problem, *Eur. Water Poll.Cont.*, 5, 26, 1995.
2. Papaiacovou, I., Case study—Wastewater reuse in limassol as an alternative water source, *Desalination*, 138, 55, 2001.
3. Sedlak, D.L., Gray, J.L., and Pinkston, K.E., Understanding microcontaminants in recycled water, *Environ. Sci. Technol.*, 34, 509A, 2000.

4. Glassmeyer, S.T. et al., Transport of chemical and microbial compounds from known wastewater discharges: potential for use as indicators of human fecal contamination, *Environ. Sci. Technol.*, 39, 5157, 2005.
5. Kolpin, D.W. et al., Pharmaceuticals, hormones, and other organic wastewater contaminants in U.S. streams, 1999–2000: A national reconnaissance, *Environ. Sci. Technol.*, 36, 1202, 2002.
6. de Mes, T., Zeeman, G., and Lettinga, G., Occurrence and fate of estrone, 17-estradiol and 17-ethynylestradiol in STPs for domestic wastewater, *Rev. Environ. Sci. Biotechnol.*, 4, 275, 2005.
7. Sumpter, J.P. and Johnson, A.C., Response to comment on "lessons from endocrine disruption and their application to other issues concerning trace organics in the aquatic environment," *Environ. Sci. Technol.*, 40, 1086, 2006.
8. Walter, M.V. and Vennes, J.W., Occurrence of multiple-antibiotic enteric bacteria in domestic sewage and oxidation lagoons, *Applied Environ. Micro.*, 930, 1985.
9. Guardabassi, L. et al. Antibiotic resistance in *Acinetobacter spp.* isolated from sewers receiving waste effluent from a hospital and a pharmaceutical plant, *Applied Environ. Micro.*, 3499, 1998.
10. Kim, S. et al., Tetracycline as a selector for resistant bacteria in activated sludge, *Chemosphere*, 66, 1643, 2007.
11. Hirsch, R. et al., Occurrence of antibiotics in the aquatic environment, *Sci. Total Environ.*, 225, 109, 1999.
12. Carballa, M. et al., Behavior of pharmaceuticals, cosmetics and hormones in a sewage treatment plant, *Water Res.*, 38, 2918, 2004.
13. Halling-Sørensen, B. et al., Occurrence, fate and effects of pharmaceutical substances in the environment—A review, *Chemosphere*, 36, 357, 1998.
14. Grady, C.P.L., Daigger, G.T., and Lim, H.C., *Biological Wastewater Treatment*, 2nd ed., Marcel Dekker, New York, 1999.
15. Daughton, C.G. and Ternes, T.A., Pharmaceuticals and personal care products in the environment: agents of subtle change?, *Environ. Health Perspec. Suppl.*, 107, 907, 1999.
16. Stumpf, M. et al., Polar drug residues in sewage and natural waters in the state of Rio de Janeiro, Brazil, *Sci. Total Environ.*, 225, 135, 1999.
17. Jorgensen, S.E. and Halling-Sorensen, B., Drugs in the environment, *Chemosphere*, 40, 691, 2000.
18. Jones, O.A.H., Voulvoulis, N., and Lester, J.N., Human pharmaceuticals in the aquatic environment—A review, *Environ. Technol.*, 22, 1383, 2001.
19. Kümmerer, K., Drugs in the environment: emission of drugs, diagnostic aids and disinfectants into wastewater by hospitals in relation to other sources—A review, *Chemosphere*, 45, 957, 2001.
20. Giger, W. et al., Occurrence and fate of antibiotics as trace contaminants in wastewaters, sewage sludges, and surface waters, *Chimia Int. J. Chem.*, 57, 485, 2003.
21. Gobel, A. et al., Occurrence and sorption behavior of sulfonamides, macrolides, and trimethoprim in activated sludge treatment, *Environ. Sci. Technol.*, 39, 3981, 2005.
22. Lindberg, R.H. et al., Screening of human antibiotic substances and determination of weekly mass flows in five sewage treatment plants in Sweden, *Environ. Sci. Technol.*, 39, 3421, 2005.
23. Ternes, T.A., Joss, A., and Siegrist, H., Scrutinizing pharmaceuticals and personal care products in wastewater treatment, *Environ. Sci. Technol.*, 38, 392A, 2004.
24. Jones, O.A.H., Lester, J.N., and Voulvoulis, N., Pharmaceuticals: a threat to drinking water? *Trends Biotech.*, 23, 163, 2005.

25. Thomas, P.M. and Foster, G.D., Tracking acidic pharmaceuticals, caffeine, and triclosan through the wastewater treatment process, *Environ. Toxic. Chem.*, 24, 25, 2005.

26. Batt, A.L., Bruce, I.B., and Aga, D.S., Evaluating the vulnerability of surface waters to antibiotic contamination from varying wastewater treatment plant discharges, *Environ. Poll.*, 142, 295, 2006.

27. Golet, E.M., Alder, A.C., and Giger, W., Environmental exposure and risk assessment of fluoroquinolone antibacterial agents in wastewater and river water of the Glatt Valley watershed, Switzerland. *Environ. Sci. Technol.*, 36, 3645, 2002.

28. Metcalfe, C.D. et al., Distribution of acidic and neutral drugs in surface waters near sewage treatment plants in the lower great lakes, Canada, *Environ. Toxic. Chem.*, 22, 2881, 2003.

29. Petrovic, M., Gonzalez, S., and Barcelo, D., Analysis and removal of emerging contaminants in wastewater and drinking water. *TrAC, Trends Anal. Chem.*, 22, 685, 2003.

30. Ashton, D., Hilton, M., and Thomas, K.V., Investigating the environmental transport of human pharmaceuticals to streams in the United Kingdom, *Sci. Total Environ.*, 333, 167, 2004.

31. Ingerslev, F. and Halling-Sorensen, B., Biodegradability properties of sulfonamides in activated sludge, *Environ. Toxic. Chem.*, 19, 2467, 2000.

32. Ternes, T.A. et al. Removal of pharmaceuticals during drinking water treatment, *Environ. Sci. Technol.*, 36, 3855, 2002.

33. Perez, S., Eichhorn, P., and Aga, D.S, Evaluating the biodegradability of sulfamethazine, sulfamethoxazole, sulfathiazole, and trimethoprim at different stages of sewage treatment, *Environ. Toxic. Chem.*, 24, 1361, 2005.

34. Joss, A. et al., Biological degradation of pharmaceuticals in municipal wastewater treatment: Proposing a classification scheme, *Water Res.*, 40, 1686, 2006.

35. Ternes, T.A. et al., Assessment of technologies for the removal of pharmaceuticals and personal care products in sewage and drinking water facilities to improve the indirect potable water reuse, Report EU-POSEIDON project, EVK1-CT-2000-00047, Germany, 2004.

36. Kümmerer, K., Significance of antibiotics in the environment, *J. Antimicrob. Chemo.*, 52, 5, 2003.

37. Eichhorn, P. et al., Application of ion trap-ms with h/d exchange and qqtof-ms in the identification of microbial degradates of trimethoprim in nitrifying activated sludge, *Anal. Chem.*, 77, 4176, 2005.

38. Golet, E.M. et al., Environmental exposure assessment of fluoroquinolone antibacterial agents from sewage to soil, *Environ. Sci. Technol.*, 37, 3243, 2003.

39. Kim, S. et al., Removal of antibiotics in wastewater: effect of hydraulic and solids retention times on the fate of tetracycline in the activated sludge process, *Environ. Sci. Technol.*, 39, 5816, 2005.

40. Stuer-Lauridsen, F. et al., Environmental risk assessment of human pharmaceuticals in Denmark after normal therapeutic use, *Chemosphere*, 40, 783, 2000.

41. Joss, A. et al., Removal of pharmaceuticals and fragrances in biological wastewater treatment, *Water Res.*, 39, 3139, 2005.

42. Drewes, J.E., Heberer, T., and Reddersen, K., Fate of pharmaceuticals during indirect potable reuse, *Water Sci. Technol.*, 46, 73, 2002.

43. ES&T, http://pubs.acs.org/subscribe/journals/esthag-w/2006/dec/science/kc_remove_ppcp.html?sa_campaign=rss/cen_mag/estnews/2006-12-27/kc_remove_ppcp (accessed August 8, 2007).

44. Liao, B.Q. et al., Surface properties of sludge and their role in bioflocculation and settleability, *Water Res.*, 35, 339, 2001.

45. Lee, W.T., Kang, S., and Shin, H., Sludge characteristics and their contribution to microfiltration in submerged membrane bioreactors, *J. Memb. Sci.*, 216, 217, 2003.

46. Harper, Jr. W.F. and Yi, T., Sorption of ethinyl estradiol in activated sludge processes, Proceedings of the 25th Annual Alabama Water Environ. Assoc. Con., 2004.

47. Sithole, B.B. and Guy, R.D., Models for tetracycline in aquatic environments: i. interaction with bentonite clay systems, *Water Air Soil Poll.*, 32, 303, 1987.

48. Batt, A.L., Kim, S., and Aga, D.S., Comparison of the occurrence of antibiotics in four full-scale wastewater treatment plants with varying designs and operations. *Chemosphere*, 68, 428, 2007.

49. Rasche, M.E., Hyman, M.R., and Arp, D.J., Factors limiting aliphatic chlorocarbon degradation by Nitrosomonas europaea: cometabolic inactivation of ammonia monooxygenase and substrate specificity, *Applied Environ. Micro.*, 57, 2986, 1991.

50. Keener, W.K. and Arp, D.J., Transformations of aromatic compounds by *Nitrosomonas europaea*, *Applied Environ. Micro.*, 60, 1914, 1994.

51. Batt, A.L., Kim, S. and Aga, D. S., Enhanced biodegradation of iopromide and trimethoprim in nitrifying activated sludge, *Environ. Sci. Technol.*, 40, 7367, 2006.

52. Köhler, A. et al., Organic pollutant removal versus toxicity reduction in industrial wastewater treatment: the example of wastewater from fluorescent whitening agent production, *Environ. Sci. Technol.*, 40, 3395, 2006.

53. Isidori, M. et al., A multispecies study to assess the toxic and genotoxic effect of pharmaceuticals: furosemide and its photoproduct, *Chemosphere,* 63, 785, 2006.

16 Chemical Processes during Biological Wastewater Treatment

Willie F. Harper, Jr.,
Tamara Floyd-Smith, and Taewoo Yi

Contents

16.1   INTRODUCTION

Removing pharmaceutical and personal-care products (PPCPs) during biological wastewater treatment is important for preventing the rapid accumulation of these chemicals in our environment. Accordingly, the wastewater treatment community has responded to these concerns with a great deal of applied research. Analytical methods are now available for low-level detection of PPCPs in wastewater, and surveys of wastewater influent and effluent streams have revealed the broad classes of micropollutants present in municipal wastewater (see Chapter 1 and Chapter 5 for a review of the various classes of PPCPs and veterinary medicines present in wastewater). This chapter presents experimental findings related to the sorption and biodegradation of various classes of PPCPs in biological wastewater treatment systems. This includes a review of full-scale PPCP removal performance and looks at key issues related to both sorption and biodegradation. There also is discussion related to the possible effects of antibiotic compounds on the spread of antimicrobial-resistant microorganisms via the activated sludge process.

16.2 THE ACTIVATED SLUDGE PROCESS: A BRIEF OVERVIEW

The activated sludge process is used to treat both municipal and industrial wastewater before the water is returned to the environment (or reused). In the activated sludge process, microorganisms remove soluble organic constituents from wastewater. A conventional municipal activated sludge wastewater treatment plant (WWTP) schematic is shown in Figure 16.1. Influent wastewater is screened for removal of wastewater large debris (e.g., rags, glass, rocks) and then it is fed to a primary clarifier for removal of settleable particulate matter. The primary effluent is then fed to an aeration basin where particulate and dissolved organics and nutrients are removed by a flocculent biomass. It is in this basin that actively growing microorganisms may take part in chemical reactions that remove and perhaps transform PPCPs. The wastewater is then routed to a secondary clarifier for biomass recycle and for solids separation to produce a clarified secondary effluent. In many wastewater treatment facilities, secondary effluent is further treated with granular filtration for removal of nonsettleable material or disinfection to destroy pathogens.

16.3 FULL-SCALE STUDIES

Numerous reports have explored the removal of various classes of PPCPs at full scale, generally attempting to evaluate whether municipal WWTPs are acting as persistent point sources for PPCP discharge to the environment. Ternes[1] showed that the removal efficiencies ranged from 10 to 90% in WWTPs in Germany, and Ternes et al.[2] showed that removal efficiencies for polar PPCPs varied from 12 to 90% for WWTPs in Brazil. Gomez et al.[3] conducted a 1-year monitoring study at a sewage treatment plant in Spain, and they found that the removal efficiencies for 14 organic micropollutants varied from 20% (carbamazepine) to 99% (acetaminophen). Joss et al.[4] showed that only 4 out of 35 compounds are 90% removed using state-of-the-art biological treatment systems, and 17 out of 35 are removed at less than 50% efficiency. These studies are in addition to others that present high removal efficiencies. Oppenheimer and Stephenson[5] found that removal efficiencies for frequently detected PPCPs were generally high (>80%), and another study by Jones et al.[6]

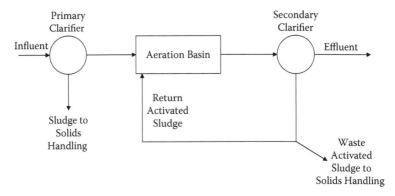

FIGURE 16.1 Conventional activated sludge wastewater treatment plant schematic.

found that ibuprofen, paracetamol, salbutamol, and mefenamic acid were removed at approximately 90% within a large sewage treatment plant in England. Overall, these efforts have shown that the removal efficiencies vary greatly.

That conclusion that PPCP removal in full-scale systems varies considerably is further supported by Lishman et al.,[7] who investigated the presence of selected acidic drugs, triclosan, polycyclic musks, and selected estrogens in WWTP influent and effluent at sites in Canada. They found that three analytes were never detected during the survey (clofibric acid, fenoprofen, fenofibrate) and three analytes were always removed at high efficiency for all treatment configurations (ibuprofen, naproxen, triclosan). Two analytes were removed at low efficiencies (gemfibrozil, diclofenac), but better removals were observed for treatment configurations with higher solid retention times. Five polycyclic musks were surveyed; general conclusions could not be reached because of the small dataset and because of numerous nonquantifiable results, but removal efficiencies generally were variable. E2 and E1 were both removed at high efficiency for all treatment systems. As shown in Figure 16.2, even where conventional WWTPs are concerned, removal efficiencies for different PPCPs can vary significantly. Diclofenac removal efficiency is negative in Figure 16.2, suggesting that diclofenac may be deconjugated during the treatment process. Generally, these full-scale studies have not collected the type and amount of data necessary to organize mass balances for specific PPCPs, so that a clear articulation of the relative roles of sorption and biodegradation in the full-scale process is generally unavailable. Some studies have complemented full-scale studies with batch experiments, so that the potential for sorption or biodegradation at full-scale can be assessed.

Removal efficiencies can vary as a function of the type of compound. Carballa et al.[8] surveyed two cosmetic ingredients (galaxolide, tonalide), eight pharmaceuticals (carbamazepine, diazepam, diclofenac, ibuprofen, naproxen, roxithromycin, sulfamethoxazole, and iopromide), and three hormones (estrone, 17β-estradiol, and 17α-

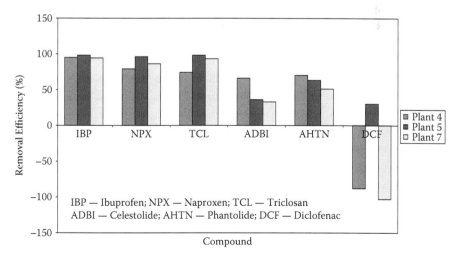

FIGURE 16.2 Removal efficiencies for selected PPCPs for three conventional activated sludge systems as reported by Lishman et al., 2006. (Image created by W.F. Harper, Jr.)

ethinylestradiol) at municipal WWTPs in Spain. They found that the overall removal efficiencies ranged between 70 and 90% for the fragrances, 40 and 65% for the anti-inflammatories, approximately 65% for 17β-estradiol, and 60% for sulfamethoxazole. However, the concentration of estrone increased along the treatment due to the partial oxidation of 17β-estradiol in the aeration tank. Nakada et al.[9] measured a host of compounds, including six acidic analgesics or anti-inflammatories (aspirin, ibuprofen, naproxen, ketoprofen, fenoprofen, mefenamic acid), two phenolic antiseptics (thymol, triclosan), four amide pharmaceuticals (propyphenazone, crotamiton, carbamazepine, diethyltoluamide), three phenolic endocrine disrupting chemicals (nonylphenol, octylphenol, bisphenol A), and three natural estrogens (17β-estradiol, estrone, estriol) in 24-h composite samples of influents and secondary effluents from municipal WWTPs in Tokyo. They found that aspirin, ibuprofen, and thymol were removed efficiently during secondary treatment (>90% efficiency). They also found that amide-type pharmaceuticals, ketoprofen, and naproxen showed poor removal (<50% efficiency), probably because of their lower hydrophobicity (log K_{ow} < 3). This study was also the first to report the presence of crotamiton (a topical treatment for *scabies*), and to show that it is persistent during secondary treatment. Overall, these results reinforce the conclusion that removal efficiencies vary for the various PPCPs and suggest that chemical characteristics also may play an important role in determining the fate of each compound in biological wastewater treatment.

Removal efficiencies also can vary as a function of the sludge retention time (SRT). Oppenheimer and Stephenson[5] studied the removal of 20 PPCPs in full-scale and pilot-scale WWTPs in the United States, and they organized their data using a bin assignment system, which assigned each detected compound into a category related to the frequency of detection (i.e., infrequent, variable, and frequent) and into another category related to the removal efficiencies (excellent removal, moderate removal, poor removal). They found that half of the PPCPs were frequently detected and were removed at less than 80% efficiency at an SRT of 5 days or less. Caffeine and ibuprofen were among nine compounds that were both frequently detected and removed well for all the systems in the study. Galaxolide and musk ketone were also frequently detected but removed at 80% only when the SRT exceeded 25 days.

Membrane bioreactor systems (MBRs) have been evaluated as possibly better technology for removing PPCPs. MBRs use a suspended growth bioreactor, like in conventional activated sludge, but replaces gravity sedimentation with micro- or ultrafiltration (Figure 16.3). The MBR is an attractive treatment configuration because it eliminates the need for secondary clarification, which in turn allows the overall treatment process to be sited on a much smaller footprint. Kim et al.[10] found that the MBR system was efficient for hormones (e.g., estriol, testosterone, androstenedione) and certain pharmaceuticals (e.g., acetaminophen, ibuprofen, and caffeine) with approximately 99% removal, but MBR treatment did not decrease the concentration of erythromycin, trimethoprim, naproxen, diclofenac, and carbamazepine. Oppenheimer and Stephenson[5] used a limited dataset to suggest that MBR provided no additional PPCP removal, when compared to similarly operated conventional systems. Kimura et al.[11] found that MBRs exhibited much better removal regarding ketoprofen and naproxen, but with respect to the other compounds, comparable removal was observed between the MBRs and conventional systems. These data

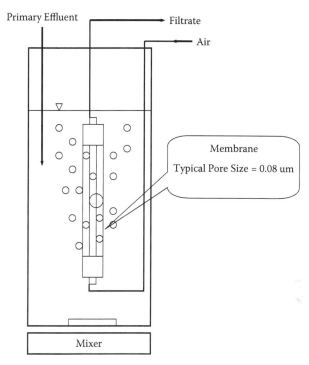

Primary Effluent → Filtrate
Air

Membrane
Typical Pore Size = 0.08 um

Mixer

FIGURE 16.3 Membrane bioreactor schematic diagram.

suggest that MBRs likely offer no inherent advantage over conventional systems for removing PPCPs, but because MBRs are operated at long solid retention times and at high mixed liquor suspended solids (MLSS) concentrations, those operational factors are likely the cause of any measured differences in PPCP removal efficiencies.

Finally, there remains a need to continue to conduct full-scale studies, with the goal of organizing accurate mass balance and fate data. To accomplish this, rigorous wastewater sampling methods must be employed. For example, these full-scale studies collected data using time-weighted composite sampling using automatic samplers, equipped with sample storage in cooled compartments. This strategy allowed the reports to collect data that are likely to represent a reasonable estimate of the PPCP concentrations of interest, as well as the inherent variability; but this approach is not infallible. Many of the PPCPs of interest are biodegradable and may be transformed while the samples remain stored in the collection container. Still other compounds are highly hydrophobic and sorb strongly to biomass solids and colloidal materials that are also present in the original sample. In these cases, it is possible to underestimate the concentrations of interest, either because the solids are not properly resuspended before sample analysis, or because of inadequate extraction techniques. Finally, time-weighted sampling collects a given wastewater volume at given time intervals, even if the wastewater flow is low. This means that time-weighted sampling may cause low-flow PPCP concentrations to be overrepresented in the composite sample. For these reasons, future sampling campaigns should consider the use of flow-weighted sampling in combination with frequent grab sampling to minimize the error associated with sample collection. Each collected sample should also be mixed vigorously to resuspend settled material, and PPCP analysis should be carried out on both the filtered and unfiltered samples. Improvements in sample collection methodology will strengthen the reliability of the data, which in turn will no doubt be the basis for future treatment plant optimization and regulatory action.

16.4 SORPTION

In general, the partitioning of organic compounds from water onto activated sludge biomass is referred to as adsorption, although it may be more appropriate to refer to this as sorption because there may be some uncertainty as to whether the compound is on the surface (adsorption) or partitioning into another phase (absorption). When sorption is of interest, it is important to establish a relationship between what is on the surface and what is in the aqueous phase, a relationship generally referred to as a sorption isotherm. The term *isotherm* comes from the idea that the equilibrium is reached at a constant temperature to distinguish this type of partitioning from condensation. These relationships are determined experimentally and then the data are used to determine a partitioning coefficient, which is a measure for the affinity of a given compound for the activated sludge biomass.

Partitioning coefficients (K_d) have been determined in a number of studies to investigate PPCP sorption to activated sludge. Ternes et al.[12] conducted a series of batch tests with primary and secondary sludge slurries to determine partitioning coefficients for a number of target PPCPs. They found that the K_d values of pharmaceuticals ranged from <1 to 500 L kg^{-1}, while that of the polycyclic musk fragrances acetyl hexamethyl tetrahydronaphthalene (AHTN) and hexahydrohexamethylcyclopentabenzopyran (HHCB) proved to be much higher and up to 5300 and 4900 L kg^{-1}, respectively. They also found significant differences between the K_d values obtained between primary sludge and secondary sludge. For acidic pharmaceuticals and musk fragrances, the K_d values were higher when measured with primary sludge; the opposite was true with neutral pharmaceuticals, iopromide, and ethinylestradiol.

The sorption equilibrium partitioning coefficients determined for steroid estrogens with activated sludge show some limited variability, but they are generally in good agreement (Figure 16.4). Clara et al.[13] found that the log (K_d) for steroid estrogens was 2.84 (2.64 to 2.97) and 2.84 (2.71 to 3.00) for E2 and EE2, respectively. In the work by Ternes et al.[12] the log (K_d) for EE2 was determined to be 2.54 (2.49 to 2.58). Yi et al.[14] found that the log K_d for EE2 was 2.7 for membrane bioreactor sludge and 2.3 when the sludge was taken from a sequencing batch reactor, since the MBR particle sizes were significantly smaller than the SBR particles. This result suggested that particle size may explain some of the variability that is reported for steroid estrogen partitioning coefficients. Andersen et al.[15] determined distribution coefficients (K_d) with activated sludge biomass for the steroid estrogens estrone (E1), 17β-estradiol (E2), and 17α-ethinylestradiol (EE2) in batch experiments, and they determined log K_d values for steroid estrogens of 2.6, 2.7, and 2.8, respectively. When Andersen et al.[15] corrected their log K_d values to account for the organic carbon content of the sludge, they found that the log K_d values were 3.16, 3.24, and 3.32, respectively. These values were remarkably consistent with the sorption partitioning coefficients

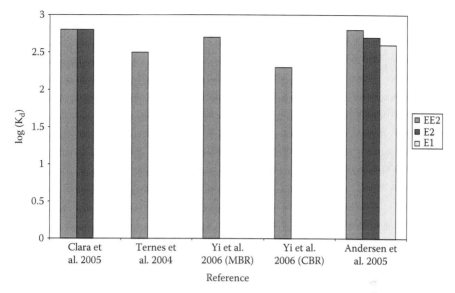

FIGURE 16.4 Partitioning coefficients determined for steroid estrogen sorption to activated sludge.

determined where soil is used as the sorbate.[16–19] Taken together, these partitioning coefficients enable practitioners to model PPCP sorption in activated sludge processes and numerically evaluate the importance of sorption as a removal mechanism.

Sorption is not always an important removal mechanism. Ternes et al.[12] found that, for compounds with the K_d values less than 500 L Kg[-1], only 20% of the target compound mass was associated with the sludge solids, which showed that the majority of the mass of the target compounds remained in solution. This result supported the idea that sorption is not an important removal mechanism for many pharmaceutical compounds. Yu et al.[20] conducted aerobic batch biodegradation (using activated sludge as microbial inocula) experiments to evaluate the biodegradation behavior of 18 target PPCPs at initial concentrations of 50, 10, and 1 µg L[-1]. The target compounds included a number of antiseptics, barbiturates, and anticonvulsants. Their sterile control studies showed no loss of target PPCPs during the entire incubation period, and sorption to the biomass was found to be negligible for all testing conditions. Urase and Kikuta[21] conducted a series batch experiment to examine the removal of 3 steroid estrogens (i.e., 17β-estradiol), 2 endocrine disruptors (i.e., bisphenol A), and 10 pharmaceutical substances by activated sludge. Many of the target PPCPs in this study were hydrophilic, had lower water–sludge partition coefficients than the steroid estrogens, and remained in the aqueous phase, with only a small fraction partitioning to the activated sludge.

When sorption is important, there is a sorption/desorption cycle that should be investigated experimentally. In some cases desorption fails to restore the full capacity of the sorbent, and when this happens some of the sorption sites remain occupied. This is referred to as sorption hysteresis, and this has been reported for many organic compounds where either soil or sludge acts as the sorbent.[22–24] Hysteresis has thus far received little attention where PPCP sorption to sludge is concerned. Recently,

Kim et al.[23] showed sorption hysteresis in the case of tetracycline sorption/desorption with activated sludge, but this is probably because tetracycline forms strong complexes with Ca (II) and other divalent cations known to be important for floc stability.[25,26] PPCP sorption hysteresis is a basic and relevant process that has not received great attention to date.

One cause of sorption hysteresis may be related to particle characteristics (e.g., size), and there is a need to study the possible fundamental connections. Yi and Harper[27] hypothesized that sorption hysteresis is more pronounced as the biomass particle size distribution shifts toward larger sizes. The rationale for this was that smaller flocs are more dense and less permeable than larger floc,[28,29] therefore allowing for much less intraparticle entrapment of PPCPs. In general, activated sludge particles in conventional processes are typically 80 to 300 μm in diameter,[30] and this structure typically consists of smaller microcolonies (approximately 8 to 15 μm) connected by exocellular polymeric and inorganic material, and with a few large flow channels that facilitate transport.[28,29] Smaller activated sludge particles can be found in bioreactors like MBRs,[14,31] and smaller particles have less internal polymer, a higher number of cells per unit volume,[28] and they do not have the large flow channels that facilitate transport.

Yi and Harper[27] investigated this hypothesis by operating two laboratory-scale bioreactor systems and an MBR and a conventional bioreactor (CBR), both operated in continuous flow mode. The experimental strategy was to harvest biomass from the bioreactors for use in a series of sorption/desorption batch tests. The data retrieved from the batch tests were used to determine sorption and desorption isotherms, from which the partitioning coefficients (K_d and K_{ds}, respectively) and sorption hysteresis (HI) index values were calculated. Sorption HI was calculated as follows:

$$HI = \frac{K_{ds} - K_d}{K_d} \tag{16.1}$$

The subscript T (23°C) and C_r (C_r level is 0.5) refer to specific conditions of constant temperature and residual solution phase concentration ratio, respectively. The partitioning coefficient determined from the sorption experiments is K_d, and the partitioning coefficient determined from the desorption experiments is K_{ds}. Samples were also collected for biomass particle size analysis.

A typical sorption/desorption result is shown in Figure 16.5 for the two different biomass floc suspensions. The suspension taken from the MBR had a mean particle size of 10 μm, while that of the CBR had a mean particle size of 120 μm. In this example the sorption/desorption experiment yielded K_d and K_{ds} values of 0.47 L/g and 0.56 L/g for the MBR biomass, and 0.32 L/g and 0.61 L/g for the CBR biomass, respectively. Using these values, the hysteresis index values for the MBR and CBR were 0.19 and 0.89, respectively. Results such as these suggest that the particle size influenced the hysteresis index for EE2 sorption. Yi and Harper[27] found that as the mean particle size increased from 10 to 230 μm, the HI increased nonlinearly from approximately 0.2 to 0.9. This result showed that the biomass particle size can have a

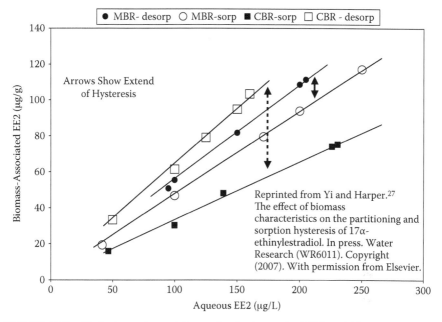

FIGURE 16.5 Typical sorption and desorption isotherms for MBR and CBR biomass.

dramatic effect on the entrapment of EE2 within activated sludge floc, which in turn may affect the ultimate fate of EE2.

16.5 BIODEGRADATION

Biodegradation is likely due to cometabolic activity because PPCPs are not present in high enough concentration to support substantial biomass growth. This means that PPCP transformation is most likely to occur during exponential growth stages and during active degradation of the primary substrates present in wastewater. The published reports of cometabolism of PPCP are currently limited. Most of the published reports that concern cometabolism focus on the removal of xenobiotics that are produced as a result of industrial and military activity (e.g., chlorinated solvents such as trichloroethylene, nitroaromatic compounds, explosives, dyes, polyurethane foams). These compounds may be present in the environment at much higher concentrations than PPCPs are, but many industrial pollutants and PPCPs share some of the same structural features (i.e., polyaromatic rings), so there may be common reaction mechanisms. It is also known that cometabolism is often an initiating reaction, producing intermediates that may be more biodegradable (and therefore would participate in the central metabolic pathways), or that may be susceptible to adsorption or polymerization reactions and rendered nonbioavailable (i.e., dead-end product). Quintana et al.[32] observed the cometabolic transformation of four acidic pharmaceuticals in laboratory-scale experiments. Although cometabolism is likely when biodegradation is occurring, there is only limited information that clearly connects cometabolism with the removal of PPCPs. One interesting example comes from Alexy et al.,[33] who found that each of 18 antibiotics was not biodegraded, but some partial biodegradation was observed when sodium acetate also was present. This suggests that when sodium acetate is available as a primary substrate, the antibiotics may be subject to cometabolism. Biodegradation may sometimes result in the formation of a stable byproduct. Haib and Kummerer[34] found that diatrizoate (found in X-ray contrast media) was biodegraded aerobically to 3,5-diamino-2,4,6-triodobenzoic acid, which was not further degraded by bacteria. Quintana et al.[32] also found that biotransformation of ketoprofen and bezafibrate produced more stable metabolites.

A wide variety of mono- and dioxygenase enzymes can transform xenobiotics during exponential growth conditions,[35] but biotransformation of pollutants in the absence of bacterial growth also may occur as a result of enzymes previously produced by dead (nonviable) bacteria and as a result of extracellular enzymes excreted by viable bacteria.[36,37] Activated sludge communities are diverse and known to house a wide variety of nonspecific mono- and dioxygenase enzymes associated with both heterotrophic and autotrophic microorganisms.[38,39]

There is circumstantial evidence linking nitrifiers to a unique capability to biologically (perhaps cometabolically) transform steroid estrogens such as EE2. Surveys of municipal WWTPs indicated that nitrifying sludges remove EE2 more efficiently than those that do not nitrify.[40] Numerous experimental results further supported this contention: Vader et al.[41] degraded EE2 using nitrifying activated sludge, and they noted the presence of unidentified hydrophilic daughter products. Several groups[14,42–44] also biologically degraded EE2 using nitrifying mixed cultures. These combined results suggest that EE2 and NH_4 transformation rates are linked. A specific EE2 transformation mechanism may involve ammonium monooxygenase (AMO), the key enzyme that catalyzes the conversion of ammonia to nitrite in nitrifying organisms. For example, AMO is also capable of cometabolically oxidizing polycyclic aromatic rings.[45,46] The active site of AMO is buried in the core of the protein, where four neighboring α-helices provide two histidine and four glutamic acids as iron ligands.[47,48] One face of the di-iron site contains a hydrophobic pocket and may be well suited for organic substrates like EE2. Yi et al.[14] showed that EE2 and NH_4^+ are simultaneously degraded in an AMO-containing extract.

Yi and Harper[43] proposed a conceptual picture linking EE2 removal and NH_4^+ removal (Figure 16.6). AMO converts NH_3 to NH_2OH in the presence of oxygen. This step requires reducing power that is regenerated as NH_2OH is oxidized to NO_2 by hydroxylamine oxidoreductase. Electrons then enter a catalytic cycle involving a binuclear copper site located at the AMO active site. Oxygen reacts to convert the Cu(I) to Cu(II), but the oxygen remains bound as peroxide ion (O_2^-). This oxygenated form of the enzyme then reacts with organic substrates to produce the Cu(II) form.

Yi and Harper[43] evaluated the conceptual model shown in Figure 16.6 using an enzyme extract taken from an *enriched* (not pure) culture of nitrifiers. They determined the ratio of EE2/NADH removed by incubating EE2, NADH, and other components in the presence of an AMO-containing enzyme extract; the molar ratio of NADH/ EE2 determined during the incubation was 2.2, which is consistent with the action of monooxygenase-mediated biotransformation shown in Figure 16.6. This result shows that the cometabolic biotransformation of EE2 was monooxygenase mediated, as opposed to being dioxygenase mediated, because the NADH/EE2

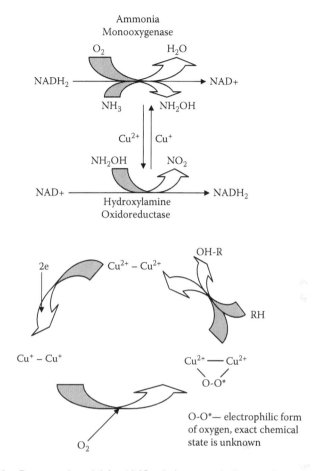

FIGURE 16.6 Conceptual model for AMO role in cometabolic transformation.

molar ratio of the later is 1:1 (as opposed to 2:1). This result demonstrates the potential for monooxygenase-mediated EE2 biotransformation *in vitro*.

Yi and Harper[43] investigated the relationship between the measured NH_4 rates and the measured EE2 biotransformation rate using an enriched culture of nitrifying bacteria. Figure 16.7 shows their data, along with that of Shi et al.[42] and Vader et al.[41] The results showed a linear relationship between nitrification and EE2 biodegradation rates over the range of NH_4 and EE2 biotransformation rates tested. The EE2 biotransformation rate increased from 1.1 to 4.1 μmol EE2/g VSS/h, while the NH_4 biotransformation rate increased from 0.3 to 3.1 mmol NH_4/g VSS/h. These data taken together strongly show a linear link between nitrification and EE2 removal in enriched nitrifying cultures and therefore support the notion that EE2 biotransformation can be cometabolically mediated under the operating conditions that allow for enrichment of nitrifiers.

These results support the conclusion that nitrifying activated sludge cultures may play a role in biotransforming pharmaceuticals in biological WWTPs, but heterotrophic organisms also likely play a key role. Furthermore, these results show

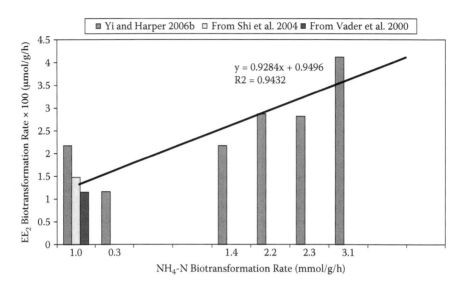

FIGURE 16.7 The relationship between NH_4-N and EE2 biotransformation.

that nitrifiers may, at a minimum, provide for initial degradation of PPCPs like EE2 into an intermediate that can then be degraded further by heterotrophic organisms. The work of Shi et al.[42] also supports this idea. They conducted EE2 biodegradation experiments with a nitrifying pure culture and a nitrifying mixed culture. They detected daughter products in the pure-culture experiments but not in the mixed-culture experiments, perhaps because the heterotrophs completely degraded the daughter products produced by the nitrifiers. At this point it is not clear if AMO is kinetically dominant in full-scale WWTPs among all enzymes that might be capable of transforming pharmaceuticals, especially if the enzymes are present in fast-growing heterotrophic organisms. Proof that nitrifiers are responsible for transformation of steroids in full-scale systems has not been shown definitively. It may be that nitrifiers will cometabolically transform pharmaceuticals containing aromatic structures when they are present in low organic carbon, ammonium-enriched environments through the enzyme AMO. However, heterotrophic cultures also may contribute to and, in fact, may predominate these biotransformations if the sewage also contains mono- and dioxygenase inducers that function in heterotrophic bacteria. Other scavenging, biodegradative mechanisms are likely to exist and function among the complex collection of heterotrophic bacteria present in the low organic carbon environments found during the nitrification phase of bioreactors. Questions related to the relative importance and the potentially synergistic interplay between nitrifiers and heterotrophs need to be further elucidated to clarify this issue.

Nyholm et al.[49] suggested that biodegradation can be enhanced by operating at longer SRT. They operated laboratory-scale bioreactors over a range of SRT values (1 to 32 days) and sludge loadings (0.1 to 0.9 mgBOD5/mg MLSS/d), and they spiked five organic micropollutants (2,4-dichlorophenoxy acetic acid (2,4-D); 2,4,6-trichlorophenol (TCP); pentachlorophenol (PCP); 4-nitrophenol (4-NP) and lindane) into the influent. They found that adaptation was generally required, and that removal by biodegradation in successfully adapted systems was generally within a range of about 40 to about 95% except for 4-NP, which was degraded to concentration levels below the analytical detection limit. They found that PCP, TCP, and 2,4-D were degraded best at high sludge ages.

Biofilm experiments also have offered insight into the biodegradability of selected PPCPs. For example, Boyd et al.[50] investigated removal of naproxen and its chlorination products using a laboratory-scale biofilm bioreactor process. The bioreactor was a plug-flow bioreactor, and it used 31 m of polypropylene tubing as the support matrix for the biofilm. The bioreactor was fed a naproxen solution and then fed a solution at the same naproxen concentration following contact with free chlorine. Naproxen was not degraded biologically, and the naproxen solution containing products of chlorination caused biomass sloughing and discharge from the bioreactor. Zwiener and Frimmel[51] investigated the biodegradation of three active compounds of pharmaceuticals (clofibric acid, ibuprofen, and diclofenac) in short-term tests with a miniaturized upflow biofilm bioreactor with an oxic/anoxic configuration. The biofilm reactor removed 85% of the applied dissolved organic carbon (DOC), but clofibric acid and diclofenac were not eliminated and were discharged at a level of approximately 95% of their initial concentration; they did find, however, that the elimination in the anoxic region of the biofilm reactor improved the removal efficiencies of clofibric acid and diclofenac to values between 60 and 80% of their initial concentration. Winkler et al.[52] found that ibuprofen (as well as its hydroxylated and carboxylated metabolites) was biodegraded in a river biofilm reactor, but clofibric acid was not.

Synthetic antibiotics, which do not appear to be readily biodegradable, deserve special attention. Ingerslev et al.[53] studied the primary aerobic and anaerobic biodegradability of the antibiotics olaquindox (OLA), metronidazole (MET), tylosin (TYL), and oxytetracycline (OTC). They conducted batch experiments at intermediate concentrations (50 to 5000 µg/L) using shake flasks inoculated with C14-labeled antibiotic compounds and mixed with sediment or activated sludge. They found that these compounds were slowly biodegradable during aerobic conditions, with half-life values that were typically between 1 and 5 weeks. During anaerobic conditions the biodegradation rates were slower, with half-life values of up to 12 weeks. Alexy et al.[33] studied the biodegradability of 18 clinically important antibiotics, and in addition to finding that none of them were readily biodegradable, they also found that half of the antibiotics tested inhibited biological activity when present at parts-per-billion levels. A study by Kummerer et al.[54] also revealed that none of the test antibiotic compounds (ciprofloxacin, ofloxacin, metronidazole) were biodegraded and that, in addition, the genotoxicity was not eliminated during batch experiments. Zhou et al.[55] treated a high-strength pharmaceutical wastewater with a pilot-scale system composed of an anaerobic baffled reactor followed by a biofilm airlift suspension reactor. They found that ampicillin and aureomycin, with influent concentrations of 3.2 and 1.0 mg/L, respectively, could only be partially degraded, with overall removal efficiencies of less than 10% at steady state. These results imply that biodegradation is not likely to play a large role in determining the ultimate fate of synthetic antibiotics in conventional biological wastewater treatment systems.

Although a number of elucidating studies concerning biodegradation of PPCPs have been conducted, research on the biodegradation of PPCPs should continue, with particular attention to the identification of daughter products and the application of molecular methods to identify the important microorganisms and mechanisms. Currently, there are numerous examples in the literature reporting on the biodegradation of PPCPs in biological treatment systems but without any direct evidence of biotransformation (e.g., metabolites). This is a weakness that currently exists in the literature, and it does not serve to clarify the dialogue concerning the fate of PPCPs. There are also examples[41] of reports that show unidentified "daughter products"; these reports will be strengthened with clear identification of metabolites, which can be readily accomplished by combining the latest tools in high performance liquid chromatography tandem mass spectrometry (HPLC/MS/MS) technology with well-established methods such as thin layer chromatography and NMR.

16.6   ANTIBIOTIC-RESISTANT MICROORGANISMS AND THE ACTIVATED SLUDGE PROCESS

There is a notion that asserts that the sludge in WWTPs could be where bacteria obtain the ability to resist antibiotics through the exchange of genetic fragments. This idea leads to concerns about whether biological WWTPs are an important source of antibiotic-resistant microorganisms and resistance genes. This is a major public health issue, and it has become more important as water quality surveys have revealed the broad range of antibiotics present in WWTPs effluents.[56] Evaluating biological WWTPs as a point source for antibiotic-resistant microorganisms is a key task toward the overall goal of understanding the spread of antibiotic resistance microorganisms in the aquatic environment. As a result, a number of studies have been conducted to evaluate antibiotic-resistant microorganisms and determine whether or not the activated sludge process is an important source of antibiotic-resistant microorganisms in the environment.

There is evidence that antibiotic-resistant microorganisms are present in activated sludge treatment plants. For example, Schwartz et al.[57] cultivated heterotrophic bacteria resistant to vancomycin, ceftazidime, cefazolin, and penicillin G from municipal activated sludge. They also estimated that the amount of vancomycin-resistant enterococci in activated sludge to be approximately 16% (as a percentage of cultivable species), and the amount of cefazolin-resistant enterobacteriaceae in activated sludge biofilms to be approximately 19% (approximately 11% in the WWTP discharge). The vancomycin-resistant heterotrophic bacteria were also completely resistant to tetracycline and erythromycin. In an early example, Mach and Grimes[58] examined enteric bacteria for their ability to transfer antibiotic resistance. They isolated resistant *Salmonella enteritidis*, *Proteus mirabilis*, and *Escherichia coli* from primary sewage effluent, and they demonstrated resistance to ampicillin, chloramphenicol, streptomycin, sulfadiazine, and tetracycline by spread plate and tube dilution techniques. Each donor they isolated was mated with susceptible *E. coli* and *Shigella sonnei* species, and they found that the donors transferred the genes of interest at transfer frequencies (given as the ratio of resistant recipient bacteria per resistant donor) of $2.1 \times 10^{(-3)}$ and *in situ* transfer frequencies of $4.9 \times 10^{(-5)}$ to $7.5 \times 10^{(-5)}$. These transfer frequencies suggested that a significant level of resistance transfer occurs in WWTPs, even in the absence of antibiotics as selective agents. More recent efforts have complemented these culture-dependent results with molecularly based culture-independent approaches. For example, Volkmann et al.[59] used real-time polymerase chain reaction (PCR) assays to quantify the presence of antibiotic-resistance genes for vancomycin (vanA) and ampicillin (ampC), and they found that, in municipal wastewater, the resistance gene vanA was detected in 21% of the samples, and ampC in 78%.

The presence of hospital wastewater also may affect antibiotic resistance rates. Reinthaler et al.[60] evaluated the resistance patterns of *E. coli* in WWTPs and found that the highest resistance rates were found in *E. coli* strains of a sewage treatment plant that treats not only municipal sewage but also sewage from a hospital. They also found that, among the antimicrobial agents tested, the highest resistance rates in the penicillin group were found for ampicillin (up to 18%) and piperacillin (up to 12%); in the cephalosporin group for cefalothin (up to 35%) and cefuroxime-axetil (up to 11%); in the group of quinolones for nalidixic acid (up to 15%); and for trimethoprime/sulfamethoxazole (up to 13%) and for tetracycline (57%). They determined that more than 102 CFU *E. coli*/mL reached the receiving water, and thus sewage treatment processes contribute to the dissemination of resistant bacteria in the environment.

Although antibiotic-resistant microorganisms are present in activated sludge, there are indications that the biological treatment process can reduce the volumetric concentration of antibiotic-resistant microbial species. Auerbach et al.[61] found that the activated sludge process reduced the volumetric concentration (expressed as gene copies per mL) of two genes that confer tetracycline resistance (tetQ and tetG); they also found that the fraction of bacterial species carrying tetQ and tetG was sometimes higher in the effluent than in the influent. This shows that although the overall concentration of tetG and tetQ carrying species was decreased, the fraction of species carrying these genes may not be attenuated. Andersen[62] used multiple antibiotic-resistance indexing to show that resistance levels of *E. coli* decreased during wastewater treatment. Many more studies are required to better understand how well these antibiotic-resistant microorganisms are removed. Future studies must include the full complement of antibiotic-resistance genes and must employ the quantitative molecular approach of Auerbach et al.,[61] which allowed for the determination of the volumetric concentration of antibiotic-resistance genes as well as the fraction of bacterial species carrying the antibiotic-resistance genes of interest.

One possible mechanism for antibiotic-resistant gene exchange in activated sludge could be related to a natural genetic engineering device called the integron, which allows diverse species of gram-negative bacteria to exchange and accumulate entire libraries of useful genes. Integrons may be a key component in the spread of antibiotic resistance. Biological WWTPs, where different bacteria and antibiotics aggregate, could be where the exchange is occurring. Szczepanowski et al.[63] found that bacteria residing in the sludge of a municipal water treatment plant contain integron-specific DNA sequences, so that their presence in this environment indicates that sludge is a specific location where genes are coming together and being distributed. Tennstedt et al.[64] found that 12% of the plasmids isolated from municipal sludge contained class 1 integron-specific sequences, and that these sequences contained genes that code for two chloramphenicol-resistance proteins.

16.7 CONCLUSION

Overall, the removal of various classes of PPCPs depends of the chemical characteristics of the PPCPs in question, that of the sludge, and operating conditions of the WWTP. Removal efficiencies at full scale vary considerably across the numerous classes of compounds of concern. Operating at higher SRT has been proposed as a strategy for improving removal efficiencies for numerous classes of micropollutants. In general, both sorption and biodegradation can play a role in the removal of many PPCPs, but the relative importance of one or the other must be determined carefully for the system and pollutants of interest. When biodegradation occurs, cometabolism likely plays a role for many PPCPs, because the concentrations of these pollutants are too low to support substantial biomass growth. When sorption is an important removal mechanism, it is possible to observe sorption hysteresis, which could result in the entrapment of target compounds within the floc and a reduction in the long-term sorption capacity. The presence of antibiotics in wastewater streams has raised concerns related to the proliferation of antibiotic-resistant microorganisms in activated sludge processes. It appears that the activated sludge process can reduce the numbers of antibiotic-resistant microorganisms, but there is potential to discharge antibiotic-resistant organisms into the aquatic environment.

REFERENCES

1. Ternes, T.A. Occurrence of drugs in German sewage treatment plants and rivers. *Water Research* 1998, 32, 3245–3260.
2. Ternes, T.A., Kreckel, P., and Mueller, J. Behaviour and occurrence of estrogens in municipal sewage treatment plans—II. Aerobic batch experiments with activated sludge. *Science of the Total Environment* 1999, 225, 91–99.
3. Gomez, M.J., Martinez Bueno, M.J., Lacorte, S., Fernandez-Alba, A R., and Aguera, A. Pilot survey monitoring pharmaceuticals and related compounds in sewage treatment plant located on the Mediterranean coast. *Chemosphere* 2006, In press.
4. Joss, A., Zabczynski, S., Gobel, A., Hoffman, B., Loffler, D., McArdell, C.S., Ternes, T., Thomsen, A., and Siegrist, H. Biological degradation of pharmaceuticals in municipal wastewater treatment: proposing a classification scheme. *Water Research* 2006, 40, 1686–1696.
5. Oppenheimer, J. and Stephenson, R. Characterizing the passage of personal care products through wastewater treatment processes 2006, 79th Annual Water Environment Federation Technical Exposition and Conference, Dallas, TX.
6. Jones, O.A.H., Voulvoulis, N., and Lester, J.N. The occurrence and removal of selected pharmaceutical compounds in a sewage treatment works utilising activated sludge treatment. *Environmental Pollution* 2006, In press.

7. Lishman, L., Smyth, S.A., Sarafin, K., Kleywegt, S., Toito, J., Peart, T., Lee, B., Servos, M., Beland, M., and Seto, P. Occurrence and reductions of pharmaceuticals and personal care products and estrogens by municipal wastewater treatment plants in Ontario, Canada, *Science of the Total Environment* 2006, 367, 544–558.

8. Carballa, M., Omil, F., Juan, J.M., Lema, M., Llompart, M., Garcia-Jares, C., Rodriguez, I., Gomez, M, and Ternes, T. Behavior of pharmaceuticals, cosmetics and hormones in a sewage treatment plant. *Water Research* 2004, 38, 2918–2926.

9. Nakada, N., Tanishima, T., Shinohara, H., Kiri, K., and Takada, H. Pharmaceutical chemicals and endocrine disrupters in municipal wastewater in Tokyo and their removal during activated sludge treatment. *Water Research* 2006, 40, 3297–3303.

10. Kim, S.D., Cho, J., Kim, I.S., Vanderford, B.J., and Snyder, S.A. Occurrence and removal of pharmaceuticals and endocrine disruptors in South Korean surface, drinking, and waste waters. *Water Research* 2006, In press.

11. Kimura, K., Hara, H., and Watanabe, Y. Removal of pharmaceutical compounds by submerged membrane bioreactors (MBRs). *Desalination* 2005, 178, 135–140.

12. Ternes, T., Herrmann, N., Bonerz, M., Knacker, T., Siegrist, H., and Joss, A. A rapid method to measure the solid-water distribution coefficient (Kd) for pharmaceuticals and musk fragrances in sewage sludge. *Water Research* 2004, 38, 4075–4084.

13. Clara, M., Strenn, E., Saracevic, E., and Kreuzinger, N. Adsorption of bisphenol-A, 17⊠ estradiol and 17β-ethinylestradiol to sewage sludge. *Chemosphere* 2004, 56, 843–851.

14. Yi, T., Harper, W.F., Holbrook, R.D., and Love, N.G. The role of particle characteristics and ammonium monooxygenase in removal of 17α-ethinylestradiol in bioreactors. *ASCE J. Env. Eng.* 2006, 132, 1527–1529.

15. Andersen, H., Hansen, M., Kjolholt, J., Stuer-Lauridsen, F., Ternes, T., and Halling-Sorensen, B. Assessment of the importance of sorption for steroid estrogens removal during activated sludge treatment. *Chemosphere* 2005, 61, 139–146.

16. Holthaus, K.I.E., Johnson, A.C., Jurgens, M.D., Williams, R.J., Smith, J.J.L., and Carter, J.E. The potential for estradiol and ethinylestradiol to sorb to suspended and bed sediments in some English rivers. *Environ. Toxicol. Chem.* 2002, 21, 2526–2535.

17. Bowman, J.C., Readman, J.W., and Zhou, J.L. Sorption of the natural endocrine disruptors, oestrone and 17β-oestradiol in the aquatic environment. *Environ. Geochem. Health* 2003, 25, 63–67.

18. Casey, F., Larsen, G.L., Hakk, H., and Simunek, H. Fate and transport of 17-estradiol in soil-water systems. *Environ. Sci. Technol.* 2003, 37, 2400–2409.

19. Ying, G.G., Kookana, R.S., and Dillon, P. Sorption and degradation of selected five endocrine disrupting chemicals in aquifer materials. *Water Research* 2003, 37, 3785–3791.

20. Yu, J.T., Bouwer, E.J., and Coelhan, M. Occurrence and biodegradability studies of selected pharmaceuticals and personal care products in sewage effluent. *Agricultural Water Management* 2006, In press.

21. Urase, T. and Kikuta, T. Separate estimation of adsorption and degradation of pharmaceutical substances and estrogens in the activated sludge process. *Water Research* 2005, 39, 1289–1300.

22. Conrad, A., Codoret, P., Corteel, P., Leroy, J., and Block, C. Adsorption/desorption of linear alkybenzensulfonate (LAS) and azoproteins by/from activated sludge flocs. *Chemosphere* 2005, In press.

23. Kim, S., Eichhorn, P., Jensen, J., Weber, S., and Aga, D. removal of antibiotics in wastewater: effect of hydraulic and solids retention times on the fate of tetracycline in the activated sludge process. *Environmental Science and Technology* 2005, 39, 5816–5823.

24. Huang, W., Peng, P., Yu, Z., and Fu, J. Effects of organic matter heterogeneity on sorption and desorption of organic contaminants by soils and sediments. *Applied Geochemistry* 2003, 18, 955–972.

25. Martin, S.R. Equilibrium and kinetic studies on the interaction of tetracyclines with calcium and magnesium. *Biophysical Chemistry*. 1979, 10, 319–326.
26. Sobeck, D. and Higgins, M. Examination of three theories for mechanisms of cation-induced bioflocculation. *Water Research* 2002, 36, 527–538.
27. Yi, T. and Harper, W.F. The effect of biomass particle characteristic on the partitioning and sorption hysteresis of 17α-ethinylestradiol. *Water Research* 2007, 41, 1543–1553.
28. Snidaro, D., Zartarian, R., Jorand, F., Bottero, J., Block, J., and Manem, J. Characterization of activated sludge floc structure. *Water Science and Technology* 1997, 36, 313–320.
29. Chu, C., Lee, D., and Tay, J. Floc model and intrafloc flow. *Chemical Engineering Science* 2005, 60, 565–575.
30. Eddy, M.A. *Wastewater Engineering: Treatment and Reuse*, 2003, New York: McGraw-Hill Publishers.
31. Ng, H. and Hermanowicz, S. Membrane bioreactor operation and short solids retention times: performance and biomass characteristics. *Water Research* 2005, 39, 981–992.
32. Quintana, J.B., Weiss, S., and Reemtsma, T. Pathways and metabolites of microbial degradation of selected acidic pharmaceutical and their occurrence in municipal wastewater treated by a membrane reactor. *Water Research* 2005, 39, 2654–2664.
33. Alexy, R., Kumpel, T., and Kummerer, K. Assessment of degradation of 18 antibiotics in the closed bottle test. *Chemosphere* 2005, 62, 294–302.
34. Haib, A. and Kummerer, K. Biodegradability of X-ray contrast compound diatrizoic acid, identification of aerobic degradation products and effects against sewage sludge micro-organisms. *Chemosphere* 2006, 62, 294–302.
35. Schwarzenbach, R., Gschwend, P., Imboden, P.M., and Dieter, M. Environmental Organic Chemistry 2003, New York: John Wiley & Sons.
36. Madigan, M. and Parker, B. *Biology of Microorganisms*, 1997, Prentice Hall.
37. Kragelund, C., Nielsen, J.L., Rolighed, T., and Nielsen, P.H. Ecophysiology of the filamentous alphaproteobacterium maganema perideroedes in activated sludge. *FEMS Microbiology Ecology* 2005, 54, 111–122.
38. Gessesse, A., Dueholm, T., Petersen, S.B., and Nielsen, P.H. Lipase and protease extraction from activated sludge. *Water Research* 2003, 37, 3652–3657.
39. Cadoret, A., Conrad, A., and Block, J. Availability of low and high molecular weight substrates to extracellular enzymes in whole and dispersed activated sludges. *Enzyme and Microbial Technology* 2002, 31, 179–186.
40. Servos, M., Bennie, D., Burnison, B., Jurkovic, A., McInnis, R., Neheli, T., Schnell, A., Seto, P., Smyth, S., and Ternes, T. Distribution of estrogens, 17-estradiol and estrone in Canadian municipal wastewater treatment plants. *Science of the Total Environment* 2004, 336, 155–170.
41. Vader, J., van Ginkel, C., Sperling, F., de Jong, F., de Boer, W., de Graaf, J., van der Most, M., and Stockman, P.G.W. Degradation of ethinyl estradiol by nitrifying activated sludge. *Chemosphere* 2000, 41, 1239–1243.
42. Shi, J., Fujisawa, S., Nakai, S., and Hosomi, M. Biodegradation of natural and synthetic estrogen by nitrifying activated sludge and ammonia-oxidizing bacterium *Nitrosomonas europea*. *Water Research* 2004, 38, 2323–2330.
43. Yi, T. and Harper, W.F. The link between nitrification and biotransformation of 17α-ethinylestradiol. *Environmental Science and Technology* 2007, 41, 4311–4316.
44. Dytczak, M.A., Londry, K.L., and Oleszkiewwicz, J.A. Transformation of estrogens in nitrifying sludge under aerobic and alternating anoxic/aerobic conditions 2006, 79th Annual Water Environment Federation Technical Exposition and Conference, Dallas, TX.

45. Chang, S., Hyman, M., and Williamson, K. Cooxidation of napthalene and other poly-cyclic aromatic hydrocarbons of the nitrifying bacterium, *Nitrosomonas europaea*. *Biodegradation* 2003, 13, 373–381.

46. Vannelli, T. and Hooper, A. NIH shift in the hydroxylation of aromatic compounds by the ammonia-oxidizing bacterium *Nitrosomonas europaea*. Evidence against an arene oxide intermediate. *Biochemistry* 1995, 34, 11743–11749.

47. Zahn, J., Arciero, D., Hooper, A., and Dispirito, A. Evidence for an iron center in the ammonia monoxygenase from *Nitrosomonas europaea*. *FEBS Letters* 1996, 397, 35–38.

48. Siegbahm, P., Crabtree, R., and Nordlund, P. Mechanism of methane monooxygen-ase—A structural and quantum chemical perspective. *JBIC* 1998, 3, 314–317.

49. Nyholm, N., Ingerslev, F., Berg, U.T., Pedersen, J.P., and Frimer-Larsen, H. Estimation of kinetic rate constants for biodegradation of chemical in activated sludge wastewater treatment plants using short term batch experiments and μg/L range spiked concentra-tions. *Chemosphere* 1996, 33, 851–864.

50. Boyd, G.R., Zhang, S., and Grimm, D.A. Naproxen removal from water by chlorination and biofilm processes. *Water Research* 2005, 39, 668–676.

51. Zwiener, C. and Frimmel, F.H. Short-term tests with a pilot sewage plant and biofilm reactors for the biological degradation of the pharmaceutical compounds clofibric acid, ibuprofen, and diclofenac. *Science of the Total Environment* 2003, 309, 201–211.

52. Winkler, M., Lawrence, J.R., and Neu, T.R. Selective degradation of ibuprofen and clo-fibric acid in two model river biofilm systems. *Water Research* 2001, 35, 3197–3205.

53. Ingerslev, F., Torang, L., Loke, M., Halling-Sorensen, B., and Nyholm, N. Primary bio-degradation of veterinary antibiotics in aerobic and anaerobic surface water simulation systems. *Chemosphere* 2001, 44, 865–872.

54. Kummerer, K., Al-Ahmad, A., and Mersch-Sundermann, V. Biodegradability of some antibiotics, elimination of the genotoxicity and affection of wastewater bacteria in a simple test. *Chemosphere* 2000, 40, 701–710.

55. Zhou, P., Su, C., Li, B., and Qian, Y. Treatment of high-strength pharmaceutical waste-water and removal of antibiotics in anaerobic and aerobic biological treatment pro-cesses. *J. Envir. Eng.* 2006, 132, 129–136.

56. Kolpin, D., Furlong, E., Meyer, M., Thurman, E., Zaugg, S., Barber, L., and Buxton, H. Pharmaceuticals, hormones, and other organic wastewater contaminants in U.S. streams, 1999–2000: A national reconnaissance. *Environmental Science and Technol-ogy* 2002, 36, 1202–1211.

57. Schwartz, T., Kohnen, W., Jansen, B., and Obst, U. Detection of antibiotic-resistant bacteria and their resistance genes in wastewater, surface water, and drinking water biofilms. *FEMS Microbiology Ecology* 2003, 43, 325–335.

58. Mach, P.A. and Grimes, D.J.R. Plasmid transfer in a wastewater treatment plant. *Appl Environ Microbiol* 1982, 44, 1395–1403.

59. Volkmann, H., Schwartz, T., Bischoff, P., Kirchen, S., and Obst, U. Detection of clini-cally relevant antibiotic-resistance genes in municipal wastewater using real-time PCR (Taqman). *Journal of Microbiological Methods* 2004, 56, 277–286.

60. Reinthaler, F.F., Posch, J., Feierl, G., Wust, G., Haas, D., Ruckenbauer, G., Mascher, F., and Marth, E. Antibiotic resistance of E. coli in sewage and sludge. *Water Research* 2003, 37, 1685–1690.

61. Auerbach, E.A., Seyfried, E.E,. and McMahon, K.D. Tetracycline resistance genes in activated sludge wastewater treatment plants, 2006, 79th Annual Water Environment Federation Technical Exposition and Conference, Dallas, TX.

62. Andersen, S.R. Effects of waste water treatment on the species composition and antibi-otic resistance of coliform bacteria. *Current Microbiology* 1993, 26, 97–103.

63. Szczepanowski, R., Braun, S., Riedel, V., Schneiker, S., Krahn, I., Puhler, A., and Schluter, A. The 120 592 bp IncF plasmid pRSB107 isolated from a sewage-treatment plant encodes nine different antibiotic-resistance determinants, two iron-acquisition systems and other putative virulence-associated functions. *Microbiology* 2005, 151, 1095–1111.
64. Tennstedt, T., Szczepanowski, R., Braun, S., Puhler, A., and Schluter, A. Occurrence of integron-associated resistance gene cassettes located on antibiotic resistance plasmids isolated from a wastewater treatment plant. *FEMS Microbiology Ecology* 2003, 45, 239–252.

Index

A

AC; *see* activated carbon
accelerated solvent extraction (ASE), 87, 184
acetylhexamethyl tetrahydronaphthalene
(AHTN), 368
activated carbon (AC), 242
adsorption, activated sludge biomass and, 368
advanced oxidation processes (AOPs), 222, 224
AEC; *see* anion exchange capacity
AF; *see* assessment factor
AFOs; *see* animal feeding operations
AHTN; *see* acetylhexamethyl
tetrahydronaphthalene
Allium cepa L., 130
4-aminophenyl methyl sulfone (APMS), 270
ammonia oxidizing bacteria (AOB), 316, 357
ammonium monooxygenase (AMO), 372, 374
AMO; *see* ammonium monooxygenase
analgesics, 34, 67
androgens
biosynthesis of, 302
degradation of, 304
excretion of, 303–304
metabolism of, 302–303
natural, 302
animal feeding operations (AFOs), 292
animal feeding operations, hormones in waste
from, 291–329
background, 292–295
concentrated animal feeding operations in
United States, 292–293
hormones and CAFO, 293–295
best management practices to reduce
hormone loss, 317–319
buffer strips, 319
constructed wetlands, 318–319
controlled stream access, 319
fate of hormones during manure storage,
treatment, and land application,
313–317
aeration of dairy manure, 316–317
anaerobic digestion, 315
chemical and biological phosphorus
removal, 316
composting, 315
conventional manure storage and
treatment systems, 313–314
innovative manure treatment systems, 315

nitrification and denitrification, 315–316
hormone growth promoters, 306–310
naturally occurring, 307
synthetic, 307–310
hormones in CAFOs, 295–301
biosynthesis of estrogens, 296–297
degradation of estrogens, 300
excretion of estrogens, 298–300
metabolism of estrogens, 297–298
natural estrogens, 295–296
manure available for land application, 294
natural androgens, 302–304
biosynthesis, 302
degradation, 304
excretion, 303–304
metabolism, 302–303
natural progestagens, 304–306
biosynthesis of progesterone, 304–305
degradation of progesterone, 306
excretion of progesterone, 306
metabolism of progesterone, 305–306
number of U.S. CAFOs, 294
research needs, 320
routes of hormone loss from CAFOs, 310–313
groundwater, 312–313
soil and runoff, 311–312
streams and rivers, 312
size threshold of CAFOs, 293
anion exchange capacity (AEC), 141
antibacterial agents, reaction and transformation
of with aqueous chlorine under
relevant water treatment conditions,
261–289
background, 263–268
agents of investigation, 263–267
chemical oxidation by aqueous chlorine,
267
prior work on reaction of pharmaceuticals
with chlorine, 267–268
kinetic models, 274
materials and methods, 268–271
chemical reagents, 268–269
reaction setup and monitoring, 269–271
surface water and wastewater samples,
269
results and discussion, 271–285
identification of reactive functional
groups, 275–276
reaction kinetics and modeling, 271–275